AUDIO PRODUCTION WORKTEXT
Concepts, Techniques, and Equipment

The eighth edition of *Audio Production Worktext* gives readers an expansive introduction to the modern radio production studio, the equipment found in that studio, and the basic techniques needed to accomplish radio production work. This new edition is updated throughout and features new sections on mobile technology, audio editing apps and software, and digital editing. It includes updated graphics and expanded content on portable digital audio and features a worktext/website format tailored for both students and teachers. *Audio Production Worktext, Eighth Edition* offers a solid foundation for anyone who wishes to know more about audio equipment and modern production techniques.

Key features include:

- Updated graphics and artwork throughout that reflect the latest in the audio production world
- Production Tips – brief tips that provide interested notes relevant to various audio production topics
- Worktext format that features self-study questions, hands-on projects, an appendix on analog audio equipment for useful background information, and an updated Glossary

- Updated companion website that includes a sample syllabus, test bank, PowerPoint images, and chapter Self-Study questions and projects.

Samuel J. Sauls, Ph.D., retired as an Associate Professor from the University of North Texas in August 2013 after 29 years on the faculty in the Department of Radio, Television and Film. From 1984 to 1994 he was the Station Manager of the University radio station, KNTU-FM. During his tenure, he also held the roles of Associate Chair and Director of Graduate Studies in the department. He has a combined total of 15 years' experience in commercial and noncommercial radio and also worked for four years in Saudi Arabia producing soundtracks in English, Arabic, and French. He has served on the Broadcast Education Association board of directors, was BEA President in 2011–2012, and is a former President of the Texas Association of Broadcast Educators. His first book, *The Culture of American College Radio*, was published in 2000 and his most recent co-authored book, *The Sump'n Else Show*, was published in 2014. Dr. Sauls has published some 16 academic articles and has made over 65 conference paper and panel presentations, as well as reviewing over 20 journal articles and 15 books for publication consideration. He is an Adjunct Professor in the Department of Communications at Susquehanna University.

Craig A. Stark, Ph.D., is an Associate Professor of Communications at Susquehanna University, where he teaches courses in media law, writing for new media, gaming and interactive media, and critical issues in emerging media. He also serves as the faculty advisor for the school's radio station, WQSU FM. He has worked and taught in educational broadcasting for over twenty years. Prior to entering the academic field, he worked in commercial radio and video production as an on-air announcer, sports reporter, and broadcast engineer. He is active in several professional and academic organizations including the Broadcast Educators Association, and is the author of several articles related to media studies.

AUDIO PRODUCTION WORKTEXT

Concepts, Techniques, and Equipment

8th Edition

SAMUEL J. SAULS AND
CRAIG A. STARK

Routledge
Taylor & Francis Group

NEW YORK AND LONDON

Please visit the companion website at
www.routledge.com/cw/Sauls

This edition published 2016
by Routledge
711 Third Avenue, New York, NY 10017

and by Routledge
2 Park Square, Milton Park, Abingdon, Oxon, OX14 4RN

Routledge is an imprint of the Taylor & Francis Group, an informa business

© 2016 Taylor & Francis

The right of Samuel J. Sauls and Craig A. Stark to be identified as authors of this work has been asserted by them in accordance with sections 77 and 78 of the Copyright, Designs and Patents Act 1988.

Seventh edition first published 2013
By Focal Press

Library of Congress Cataloging-in-Publication Data
Names: Sauls, Samuel J., author. | Stark, Craig A., author.
Title: Audio production worktext : concepts, techniques, and equipment /
    Samuel J. Sauls and Craig A. Stark.
Description: Eighth edition. | New York, NY : Routledge, 2016. | Includes index.
Identifiers: LCCN 2015040242| ISBN 9781138839458 (hardback) |
    ISBN 9781138839465 (pbk.) | ISBN 9781315733418 (ebook)
Subjects: LCSH: Sound studios—Textbooks. | Radio stations—Equipment and supplies—Textbooks. |
    Radio—Production and direction—Textbooks. | Sound—Recording and reproducing—Equipment
    and supplies—Textbooks.
Classification: LCC TK6557.5 .S28 2016 | DDC 791.4402/32—dc23
LC record available at http://lccn.loc.gov/2015040242

ISBN: 978-1-138-83945-8 (hbk)
ISBN: 978-1-138-83946-5 (pbk)
ISBN: 978-1-315-73341-8 (ebk)

Typeset in Electra LT Std Regular
by Apex CoVantage, LLC

CONTENTS

Preface   *xiii*
Acknowledgments   *xv*

CHAPTER 1   PRODUCTION PLANNING

1.1   Introduction   1
1.2   The Idea   2
1.3   Goals and Objectives   2
1.4   The Target Audience   3
1.5   Style   3
1.6   Production Personnel   4
1.7   Production Elements   5

Production Tip 1A—Pick Any Two   6
1.8   The Script   6
1.9   Paperwork   7
1.10 The Importance of Voice   7

Production Tip 1B—Copy Marking   10
1.11 Equipment and Facilities   11
1.12 Laws   11
1.13 Ethics   12
1.14 Conclusion   14

Self-Study   15
Questions   15
Answers   16

Projects   18
Project 1—Undertake Production Planning for a 15-Minute Interview Show   18
Project 2—Assess Your Skills   18

CHAPTER 2   THE STUDIO ENVIRONMENT

2.1   Introduction   21
2.2   The Audio Chain   22
2.3   The Studio Layout   22
2.4   Production Studio Furniture   23
2.5   Studio Sound Considerations   24
2.6   Studio Construction Materials   25

2.7   Studio Size and Shape   26
2.8   Studio Aesthetics   26

Production Tip 2A—Static Electricity   *27*
2.9   On-Air/Recording Lights   27
2.10   Hand Signals   27
2.11   Noise and Distortion   28
2.12   Is it a Sound Signal or an Audio Signal?   29
2.13   Sound Defined   29
2.14   Key Characteristics of Sound Waves   29
2.15   Frequency Response   31
2.16   Conclusion   31

Self-Study   *32*
Questions   32
Answers   35

Projects   *38*
Project 1—Tour an Audio Facility and Write a Report Describing It   38
Project 2—Redesign Your Production Studio   38
Project 3—Draw an Audio Chain Flowchart for Your Production Studio   39

CHAPTER 3   DIGITAL AUDIO PRODUCTION

3.1   Introduction   41
3.2   The Analog Roots of Digital Production   41
3.3   The Digital Process   41
3.4   Reasons for Editing   43
3.5   Desktop Audio Production: The Digital Audio Editor   43

Production Tip 3A—Audacity   *43*
3.6   Digital Audio Workstations and Other Digital Editing Solutions   44
3.7   Strong Points and Weak Points of Digital Production   45

Production Tip 3B—Maintaining Digital Equipment   *46*
3.8   Audio Synchronization   46
3.9   Latency Issues   47
3.10   Digital Audio Editing   47

Production Tip 3C—What's up with Adobe® Creative Cloud®?   *48*
3.11   Multitrack Editing Techniques   50
3.12   Multitrack Voice Effects   51
3.13   Track Sheets   52
3.14   The Mix Down   53
3.15   Multitrack Spot Production   54
3.16   Conclusion   54

Self-Study   *55*
Questions   55
Answers   57

Projects   *60*
Project 1—Undertake Digital Audio Editing   60
Project 2—Using a Digital Audio Editor, Build a Short Music Bed and Record a "Voice and Music Bed" Spot   60
Project 3—Write and Record a 60-Second "Concert Commercial"   61
Project 4—Write and Record a Report that Compares Various Recording/Editing Software Programs   62
Project 5—Record a Public Service Announcement that Uses a Sound Effect   63

CHAPTER 4   MICROPHONES

4.1   Introduction   65
4.2   Classifying Microphones   66

4.3   Dynamic Microphones   66
4.4   Condenser Microphones   66
4.5   Microphone Pickup Patterns   67
4.6   The Omnidirectional Pickup Pattern   67
4.7   The Cardioid Pickup Pattern   68
4.8   Polar Response Patterns   68
4.9   Impedance of Microphones   69
4.10   Sensitivity of Microphones   69
4.11   Proximity Effect and Bass Roll-Off   69
4.12   Microphone Feedback   70
4.13   Multiple-Microphone Interference   70
4.14   Stereo   70
4.15   Stereo Miking Techniques   70
4.16   Surround Sound   71
4.17   Special Purpose and Other Types of Microphones   72
4.18   Microphone Accessories   74
4.19   Microphone Usage   77

Production Tip 4A—Microphone-to-Mouth Relationship and Setting Levels   77
4.20   Conclusion   77

Self-Study   78
Questions   78
Answers   81

Projects   84
Project 1—Position Microphones in Various Ways to Create Different Effects   84
Project 2—With Several Other Students, Make a Recording Using Stereo Miking Techniques   85
Project 3—Compare Sound from Different Types of Microphones   85
Project 4—Diagram/Apply Miking Techniques to Various On-Campus or Local Sporting Events   86

CHAPTER 5   THE AUDIO CONSOLE

5.1   Introduction   89
5.2   The Digital Audio Console   89
5.3   Audio Console Functions   91
5.4   Computers and Audio Consoles   91
5.5   Basic Audio Console Components   91
5.6   Input Selectors   91
5.7   Input Volume Control   92
5.8   Monitoring: Speakers and Headphones   93
5.9   Cue   94
5.10   VU Meters   95
5.11   Output Selectors   95
5.12   Output Volume Control   96
5.13   Remote Starts, Clocks, and Timers   96
5.14   Equalizers and Pan Pots   96
5.15   Other Features   97

Production Tip 5A—Manipulating Faders   97
5.16   Sound Transitions and Endings   97
5.17   Conclusion   98

Self-Study   99
Questions   99
Answers   102

Projects   105
Project 1—Learn To Operate an Audio Console   105
Project 2—Diagram and Label an Audio Board   105
Project 3—Record a Two-Voice Commercial   106

CHAPTER 6   DIGITAL AUDIO PLAYERS/RECORDERS

6.1   Introduction   109
6.2   The CD Player   109
6.3   CDs and Care of CDs   110
6.4   The CD Recorder   111
6.5   Data Compression   111
6.6   Compactflash and Other Digital Recorders   112
6.7   Storage   113
6.8   MP3/Portable Audio Players   113
6.9   Digital Distribution Networks   114

Production Tip 6A   **115**
6.10   Conclusion   115

Self-Study   **116**
Questions   116
Answers   117

Projects   **119**
Project 1—Prepare a Report on a Digital Player/Recorder or Portable Audio Production Software That Is Not Discussed in This Chapter   119
Project 2—Play and Record Several CD Selections   119
Project 3—Record a 5-minute Interview With a Classmate and Edit The Interview to 3 Minutes, Using a Portable Digital Recording Device   120

CHAPTER 7   MONITOR SPEAKERS AND STUDIO ACCESSORIES

7.1   Introduction   123
7.2   Types of Speakers   123
7.3   Basic Speaker System Components   124
7.4   Speaker System Enclosure Designs   124
7.5   Speaker Sound Qualities   125
7.6   Speaker Placement   125
7.7   Phase and Channel Orientation   127
7.8   Monitor Amplifiers   127
7.9   Speaker Sensitivity   127
7.10   Headphones   127
7.11   Hardwiring and Patching   128
7.12   Common Audio Connectors   130
7.13   Other Connectors and Connector Adapters   131
7.14   Balanced and Unbalanced Lines   132
7.15   Microphone, Line, and Speaker Levels   132
7.16   Studio Timers   133
7.17   Telephone Interface   133
7.18   Conclusion   133

Self-Study   **135**
Questions   135
Answers   137

Projects   **141**
Project 1—Compare Speaker/Listener Placement   141
Project 2—Identify Common Connectors Found in the Audio Production Studio   141

CHAPTER 8   SIGNAL PROCESSING AND AUDIO PROCESSORS

8.1   Introduction   143
8.2   Why Use Signal Processing Effects?   143

8.3   Software or Black Box Signal Processing   144
8.4   Equalizers   144
8.5   The Graphic Equalizer   144
8.6   The Parametric Equalizer   146
8.7   Audio Filters   146
8.8   Noise Reduction   147
8.9   Reverb and Digital Delay   147

Production Tip 8A—World Wide Web Effects   **148**
8.10   Dynamic Range   149
8.11   Compressors, Expanders, and Noise Gates   149
8.12   Limiters   149
8.13   Other Signal Processors   150
8.14   Multi-Effects Processors   150
8.15   Conclusion   151

Self-Study   **152**
Questions   152
Answers   154

Projects   **157**
Project 1—Record a Commercial Spot that Uses a Signal Processing Effect   157
Project 2—Use Multitrack Recording to Create a Chorusing Effect   157
Project 3—Restore an Audio Clip Using Noise-Reduction Software   158

CHAPTER 9   PRODUCTION SITUATIONS

9.1   Introduction   159
9.2   Producing Commercials   159

Production Tip 9A—Music Punctuators   **160**
9.3   Enhancing Image   160
9.4   Announcing Music   161
9.5   Recording Music   162

Production Tip 9B—Miking a Guitar   **163**
9.6   Preparing and Announcing News   163
9.7   Reporting Sports, Traffic, and Weather   165
9.8   Hosting Talk Shows   165
9.9   Performing Drama and Variety   167
9.10   Conclusion   167

Self-Study   **168**
Questions   168
Answers   170

Projects   **172**
Project 1—Record an Air-Check Tape   172
Project 2—Record a 5-Minute Radio Interview Show in Which You Are the Interviewer   172

CHAPTER 10   LOCATION SOUND RECORDING

10.1   Introduction   175
10.2   Types of Field Production   175
10.3   Common Location Sound Problems   176

Production Tip 10A—How to Get Rid of a Hum   **177**
10.4   Site Planning for Location Recording   177
10.5   Using Microphones   179
10.6   Using Recorders   180

10.7   Using Mixers   180
10.8   Using Headphones   180
10.9   Getting the Signal Back to the Studio   182
10.10   Handling Vehicles   183
10.11   Providing for Your Own Needs   184

Production Tip 10B—How to Pack a Survival Bag   **184**
10.12   Postproduction Concerns for Location Recording   185
10.13   Conclusion   185

Self-Study   **186**
Questions   186
Answers   188

Projects   **190**
Project 1—Listen and Plan for Sounds   190
Project 2—Record Atmosphere Sound, Room Tone, and Walla Walla   191

CHAPTER 11   SOUND PRODUCTION FOR THE VISUAL MEDIA

11.1   Introduction   193
11.2   The Importance of Sound to a Visual Production   193
11.3   The Need to Accommodate the Picture   193
11.4   Recording Speech   194
11.5   The Boompole   194

Production Tip 11A—Holding a Boompole   **196**
11.6   The Lavaliere   196
11.7   Other Forms of Microphone Positioning   197
11.8   Continuity and Perspective   197
11.9   The Recording Procedure   198

Production Tip 11B—Recording with Your Eyes Shut   **199**
11.10   Recording Sound Effects   200
11.11   Recording Ambient Sounds   200
11.12   Recording Music   201
11.13   Recording ADR   203
11.14   Recording Foley   204
11.15   Recording Voice-Overs   204
11.16   Postproduction Considerations   205
11.17   Final Mix   205
11.18   Conclusions   206

Self-Study   **207**
Questions   207
Answers   209

Projects   **211**
Project 1—Determine the Importance of Sound and Picture   211
Project 2—Record Sound for a Video Project   211

CHAPTER 12   INTERNET RADIO AND OTHER DISTRIBUTION PLATFORMS

12.1   Introduction   213
Production Tip 12A—Ch-Ch-Ch-Changes   **213**
12.2   Web Pages   214
12.3   Overview of the Audio Process for Streaming   214
12.4   Encoders   215
12.5   Servers   216
12.6   Playback Software and Apps   216

12.7   Software Options   216

Production Tip 12B—Internet Audience and On-Air Talent Interaction   217

12.8   On-Demand Files and Podcasting   218
12.9   Building a Home Studio for Internet Audio Production   218
12.10 Copyright   219
12.11 Internet Radio Station Listing Sites   220
12.12 Other Distribution Means   220
12.13 Satellite Radio   220
12.14 Cable and Satellite TV Radio   220
12.15 Over-the-Air Broadcasting   220
12.16 HD Radio   221
12.17 Conclusion   222

Self-Study   223
Questions   223
Answers   226

Projects   229
Project 1—Report on the Differences and Similarities among Six Radio Station Websites   229
Project 2—Tour a Broadcast Radio Station Transmitting Facility   229
Project 3—See What You Have and What You Need in order to build Your Own Audio Recording
           and Editing Facility in Your Home or Dorm Room   230

GLOSSARY   231

APPENDIX   ANALOG AND DIGITAL AUDIO EQUIPMENT

A.1   Introduction   243
A.2   Turntables   243
A.3   Turntable Use   244
A.4   Reel-to-Reel Audio Tape Recorders   245
A.5   Reel-to-Reel Recorder Use   246

Production Tip A.A—Sel Sync   247
A.6   Cassette Tape Recorders   247
A.7   Cassette Recorder Use   248
A.8   Cartridge Tape Recorders   248
A.9   Cartridge Recorder Use   249
A.10   Tape-Based Digital Recorders   249
A.11   Analog Tape Editing Tools   249
A.12   Making Edits   250
A.13   Analog Audio Consoles   251
A.14   Analog Audio Console Use   251
A.15   The MD Recorder/Player   251
A.16   The MiniDisc   252
A.17   Conclusion   253

Index   255

PREFACE

As we move into the eighth edition of the *Audio Production Worktext*, which we inherited from David Reese, Lynne Gross, and Brian Gross in the last edition, we have strived to maintain an excellent audio production textbook. To that end, we have not significantly altered the layout or intent of the text. And so, instructors who use the textbook, and the students who learn from it, can be assured that we have done everything possible to uphold the standards set by previous editions of the *Audio Production Worktext*.

We hope readers of the eighth edition will find an updated, modified worktext that suits the needs of beginning audio production students, while also providing refresher material for more experienced practitioners.

As started in the seventh edition, the supporting material for the worktext is provided online. We believe this was a good move for the text, since it allows for faster access to the ancillary materials. In this edition, the instructor website (including supplemental items such as the sample syllabus, test bank, and PowerPoint images), along with the chapter Self-Study questions and Projects, have all been reviewed and appropriately updated.

One tradition we have retained from previous editions of the text is using the latest version of Adobe® Audition® to discuss and illustrate many production concepts throughout the text. As always, you can download a trial version of the program from www.adobe.com, and demo versions of other programs are also available online.

New to this edition, instructors and students can download 1 free sound effect (a $5 value) from the Pro Sound Effects Online Library featuring over 175,000 sound effects. Users can preview each sound, search by keyword, sort by category, create playlists, and download in wav or mp3 formats. To do this, enter the code **APW-PSE1** at download.prosoundeffects.com.

We have freshened and updated both the copy and artwork throughout the book in order to reflect the ever-changing world of audio production. Chapter 1,

"Production Planning," opens the text and focuses on qualities that typify a good production person; a model production planning process; and working with the basic production elements of voice, music, and sound effects. It also addresses the importance of audio production as well as career opportunities for consideration in the audio field. Providing a good introduction, Chapter 1 emphasizes the importance of audio right at the beginning. Chapter 2, "The Studio Environment," continues to introduce the reader to the layout and design of the audio production studio and briefly describes equipment that is detailed in subsequent chapters. This chapter also discusses some basic concepts regarding sound that will help the production person understand the raw material being worked with. The chapter logically explains the audio chain, followed by the studio layout, furniture, environment, materials, size, aesthetics, lights, hand signals, and the audio sound itself.

As with the seventh edition, Chapter 3, "Digital Audio Production," first provides a brief look at how audio is converted from analog to digital. After that, the chapter introduces some of the equipment used for digital production and discusses some of the tools and techniques of digital editing through the use of Adobe® Audition®. Since digital based production studios with multitrack capabilities have become the norm, this chapter also looks at the basics of multitrack production. The chapter goes into great depth about how sound works providing the basis of audio production-understanding how it all works.

After learning about some of these basic production techniques, students should be ready to work with sound sources that can serve as inputs or outputs for the console. One of these is covered in Chapter 4, "Microphones." This chapter discusses microphone basics, including pickup patterns and transducer types. Readers also learn about different types of microphones and their uses, including lavaliere mics, shotgun mics, and parabolic mics. Along with miking techniques, other accessories are also discussed.

Chapter 5, "The Audio Console," introduces a key component of audio production work. The expanded material on digital interfaces and "virtual" consoles keeps this chapter firmly in the digital age. A solid understanding of this material is important because other audio equipment often operates through a console or mixer, be it in the studio or the field. The chapter goes into details of the components of the audio console, describing each individually. Chapter 6, "Digital Audio Players/Recorders," deals with some of the more traditional audio players and recorders still in use, while also discussing some of the more recent advancements in technology. Compact discs, CompactFlash, online storage, and hard disc players and recorders are all considered here.

Chapter 7, "Monitor Speakers and Studio Accessories", looks at some of the most tried and true, yet overlooked technology in audio production. Monitor speakers and headphones are discussed, along with the connectors, cables, and accessories that complement the major pieces of equipment in the production studio, and some that are used to bring in and send sounds from and to outside sources.

In this edition, Chapter 8, "Signal Processing and Audio Processors," continues to provide an introduction to the most popular signal processing concepts employed in audio production work. Along with digital equipment, audio editing software now provides signal processing effects often faster and more easily than before. The approach to signal processing in the eighth edition is more program specific through applications, as opposed to equipment specific. Additionally, updated screen shots from the current edition of Adobe® Audition® are provided. In Chapter 9, "Production Situations," the student will learn many of the techniques and skills used for different production situations, including basic spot production, radio announcing, newscasting, interviewing, and sports play-by-play, to name a few. The goal is to inform about different production situations.

Chapter 10, "Location Sound Recording," surveys the different types of field recording, the problems associated with location recording, the tools of field recording, and basic techniques used to capture sound in the field and get that sound back to the studio. Each type of field production has its own individual pros and cons and this chapter gives the student an idea of what to expect. Chapter 11, entitled "Sound Production for the Visual Media," *examines* the way audio is used in conjunction with video and film production encompassing techniques for recording dialogue, music, sound effects, and ambient sound. Microphone types and placement are discussed, as well as editing and postproduction considerations. This chapter gives a great overview of the use of audio with visual production. The concepts of audio production are the same, but still different when the visual aspect is added and students must be aware of those differences.

Finally, Chapter 12, "Internet Radio and Other Distribution Platforms," briefly explains how to technically create an Internet radio station, while also considering other issues ranging from online hosting to content licensing. The chapter also discusses various distribution methods such as online sites, satellite broadcasting, HD Radio, and Internet broadcasting. Provided in this chapter is a unique story about the creation of an online audio production facility, and one person's move from analog to digital audio distribution. While looking at the full spectrum of audio production, students need to understand the development and application of newer distribution platforms.

The appendix, "Analog and Digital Audio Equipment," is a brief historic survey of analog equipment and some of its production techniques, as well as an admission that some facilities still have not completely abandoned the equipment and techniques of older technology. This appendix also includes a discussion of several digital audio production technologies that are not as prominently used as they once were.

Throughout the text, you'll also find Production Tips, which provide interesting notes that are relevant to various audio production topics. Key terms are listed in boldface when they appear in each chapter and are included in the Glossary. A web indicator @ throughout the book encourages the reader to visit the text website for more information and examples of concepts.

ACKNOWLEDGMENTS

We would obviously like to once again thank David Reese, Lynne Gross, and Brian Gross, who trusted us enough to continue the excellent work which we undertook in the previous edition and continue here. We are also indebted to our colleagues at Susquehanna University for their support in our endeavor.

Many of our friends in broadcasting, education, and related audio production industries provided us with valuable insight and "real world" experiences, which help to make the eighth edition of *Audio Production Worktext* as good as it is:

- Fred Ginsburg, CAS PhD MBKS
- Michael Parks, iHeartMedia-Harrisburg, PA
- Rich Gannon, CBS Sports NFL Analyst
- Mark Holland, JAM Creative Productions, Inc.
- Steve Eberhart, Eberhart Broadcasting
- Andrew Emge, Pro Sound Effects
- Brenda Jaskulske, Department of Radio, Television and Film, University of North Texas
- Thomas Thiriez, TwistedWave Software LTD
- Chris Dauray, sE Electronics
- Rob Fissel, Voice Over Talent

We offer special gratitude to the reviewers of the seventh edition who provided thorough input for this edition:

- David Nelson, University of Central Oklahoma
- Debbi Hatton, Sam Houston State University
- Sharon Stringer, Lock Haven University
- Brian Corea, Ohio University Southern

To Sherry Williford at Stephen F. Austin State University we extend a very special thanks for her insight as we were preparing our final manuscript.

We would also like to thank all of our friends at Focal Press for their guidance through the publication of this edition of the *Audio Production Worktext*. We would like to thank our former editor, Katy Morrissey, as well as Brianna Bemel, our former Editorial Assistant. They, along with Linda Bathgate, Publisher, Communication and Media at Focal Press, saw us through the initial preparation of the revision for the Eighth Edition. Our current Editor, Ross Wagenhofer, and Editorial Assistant, Nicole Salazar, at Routledge are extended a special thanks.

We are grateful to Sarah Thomas, our Production Editor at Routledge - Taylor and Francis Books in the UK and Carmen Baumann, Solutions Architect, CNS at Apex CoVantage for their expertise in the final publication process.

We look forward to working with everyone at Taylor & Francis in the years to come.

1

PRODUCTION PLANNING

1.1 INTRODUCTION

Why Audio?

To appreciate the importance of audio production, think about what happens when sound is not present. For example, if the sound portion of a television news report is lost, the report is cancelled. But what happens when the video (visual) portion of a television news report is lost? The report can still carry on with just the audio. While film relies on both sight and sound, even "silent" movies had musical accompaniment, and of course, radio is truly the "theatre of the mind," where *all* meaning must be conveyed and communicated by sound only. This concept is just as relevant considering the importance of audio on the Web, and the deployment of web-enabled audio technologies, from streaming audio to peer-to-peer file sharing, and podcasting.

Audio is arguably *the* most crucial element in radio, television, and film production, and audio production, in its many applications, is ever-present. Every production project (as well as every production course you take) will require audio in some form. As you grow to appreciate audio, take note that bad audio can destroy any production, no matter how good the visuals appear. At the same time good audio can help to salvage a production with weak visuals. Always remember that audio done right will support any production theme, just as audio done wrong will detract from the intended message of the production.

Audio Production Careers

Careers in audio and other media are virtually limitless. At a typical radio station, broadcast or online, those with audio knowledge are employed as board operators (as shown in Figure 1.1), air personalities (DJs), production personnel, news and sports anchors and producers.

Consider some of the opportunities available to you as an audio producer or editor, an audio studio engineer, a video or film postproduction editor, a concert sound reinforcement engineer, a music producer or editor, a production recordist/mixer, or a sound designer. The next time you watch a television program or movie, look at the credits and see all of the audio jobs listed. In these productions, dozens of audio professionals are employed as boom operators, sound recordists, sound mixers, automated dialogue replacement (ADR) personnel, sound editors, rerecording mixers, and Foley artists for sound effects generation.

With a broad knowledge and understanding of audio principles, you will be much more employable. As you learn the various applications of audio production, consider where audio is used and you will find there are numerous job opportunities in audio and audio-related areas. (See Sauls, S. J. (2007). *Basic Audio Production: Sound Applications for Radio, Television, and Film*, 2nd ed., Thomson Custom Solutions, publisher, pp. 1–1 & 1–2.)

Planning

Very little effective audio content is produced by accident; it usually involves careful planning. Some of this can be done in your head, merely by thinking through what you want to do. Other planning involves putting something down on paper so that both you and those working with you understand the needs of the project. Still other planning involves making sure that you can actually gather the elements you need for your program and that you do so legally and ethically.

What follows in this chapter is a number of steps that are often a part of audio production planning. Specific steps will differ from one project to another, but they all encompass concepts that you should consider as you start to put together an audio production.

1.2 THE IDEA

Ideas can come from anywhere—a book, a dream, a conversation with a friend, or an Internet site. One of the most common sources for ideas that end up on radio is the news of the day. It doesn't necessarily need to be major news however; it can be something tucked on the back page of a newspaper or on a blog that appears to have an interesting back-story that you want to explore. Depending on what your job position is, someone else may give you an idea and ask you to develop it. This is particularly true for reporters who are assigned story ideas by their producers or others in the newsroom. But reporters also need to be on the lookout themselves for ideas that will make meaningful stories.

With the Internet and its ever-more-powerful search engines, instead of searching for ideas, there are now ways to have them come straight to you that might be of interest to you and your audience. Google News (news.google. com), for example, allows you to personalize your own feed with stories from around the world that contain specific words or phrases. If your beat is health and wellness, for example, you might set up your Google News feed to give you stories that include the phrase "cancer treatment" or "antidepressant" or "avian flu." You can also set up Google News so that it automatically sends you email or text alerts whenever a new story with the words or phrases that you specify enters its system. Combined with a smartphone or other mobile devices, breaking news can search you out wherever you go.

In order to solidify an idea, you may need to undertake additional research. If you are preparing a story about the closing of a factory in a small town for instance, you will want to find out when the factory was built, how many people it employed at its peak, what effect its closing will have on the town, and so on. Some of this information may be on the Internet. For other bits of information however, you will need to search records in the city hall or interview people. As you do your research, verify your facts. A good rule of thumb is that you should get the same information from at least two, if not three, sources before you use it. In other words, if one person tells you that the owner of the factory embezzled money, don't include that information until you have verified it.

In today's information-overload society, being able to develop unique angles on stories can really make your career. Sometimes, a short news article might make a passing mention of a person or organization that is related to an event. Fortunately, radio—both on-air and online—lends itself more readily to somewhat lengthy interviews that provide perspectives on developing stories from individuals who are involved or affected by events. It's one thing to read about an athlete caught cheating... it's quite another if your audience can hear the story from the coach's perspective. If you are able to find willing

FIGURE 1.1 Board operator positions are a great way for you to gain experience in radio broadcasting and audio production early in your career. *(Image courtesy of Michael Parks, iHeartMedia-Harrisburg, PA.)*

interviewees who have strong opinions or personalities, all the better. But again, this approach can take some research and persistence. Some searching on the Internet and a few phone calls could yield an interesting and unique **sound bite**, if not a sizzling interview.

It is not always possible to research fast-breaking news, but you can prepare yourself for such situations by being well read and knowing the community from which you generate most of your ideas. Talk show hosts need to have (in their head) facts on a large number of issues. Disc jockeys need to research the background of artists whose music they play (see Figure 1.2). Sportscasters need to have a source for statistics so that they can craft interesting background stories during the course of a sportscast. Be sure to always do your research.

1.3 GOALS AND OBJECTIVES

Before you embark on the production of an audio project, you need to decide why you are producing it. At a very basic level, is your goal to entertain, inform, or educate, or a little of each? Knowing your motivation will affect how you put the piece together in terms of writing style, how many facts you include, and what type of music you use. At a more specific level, you can consider what action you want the listener to take. For example, let's assume that you are putting together a commercial for a new medicine that relieves tension. What do you want the person hearing it to do—buy the medicine, tell their friends

FIGURE 1.2 Disc jockeys need to undertake research so that they are well versed about the music they play. *(Image courtesy of Michael Parks, iHeartMedia-Harrisburg, PA.)*

about it, ask their doctor to prescribe it, or all of the above? If you are going to record a feature, how do you want the listener to feel while listening—happy, proud, or angry?

As you think about goals and objectives, you may want to consider some specific qualities of audio. Lacking a visual image to point the way, people create their own mental images, which often are more emotionally intense. Keep in mind, while it is possible for millions of people to listen to one radio broadcast simultaneously, audio is really a one-to-one communication process, where each listener feels a personal tie to what they are hearing. Focusing on the intense, personal communication process of audio can help you create ways to reach your goals and objectives.

1.4 THE TARGET AUDIENCE

In the modern world, where many media programs are niche-oriented, it is important to determine who you expect will listen to your production. The most common way that such **target audiences** are defined is through **demographics**—statistical information such as age, gender, income, nationality, and marital status. At an obvious level, if your audience speaks Spanish, your piece should be produced in that language. Generally, if your target audience is young, you want to make your material more fast-paced than you would if your audience is older. Surprisingly, you can sometimes use the same information

but package it differently for different demographics. A feature about the Oscars might be equally interesting to young and old—all you might need to do to get each group's attention is change the music you use in the background.

Your target audience may be defined by who you work for. For example, people who listen to National Public Radio (NPR) tend to be more liberal than people who listen to Fox News. That doesn't mean you can't present conservative views on a program for NPR, but you might need to include more background information about the topic or the guest.

Keep in mind that levels of detail are also important. For instance, if you are producing a story about the generic qualities of video games, but your intended target audience is professional gamers, your program will probably not do well because the professional gamers will be bored.

Consistency should also be a concern. Listeners who tune into a show for its light-hearted content might be turned off—for a day, or forever—if, for example, they encounter a lengthy segment about the weighty topic of self-esteem. Though listeners to this program may have issues with or even a desire to improve their self-esteem, that was not the impulse that drove them to push that button on their radios. This concept holds as true for the so-called "free-for-all" world of the Internet as it does for commercial radio. A podcaster whose topics do not have an element of consistency will soon find listeners pointing their browsers elsewhere.

1.5 STYLE

Another element to consider is style. Do you want to be serious, humorous, thought-provoking, or all of the above? Style can radically alter the appeal of content to various audiences. Content can be covered within an interview or banter between on-air personalities; it can be delivered in a dramatic skit with topical musical selections; or it could be an amalgamation of these and other forms.

What style you select will have a natural effect on the production—whether it ranges from serious to silly—and vice versa. Playing with style can create interesting and powerful effects for your audience. Delivering serious content in a lackadaisical manner can create dramatic tension on its own, even if your audience would be otherwise uninterested in the content. A quick and deliberate change in style can likewise create suspense. Of course, such breaks between content and form ought to be carefully considered and are usually appropriate only for programs that aim to entertain, rather than inform.

For some programs, such conflicts become a trademark—arguably even their very reason for existence. *Le Show*, produced by Harry Shearer, and the syndicated Phil Hendrie show are prime examples. Likewise, with shows such as

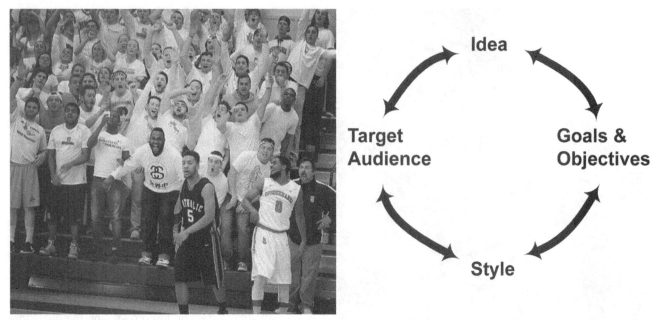

FIGURE 1.3 Study this picture long enough to think of an idea that you could use for an audio project. Then think through these questions: What would you need to research? What would be your goal for your project? Who would constitute the target audience? What style would you use? How did your idea change as you thought through the various questions?

AM radio's *Coast to Coast*, where topics can range from conspiracy theories to UFOs, it's hard to know when or whether the producers are aiming to entertain, educate, or simply titillate their audiences.

Many times, there is no clear, linear method to creating an audio production from scratch. Coming up with an idea, establishing goals and objectives, assessing the target audience, and determining style is a somewhat circular process (see Figure 1.3). Deciding that you want to make a program funny rather than serious may change your idea. Perhaps the first element you will consider is your target audience, and then you will search for an idea to appeal to that audience. The precise sequence of your thinking is not nearly as important as the need to make sure that you consider these elements before you get to the stages that cost money.

1.6 PRODUCTION PERSONNEL

The most expensive part of audio production is often the salaries of the people involved. You definitely need to consider how many people it is going to take to produce the audio product. The answer may be "one"—yourself. Audio material, because it is fairly simple and inexpensive to produce, can be completed by one person who comes up with the idea, hosts the program, engineers equipment, and records and edits.

If you possess the versatility to do it all, you will be more highly prized in the media industry than if you can only perform or only edit. The trend in the industry is to give projects to people who can handle all facets of them. Reporters, for example, not only search out the elements of the story and conduct interviews, but they operate the recording equipment and edit the package together into a meaningful whole (see Figure 1.4). As **convergence** of various media forms continues and media companies merge with each other, managers often look for people who can do all the jobs necessary to produce an audio product, and prepare information for other platforms such as newspapers, television, the Internet, and podcasting.

Many audio programs, however, are the result of teamwork. Quiz and variety shows on public radio employ more than 20 people per program. Although individual reporters contribute individual stories, a large team puts together an overall newscast or sportscast. Therefore, in addition to versatility, people who work in radio need to have good interpersonal skills so that they can work cooperatively and brainstorm with others for the good of the overall project (see Figure 1.5).

Audio producers also need to be able to write well—the main trait that professionals find lacking in their new hires. If you are a student wishing to get into audio, take every opportunity you can to improve your writing skills. If you can master the skill of writing a good **teaser** quickly, for example, you'll be in a position to wow a potential employer at a moment's notice. Radio lives and dies on teasers—they're what keep listeners tuned in through commercial breaks. Listen for teasers on your favorite radio program or podcast, then start reading news or feature stories in the newspaper or on the Internet and write some of your own. Read them out loud to yourself and

FIGURE 1.4 Radio news anchors undertake a multitude of tasks. Obviously, they read the news from a script, such as the one seen on the computer screen to the right of the anchor. They also do their own engineering, by operating an audio board and audio playback devices. They follow the list of items to be broadcast (seen on the screen to the left) and sign off that each element has in fact aired. They also look for new stories. The monitor to the far right can be used to check video feeds from television stations or news agencies and a monitor directly in front of the anchor can access databases and the Internet.

FIGURE 1.5 Brainstorming for an audio production project is a worthwhile experience that can yield beneficial improvements.

your friends. Sometimes in programming that aims to entertain, the tease can become more important than the topic itself—like a pitcher's elaborate windup that delivers a simple fastball, low and inside, sending the batter walking lazily to first base. Of course this kind of disconnect between tease and topic, if used repeatedly, can become annoying and off-putting for audiences—akin to the boy who cried "wolf."

Computer skills and the ability to troubleshoot technical problems are also high on the list of traits needed by audio people. They must also be curious and creative so that they can come up with infinite ideas, and they must be artistic enough to put the material together in a pleasing and often unique manner. Organizational skills are needed in order to proceed from idea to completion. A good sense of humor is also an asset—just as it is in most endeavors. There is some "hurry up and wait" in audio production, so people must balance their enthusiasm to get the job done with a temperament that includes patience. Another trait that has become important now that audio practitioners are expected to handle all aspects of production, is a pleasant-sounding voice and an ability to speak extemporaneously. Those wishing to get into the field should avail themselves of as much public speaking experience as they can.

Another characteristic that is common among most people who engage in audio production is passion. Be it radio, music recording, sound for visual production, or some other form, they love what they do. It's nice if some fame and fortune comes along, too, but the real motivator is a passion for audio.

1.7 PRODUCTION ELEMENTS

Once you settle on your idea, decide upon your goals and objectives, assess your target audience, determine your style, and gather together those who will be working on the project, you can get more serious about what you will need in order to put the audio piece together. There are three basic production elements that you have at your disposal: *voices*, *music*, and *sound effects*. Most productions mix several of these elements together.

Voices include the voice of the announcer or host of the program and sound bites from other sources. The announcer or host's voice is generally the most important element, because it conveys the basic information of the production (see Figure 1.6). Usually this is recorded first when putting a project together so that other elements can be recorded to fit with it. The voice will be discussed in more detail in Section 1.10.

Sound bites can come from a number of sources. You can find and interview people who have something to say that is relevant to your story and include their comments. Sometimes members of the general public come to you with newsworthy sound bites they were able to record by being in the right place at the right time. Often radio stations subscribe to services, such as the Associated Press (AP) or CNN Radio News, and have rights to use sound bites collected by those agencies. More and more media organizations have set up phone lines and social media portals where the public can provide their input and submit photos, video, and recordings for use within their programs (having to agree to the unlimited use of the materials in the process, of course).

Music can be the focus of an audio production; disc jockey shows are one example. At other times, music is in the background to help set the mood or convey additional information. For example, a sports show might feature a high-energy piece as its theme, and a Beatles track could

FIGURE 1.6 An announcer is an important part of many audio productions.

be used for a program dealing with some phenomenon of the 1960s.

Radio stations and most other audio production facilities have access to **music libraries** that include **compact discs (CDs)** of entire songs that they regularly play (or have played) on air as well as CDs of shorter pieces of music that can be used as background on productions. There are also many music library sites on the Internet from which you can download copyright-cleared music for a slight fee. Go to www.royaltyfreemusic.com for a list of music libraries. Production music, whether from a CD or an Internet site, is composed in a wide range of musical styles, including everything from symphonies to rap, and is often organized by its mood—suspense, romance, tragedy. Of course, if you are musically talented, you can compose music specifically for your production.

Sound effects also help set the scene of a production, but they are also used to augment and punctuate a point. For example, the sound of a telephone ringing or door opening creates an instant mental image. Morning radio show hosts often use unusual sound effects to increase the aural elements of their shows or to accompany some zany feature that they air on a regular basis. One way to classify sound effects is as "atmospheres" or "stingers." **Atmosphere sounds** are employed to create a natural environment, such as using seagull cries and crashing waves to set the scene at the seashore. Sound effect **stingers** are individual, short, sharp sounds designed to capture immediate attention, such as glass crashing or a gunshot.

Like production music, there are libraries of sound effects both on CD and the Internet. In addition to using these "canned" sounds, there may be times when you record natural sound live as part of the overall production or simulate a sound yourself, such as crinkling cellophane near a microphone to create a "fire" sound.

Another element you can consider using in an audio production is silence or **dramatic pause**. If you're trying to

highlight a specific copy point, don't be afraid to pause a little longer than normal before saying that phrase. Don't pause too long, however, or you will have **dead air**, which will be perceived by your listener as a technical problem rather than providing the effect that you're trying to achieve.

How you incorporate elements into your production depends on the type of production. A variety show may need live musicians, live sound effects, and people who can perform numerous voices for a mini-drama. A commercial might be composed of unusual sound effects never heard before. A sports show might have a stinger at the end of each segment. A film score might require naturally recorded sounds. Tap into your creativity to use the elements at your disposal to increase the effectiveness of your production.

PRODUCTION TIP 1A
Pick Any Two

There is a saying: "You want it fast, cheap, and good. Pick any two."

This adage definitely applies to audio production. If you do your program quickly and cheaply, it probably won't be very good. If you employ a large number of people so you can gather all the elements together quickly, you may get a good program, but it won't be cheap. If you settle for using equipment during brief periods of time when no one else is using it, you may be able to produce a good program cheaply, but it won't be fast.

The ideal, of course, is to have enough time and enough money to make the program as good as it can be. This rarely happens in the real world, but you should at least be aware of which element you are not engaging so that your life will be more bearable. Often people just breaking into the business have to work cheap (or for free) just to get a foot in the door, but that shouldn't last forever. There will certainly be times in your career when you will need to work long hours, but that should not become a way of life. If you have it in you to create quality work, you should be able to endure the fast and the cheap, especially if you realize that is what you are doing and see a way for the situation to be rectified in the not-too-distant future.

1.8 THE SCRIPT

In all probability, you will be writing a **script** at the same time you are collecting other elements for the program. You may think of an idea for the script that necessitates revising the sound effects you were planning, or you may find that you can't create a specific sound and have to change the script. In fact, as you are writing the script you may even change a few facets of your basic idea.

Basically, however, a script is a blueprint for your program. Different types of scripts have different content, but most scripts have a beginning, a middle, and an end. For a work of a dramatic nature, the beginning is usually an introduction to the characters, while the middle involves rising action and a climax. The end is a resolution to the story. A news story often starts with the voice of the reporter explaining the story, which then goes to a sound bite that elaborates and confirms major points, and ends with a wrap-up by the reporter. Many news stories are also often organized around the five Ws and an H; in other words, somewhere within the story it answers the questions who, what, where, when, why, and how. To that, some people add "so what?"—the need to explain the importance of the story. There are many books that cover the intricacies of writing scripts. Those who are seriously interested in audio production should avail themselves of this material.

There are standard forms for scripts in the industry. If you work in radio news, your news gathering organization will probably have a template that you can store in your computer and use for each story (see Figure 1.7). Although these differ from one facility to another, they generally include a **slug** (title of the story), the date, the length, and the names of the main people who need to work with the story in order to get it to air. These templates are usually set up in such a way that makes it easy for the writer to keep track of time—for example, each line equals approximately 5 seconds. Timing is very important in the radio business, and people who are on story assignments are usually given a length that their story should adhere to. Each story is typed on a separate sheet of paper or into its own computer file so that the stories can be juggled when they are put together for an entire newscast.

A talk or variety show where some elements are scripted word for word and others are spontaneous may use a **rundown sheet**, which lists the items in the order they are supposed to occur and gives an approximate timing for each (see Figure 1.8). If you are producing audio used in conjunction with video, you will probably encounter a **two-column script** with one column for audio and the other for the video that is to be shown as the audio is heard (see Figure 1.9).

1.9 PAPERWORK

Some audio productions need very little in the way of paperwork. You can probably keep in your head all the details needed for something that is short and that you do all by yourself. But other people can't read your mind, so you often need to put things on paper so that they can accomplish their jobs effectively and efficiently. If a program is complicated and involves a number of recording sessions, it is a good idea to prepare a schedule that lists what needs to be done and reminds you of any special needs the sessions may have (see Figure 1.10).

If you have guests on your show, you should have them sign **performance releases** that give you the right to use and edit their contributions (see Figure 1.11). The facility where you produce your work, including your school, may require you to fill out a form to reserve equipment, and the entity that is going to play whatever you produce (radio station, TV program, website) may request a production release that gives them permission to distribute the work.

1.10 THE IMPORTANCE OF VOICE

As noted earlier, the voice or voices heard in your audio production play a key role in conveying both the basic information and style of the piece. Although you can't control the voice of someone in a sound bite, you should give careful consideration to how your host, anchorperson, or announcer sounds. Voices that are generally considered to be "good" voices are strong in three areas. First, they tend to be in the lower pitch range and are full, resonant voices

---

SLUG: FIRE   REPORTER: JENKINS
DATE: 3/8   PRODUCER: KAMINSKI
LENGTH: 47 SEC.   ANCHOR: SANCHEZ

FIRE CREWS ARE ASSESSING THE DAMAGE AT THE BRISTOL STREET
WAREHOUSE TODAY AFTER A FIRE RAVAGED THROUGH THE FACILITY
EARLY THIS MORNING. THE FIRE APPARENTLY STARTED IN THE BASEMENT
OF THE BUILDING EITHER IN THE FURNACE AREA OR THE KITCHEN.
ALTHOUGH TWO OTHER BUILDINGS BURNED IN THIS AREA DURING THE
PAST YEAR, THE FIRE CHIEF DOES NOT THINK THE FIRES ARE RELATED.

SOT: 17 SEC. FIRE CHIEF: IN: FROM THE LOOKS OF THE FIRE...OUT:
...WE'LL SEE TOMORROW.

ARSON HAS NOT BEEN RULED OUT AS A CAUSE. TOM JENKINS REPORTING
FROM THE SCENE OF THE FIRE.

FIGURE 1.7 This is a sample script for a news story. Different news organizations use different forms, but most have something to keep their stories consistent.

TO YOUR HEALTH
"Honey and Cinnamon" – No. 35

No.	Talent	Slug	Source	Content	Segment Time	Total Time
1	Hannah	Tease	Studio	Are cinnamon and honey what you need to keep healthy? Some medical experts say "yes."	:05	:05
2	Jim Hogan recorded	Opening	Server	Music and VO	:30	:35
3	Hannah	Intro	Studio	Several new research studies have indicated that honey and cinnamon have healing and preventative powers. Today we have in our studio one of the doctors who conducted these studies, Dr. Haim Whetcome and one of the outspoken critics of this work, Dr. Theodore Axel. Welcome. Let me start with you Dr. Whetcome. What, specifically, have your studies shown?	:25	1:00
4	Hannah & Dr. Whetcome	Pro discussion	Studio	Dr. Whetcome discusses studies.	2:00	3:00
5	Hannah, Dr. Axel & Dr. Whetcome	Anti discussion	Studio	Dr. Axel, how do you react to this? Dr. Axel discusses his points of view.	2:00	5:00
6	Hannah, doctors, & listeners	Call-in questions	Phone and Studio	Questions from callers and answers by doctors	5:00	10:00
7	Commercials	Commercial break	Server	DinoAid, Preparation Hope, Xetrol, Baskin-Willow	2:00	12:00
8	Network anchors	Hourly news	Network	News provided by the network	3:00	15:00

FIGURE 1.8 A sample rundown sheet.

that are pleasing to hear. Second, your announcer should speak at a rate or pace that is easy to understand. Finally, the voice should have exceptional clarity, which usually means good articulation and pronunciation. Remember, *you* may very well be the "announcer" for your production and although a few people just have naturally good-sounding voices, almost everyone else can improve their vocal style. The most important vocal elements that you have a degree of control over include pitch, rate, tone, and volume.

Pitch is the highness or lowness of your voice and lower-pitched voices are more pleasant to listen to. Everyone has a natural pitch range, from low to high, and you should strive to be about one-quarter up from the lowest pitch at which you can speak. Male announcers, of course, have an advantage here, as their voices are typically about an octave lower than a female's. The best announcers utilize their full pitch range while emphasizing the naturally lower end of their speaking voice.

Rate is the number of words you speak in a given period of time. Typical "out loud" delivery is about 160 words per minute. This is usually not as fast as your normal conversational rate, but is slightly faster than you are used to when reading out loud. You're probably not an accurate judge of how fast you talk, so it's a good idea to record your speech and play it back to determine your rate as other people hear it. Of course, if you're dealing with a script, that may determine the speaking rate—a 60-second commercial must be read in 60 seconds, regardless of whether there are 140 words or 175 words.

A voice's **tone** is essentially the quality of sound that is made. You know that a guitar and piano can play the same notes, but you can easily tell the different tone quality between them because of the material that surrounds the strings that are vibrating to produce the sound. The human voice is similar and can take advantage of various resonators, such as the chest, larynx, or nasal cavity, to enhance the quality of sound. One of the best things an announcer

DUI: THE CONSEQUENCES

VIDEO	AUDIO
MARCIA	MARCIA: It's 2 AM. Do you know where your friends are?
SERVER: OPENING MONTAGE	MUSIC FROM SERVER: Runs :30. Ends with high note
MARCIA	MARCIA: DUIs are on the increase among teenagers. After a number of years of decline, they are once again rising. Arresting officers, who are on the front line, are in a position to have some explanations. We have two such officers with us today, Patrol Officer Wayne Polin and Patrol Officer Wendy Washington.
VARIOUS SHOTS OF MARCIA AND THE TWO OFFICERS	MARCIA AND TWO OFFICERS: Discussion of drinking and teenagers, ending with points about the importance of friends.
MARCIA	MARCIA: Let's take a look at a typical scene that demonstrates what the officers are talking about. SOUND EFFECT: Zing noise
RICHARD IN FOREGROUND WITH BAR SCENE IN BACKGROUND	RICHARD: Hey, it was Rachel's birthday. All we wanted to do was buy her a drink to help celebrate. It wasn't our fault.
ZOOM IN TO BAR SCENE	SOUND EFFECTS: Noisy bar talking MUSIC: "Revelry"
RACHEL, RICHARD, PHYLLIS, ALLEN AT TABLE	PHYLLIS (singing a little drunkenly): Happy birthday, dear Jennifer. SOUND EFFECTS: Noisy bar talking at softer level MUSIC: "Happy Birthday" at low level RACHEL (standing up unsteadily): Hey, thanks, guys, but I gotta go. I got another friend who thinks she's throwing me a surprise party. She'll be wiped out if I don't show up.
CU OF RICHARD	RICHARD: Rachel, are you sure you're OK to drive? MUSIC: "Happy Birthday" fades to "Ominous"
RACHEL	RACHEL: Sure, I've driven in far worse shape than this before. I'll be fine.

FIGURE 1.9 A sample two-column script, often used when audio is paired with video.

SCHEDULE

Title: Video Game Sounds

Director: Tony Caprizi

Date	Time	Location	Talent	Operators	Equipment	Description	Comments
9/6	9:00 AM– 11:00 AM	9th Street Arcade	Hector	Alex C. Shari K.	Shotgun mic Boompole DAT Flash recorder	Hector's intro and outro General arcade sounds	Check with manager upon arrival
9/6	12:00 PM– 2:00 PM	Electro-Arts Lab 1515 2nd St.	Hector Phil J.	Alex C. Shari K.	Shotgun mic Omni mic Boompole DAT Flash recorder	Demonstra-tion of latest sounds	Be sure to get oscillator
9/9	11:00 AM– 4:00 PM	Arts Museum 1520 East Parkway	Hector Phil J. Mary T.	Shari K.	Flash recorder	Interviews about latest sounds	
9/10	5:00 PM– 8:00 PM	Studio Edit Suite		Chris G.	Editing equipment Turntable	Edit open and close and some effects	Bring LP for "wow" effect
9/11	1:00 PM– 3:00 PM	Nolan C.'s home 926 Lake Lane	Hector Nolan C.	Shari K.	Flash recorder Lav mic Cardioid mic Boompole DAT	Interview about early sounds (and hopefully demo of equipment)	Keep mic/boompole in car until get permission to use it

FIGURE 1.10 A sample of a schedule for an audio production.

PERFORMANCE RELEASE

Note: This is not intended as an authentic legal document. It is just a representation of a performance release. Check with your organization's lawyer before using.

In consideration of my being part of the program

_____"The Philip Spear Hour"_____
    (title or subject)
and for no subsequent remuneration, I do hereby on behalf of myself, my heirs, executors, and administrators authorize

_____Sauls & Stark Productions_____
    (producer, station, or production company)
to use live or recorded my name, voice, likeness, biography, and performance for television or audio distribution throughout the world and for audiovisual and general education purposes in perpetuity.

    I further agree on behalf of myself and others as above stated that my name, likeness, biography, and performance may be used for promotion purposes and other uses. Further, I agree to indemnify, defend, and hold the producer (or station or production company) harmless for any and all claims, suits, or liabilities arising from my appearance and the use of any of my materials, name, likeness, biography, or performance.

Signature_____*Nedra Thomas*_____

Printed Name _____Nedra Thomas_____

Email Address_____Nedra@ozz.com_____

Date_____11/26/10_____

FIGURE 1.11   A performance release.

can do is to relax and let his or her body produce a natural, full sound. Using a high-quality microphone will also help produce rich tonal quality.

Volume is the loudness level of your voice and contributes to the energy or enthusiasm that your voice communicates to the listener. Another way to think of this is projection—pushing the sound out of your mouth. However, be careful of talking "at" the listener, rather than "to" the listener. If your voice is always "loud," you'll sound boring and uninterested. Good announcers use a variation in loudness and softness of their voice to help add to the interpretation of the copy they're reading. Also remember that a microphone and audio console can't compensate for a soft or weak voice. Although you can electronically turn up the volume, it will still have a lifeless quality if the proper natural volume is not being produced.

Good announcers utilize all these vocal qualities to their advantage and work to develop a pleasant, energetic voice. You will produce better audio if you look for just the right voice for your production or if you work to develop your own voice and announcing style.

PRODUCTION TIP 1B
Copy Marking

Copy marking means using a system of graphic symbols to help you interpret a script; basically, it's adding supplementary punctuation to the script. There's no universal system and many announcers develop their own, but here are a few ideas for copy marking.

Underline words or phrases you want to give extra emphasis. Double underline means more emphasis. Use a forward slash/ to indicate a pause. A double slash// means a longer pause, with each slash being about a "one-thousand-one" count. Triple periods ... and triple dashes --- are also often used to indicate a pause. A two-headed arrow >> under a word or phrases means to vocally "stretch it out," and a squiggly line ~ under a word or phrase means to read it more quickly. [Bracket a group of words] that you want to read as one complete thought. Use an arrow over a word or phrase to indicate a pitch change; bend the arrow upwards to show you should raise your pitch and downwards to lower the pitch of your voice. Words that might be difficult to pronounce are often "flagged" by putting a box around them and writing correct or phonetic pronunciation above the flagged word. Exclamation! and question? marks give an indication of how to read a sentence, but because they come at the end of the sentence, they can be missed. Put upside down punctuation at the beginning of this type of sentence to indicate how it should be read at the start of the sentence. If words or phrases are changed, draw a single line through the ~~old word~~ and write the new one above it.

Use parentheses to indicate other directions to the announcer, such as ("read with urgency.") Of course, whatever is within the parentheses is never actually spoken. Along with regular punctuation, copy marking will help you interpret a script in your own unique style.

1.11 EQUIPMENT AND FACILITIES

You will definitely need some specific equipment in order to produce your audio material and you will need a place (or places) to do the recording and editing. You will also need the skills to use the equipment and facilities so that what you produce is aesthetically pleasing and technically clean. You must, for example, be able to record material at a consistent volume level and then mix and edit material together so that it moves smoothly from one element to another and so that one element (such as music) does not drown out another element (such as an announcer's voice).

The remaining chapters in this book are devoted primarily to facilities, equipment, and aesthetics. By the time you finish reading the text material, using the worktext website, and undertaking the self-help projects, you should be ready to make your audio ideas become reality.

1.12 LAWS

There are many laws and regulations that govern the media, and if you are working for a radio station, there will probably be someone who has the overall responsibility for making sure the station abides by the law. However, everyone who produces material for radio, television, or the Internet should understand some basic legal principles. One principle that allows the media to engage in investigative reporting and the broadcast of controversial ideas resides in the First Amendment of the Constitution, which guarantees freedom of speech. This does not mean that anyone can say anything, however.

There are laws, for example, that prohibit **indecency** and **obscenity** on the airwaves. Indecency is defined as language that, in context, depicts or describes, in terms patently offensive as measured by contemporary community standards for the broadcast medium, sexual or excretory activities or organs. Obscenity, which is a more serious crime, is defined as material that contains the depiction of sexual acts in an offensive manner; appeals to prurient interests of the average person; and lacks serious literary, artistics, political, or scientific value. A number of radio and television broadcasters have been heavily fined for indecent comments and indecent lyrics in songs. One of the problems with indecency is that the definition deals with "contemporary community standards," and those change over time and place. What was considered indecent in Pennsylvania in 1990 may be commonplace in California now. One major legal issue has been to try to determine what constitutes indecency at any point in time.

Indecent material is protected speech under the First Amendment to the Constitution. Over time, however, the Federal Communications Commission (FCC) has established guidelines for the broadcasting of indecent speech. In general terms, stations are not allowed to air indecent material when there is a good possibility that children may be listening. The FCC has also recognized a "safe harbor" period, which is from 10:00 p.m. to 6:00 a.m. local time. Safe harbor refers to a time when it is safe to air indecent material. If there is any possibility that the FCC may consider the materials produced as indecent or obscene, station management and/or legal consul should be contacted in advance of the broadcast. (See Sauls, S. J. (2007). *Basic Audio Production: Sound Applications for Radio, Television, and Film*, 2nd ed., Thomson Custom Solutions, publisher, p. 9–11.)

Another legal principle that affects media is **libel**, which is saying something harmful and false about a person. Ordinary people are covered by a different standard than famous people who are considered **public figures**. An ordinary person can sue for libel if an unfavorable, false statement is made under any conditions, even if it was an innocent mistake. Public figures however, must prove that the reporter acted with **actual malice** in reporting the negative falsehood. For example, a public figure must prove that the reporter knew the report was false but broadcast it anyway.

Along the same line, there are **invasion of privacy** laws. These are state laws as opposed to national laws, so they differ from state to state. However, the laws usually allow for people to be left alone and stipulate ways that reporters can't pursue them, such as not trespassing on their private property.

There also is a law against **payola**, which is the acceptance (usually by a disc jockey) of money or gifts in exchange for favoring certain records or songs. If station employees receive money from individuals other than their employers for airing records, they must disclose that fact before they play the music under penalty of fine or imprisonment. The same principle holds true if a disc jockey promotes a business, such as a restaurant or car dealership; this is known as **plugola**. There are also laws against deceptive tricks, commonly called **hoaxes**, such as faking the report of a murder just for the fun of it.

Guiding Principles:

Journalism's obligation is to the public. Journalism places the public's interests ahead of commercial, political and personal interests. Journalism empowers viewers, listeners and readers to make more informed decisions for themselves; it does not tell people what to believe or how to feel. Ethical decision-making should occur at every step of the journalistic process, including story selection, news-gathering, production, presentation and delivery. Practitioners of ethical journalism seek diverse and even opposing opinions in order to reach better conclusions that can be clearly explained and effectively defended or, when appropriate, revisited and revised. Ethical decision-making – like writing, photography, design or anchoring – requires skills that improve with study, diligence and practice. The RTDNA Code of Ethics does not dictate what journalists should do in every ethical predicament; rather it offers resources to help journalists make better ethical decisions – on and off the job – for themselves and for the communities they serve.

Journalism is distinguished from other forms of content by these guiding principles:

Truth and accuracy above all

o The facts *should* get in the way of a good story. Journalism requires more than merely reporting remarks, claims or comments. Journalism verifies, provides relevant context, tells the rest of the story and acknowledges the absence of important additional information.
o For every story of significance, there are always more than two sides. While they may not all fit into every account, responsible reporting is clear about what it omits, as well as what it includes.
o Scarce resources, deadline pressure and relentless competition do not excuse cutting corners factually or oversimplifying complex issues.
o "Trending," "going viral" or "exploding on social media" may increase urgency, but these phenomena only heighten the need for strict standards of accuracy.
o Facts change over time. Responsible reporting includes updating stories and amending archival versions to make them more accurate and to avoid misinforming those who, through search, stumble upon outdated material.
o Deception in newsgathering, including surreptitious recording, conflicts with journalism's commitment to truth. Similarly, anonymity of sources deprives the audience of important, relevant information. Staging, dramatization and other alterations – even when labeled as such – can confuse or fool viewers, listeners and readers. These tactics are justified only when stories of great significance cannot be adequately told without distortion, and when any creative liberties taken are clearly explained.
o Journalism challenges assumptions, rejects stereotypes and illuminates – even where it cannot eliminate – ignorance.
o Ethical journalism resists false dichotomies – either/or, always/never, black/white thinking – and considers a range of alternatives between the extremes.

FIGURE 1.12 Selected provisions from the Code of Ethics and Professional Conduct of the Radio–Television Digital News Association. You might want to consider these while working on the ethics situations in Figure 1.13. The entire code can be accessed at http://rtdna.org/content/rtdna_code_of_ethics

Copyright is another legal issue that must be dealt with, particularly in relation to music. Most radio stations pay music licensing organizations for the right to use just about all the music that exists. If your production is going to be aired on radio, you probably do not need to worry about getting music copyright permission. You can also play the work within a classroom, but if you were to make a CD so that you could sell it, you would need to contact the copyright holder and secure permission—and probably pay a fee. If you use copyrighted material other than music, such as a poem, you need to secure copyright clearance for airing on radio.

There are numerous other laws that affect those producing audio material, but these are the ones you are most likely to encounter. A more detailed (but not exhaustive) look at copyright is presented in Chapter 12.

1.13 ETHICS

Whereas laws are codified and have set provisions and punishments that apply to everyone, ethics involves personal decisions regarding what is right and wrong. An ethical person, after considering the moral value of good and bad, will do what is good, even though there is no actual punishment for doing what is bad. Everyone pays lip service to being ethical. However, there are many temptations in life that lead to ethical misbehavior, in part because ethical behavior can be difficult to define. If there is a law about something (such as the payola law) then there are parameters and

punishments that are clearly spelled out. But, from an ethics point of view, what if an old high school friend, who just happens to have released a record, buys you a hamburger?

Ethics boils down to decisions individuals make, but to aid individuals who work in the media, organizations develop codes or guidelines that outline principles that a committee of respected practitioners have decided represent ethical behavior. Sometimes individual stations or networks have guidelines for their employees and often trade organizations issue codes for people within certain professions. Figure 1.12 lists some of the provisions in the code of the Radio–Television Digital News Association, an organization for electronic journalists.

But codes can't cover everything, and it is often up to individuals to make decisions—sometimes on the spur of the moment—regarding what is right and wrong. For example, if a man holding hostages calls a radio station and threatens to kill the hostages if he is not given immediate broadcast time to air his demands, what should you (as the person answering the phone) do? That's not the time to sit down and read the ethics manual or call for a meeting of top management. If you don't provide the airtime, the hostages may be killed, but if you do provide it, the demands may result in the death of more people and you will be encouraging other unbalanced people to demand airtime. It's a no-win situation and yet you must make an ethical decision.

The best preparation for making crucial ethical decisions is to have a set of values that you live by on a daily basis.

ETHICAL DECISION-MAKING

1. You and many other reporters are outside a hospital to which a famous celebrity has just been admitted for psychiatric care. A hospital employee comes over to you and confidentially tells you he has access to all patient files and can get you the complete medical records of this celebrity if you will pay him $100. In considering what you would do, you might want to think about the following options:
   - Pay him the $100 and await the files
   - Say you will pay him $100 after he delivers the files to you
   - Try to talk him down to $50
   - Suggest that he make his offer to a reporter from a competing station
   - Pay the money and get the file and then report the incident to the head of the hospital
   - Call your boss and ask what to do
   - Call the police
   - Scream
   - Pretend you don't hear him
   - Write a story about the fact that he approached you with the offer

2. You are putting together an audio production about automobile repairs and learn about a rather rare "ping" in a certain type of car that, if not fixed early, can lead to expensive damage. You manage to record this "ping" from a car that is in for repair, and plan to use the recording in the program. However, when you go to edit your program you cannot find the sound file that contains the "ping." You call the audio repair place to see if you can rerecord the sound, but the car with the "ping" has been fixed and returned to its owner. You go through a sound effects library and find a sound that is somewhat different but does sound like a "ping." Which of the following options, if any, would you consider?
   - Include the sound effects "ping"
   - Include the sound effect but give a disclaimer telling why you used a "ping" that was not authentic
   - Do the program without the "ping" sound
   - Put together the program without mentioning this "ping" problem at all
   - Accuse a co-worker of stealing the "ping" sound from your computer and tell her to find a way to rerecord it
   - Try to find another repair shop that happens to be working on a car with this "ping"

3. A committee of citizens is trying to recall the mayor from office. You don't like this mayor because he has not been at all accessible to the press, so you personally support the recall committee. You manage to catch the mayor before one of his rare news conferences and you ask him what he thinks of the recall effort. He says, "That recall committee is a bunch of thugs—I'm just kidding, of course. Don't broadcast that." What would you do under those circumstances? For example:
   - Follow the mayor's suggestion and do nothing—don't make it a story at all.
   - Broadcast the whole quote.
   - Broadcast just the first part of the quote, stopping with the word "thugs."
   - Broadcast the whole quote and mention in your report that anyone dumb enough to say such a thing to a reporter shouldn't be the mayor.
   - Give your recording to the recall committee and let them do with it as they will.

4. You are the producer for a commercial for a new restaurant in town and your station disc jockey has agreed to voice the commercial. During the recording of the commercial you overhear the restaurant manager asking the disc jockey to plug the restaurant when she is just chatting on the radio. He suggests that she particularly laud the roast beef special. The disc jockey seems somewhat reluctant, so the manager taps his wallet. What might you do at this point?
   - Nothing. This is a private matter between the disc jockey and the restaurant manager and is none of your business.
   - Nothing. No crime has been committed. Wait and see if you witness this being taken any further before acting.
   - Pull the disc jockey aside and remind her that plugging restaurants is considered payola.
   - Tell the restaurant manager that he should stop trying to get the DJ to commit a crime.
   - Call the police.
   - Call the general manager of your station and tell him what is going on.

FIGURE 1.13 Read through these situations and decide what you think would be the ethical thing to do in each instance. In some instances, you might want to do nothing or do more than one thing. Some ideas are included, but you will no doubt come up with ideas that are more comfortable for you.

When you hear about people who have had to make difficult ethical decisions, think what you would have done in their situation and have a rationale for it. If you have a good feel for what you think is right and wrong, you will find that making ethical decisions is easier. Figure 1.13 outlines some situations in need of ethical decisions. Decide what you would do in each case and talk over your decision with others.

1.14 CONCLUSION

Regardless of the type of material you are producing or the media form that will deliver it, you must do some advance thinking in order for it to be effective. Content is the most important element of audio production. You can work in a state-of-the art facility or in your garage and in all probability, your audience wouldn't know the difference. But they will know whether you have an effective idea that is well written, interestingly produced, clearly voiced, and legally and ethically based.

Self-Study

1. Where can ideas for audio projects come from?

   a) the news of the day
   b) conversations with a friend
   c) material in a book
   d) all of the above

2. In what way do ordinary people have different legal rights than public figures?

   a) Ordinary people do not have to prove actual malice in libel suits and public figures do.
   b) Ordinary people are subject to national invasion of privacy laws; public figures are subject to state invasion of privacy laws.
   c) Ordinary people don't have to obey copyright laws, and public figures do.
   d) Ordinary people are not allowed to perform hoaxes, and public figures are.

3. Which of the following is not a trait or skill that an audio production person should possess?

   a) convergence
   b) versatility
   c) equipment savvy
   d) sense of humor

4. What are short individual sounds intended to capture attention called?

   a) atmosphere sounds
   b) stingers
   c) sound bites
   d) dead air

5. Which type of script is most likely to include the five Ws and an H?

   a) news script
   b) rundown
   c) template
   d) slug

6. Which is a characteristic of a two-column script?

   a) It includes a pre-planning schedule.
   b) It gives a producer permission to use an interviewee's sound bites.
   c) It always has rising action and a climax.
   d) It is used when video and audio are recorded together.

7. Which term is defined as language that, in context, depicts or describes, in terms patently offensive as measured by contemporary community standards for the broadcast medium, sexual or excretory activities or organs?

   a) obscenity
   b) indecency
   c) libel
   d) payola

8. Which of the following would be considered a target audience?

   a) the radio network NPR
   b) people who do not fit any demographic
   c) a style that relates to theater of the mind
   d) married women in their thirties

9. Which of the following does ethical behavior involve?

   a) succumbing to life's temptations
   b) consideration of the moral value of good and bad
   c) using equipment in such a way that music does not drown out an announcer's voice
   d) codified laws

10. Which of the following refers to the number of words spoken in a given time period?

   a) pitch
   b) rate
   c) tone
   d) volume

ANSWERS

If you answered A to any of the questions:

1a. This answer is correct, but there is a better answer. (Reread 1.2.)
2a. Right. Only public figures must prove actual malice.
3a. Yes, this is the correct answer. Convergence involves media forms melding together and is not a personality trait or skill.
4a. No. Atmosphere sounds are not short or individual. (Reread 1.7.)
5a. Correct. A news script would answer who, what, where, when, why, and how.
6a. No. A schedule is not part of a script. (Reread 1.8 and 1.9.)
7a. No, but you are close. (Reread 1.12.)
8a. No. NPR might have a target audience, but NPR, itself, is not one. (Reread 1.4.)
9a. No, quite the opposite. (Reread 1.13.)
10a. No. Pitch refers to highness and lowness. (Reread 1.10.)

If you answered B to any of the questions:

1b. This answer is correct, but there is a better answer. (Reread 1.2.)
2b. No. One form of these laws does not exist. (Reread 1.12.)
3b. Wrong. Radio production people who can do many different things are prized. (Reread 1.6.)
4b. Yes. "Stingers" is the correct term.
5b. No. A rundown is a script form and may contain the five Ws and an H, but it is not the form most likely to contain them. (Reread 1.8.)
6b. No. A script is not a performance release. (Reread 1.8 and 1.9.)
7b. Correct. This is the definition for indecency.
8b. No. This is impossible. (Reread 1.4.)
9b. Yes, this is the correct answer.
10b. Yes, this is the right answer.

If you answered C to any of the questions:

1c. This answer is correct, but there is a better answer. (Reread 1.2.)
2c. No. Everyone should obey copyright laws. (Reread 1.12.)
3c. No. You need to be able to operate and troubleshoot equipment if you are involved with audio production. (Reread 1.6.)

4c. Wrong. Sound bites can capture attention, but they aren't individual sounds. (Reread 1.7.)

5c. No. A template is part of a script, but not a script. (Reread 1.8.)

6c. Wrong. Two-column scripts don't just deal with drama. (Reread 1.8.)

7c. No. You are not on the right track. (Reread 1.12.)

8c. No. This is totally unrelated. (Reread 1.3, 1.4, and 1.5.)

9c. No. This is something you should do, but it is a technical point, not an ethical one. (Reread 1.11 and 1.13.)

10c. No. Tone refers to quality. (Reread 1.10.)

If you answered D to any of the questions:

1d. Correct. Ideas can come from just about anywhere, including all of the options listed in a, b, and c.

2d. No. Hoaxes are out of bounds for anyone. (Reread 1.12.)

3d. No. A sense of humor is a good trait in most occupations. (Reread 1.6.)

4d. Wrong. Dead air is no sound at all. (Reread 1.7.)

5d. No. A slug is part of a script form but not a script itself. (Reread 1.8.)

6d. Yes. This is the correct answer.

7d. Wrong. You are not on the right track. (Reread 1.12.)

8d. Correct. This would be an example of a demographic you might use as a target audience.

9d. No. Laws are codified, but ethics aren't. (Reread 1.13.)

10d. Wrong. "Volume" refers to loudness. (Reread 1.10.)

Projects

PROJECT 1

Undertake production planning for a 15-minute interview show.

Purpose

To help you walk through the steps of preplanning with a project that is fairly simple.

Notes

1. You may not need to consider all things for your particular show. For example, it is possible that you won't need any sound effects, so you do not need to mention them.
2. You can write all your points continuously on the same sheet of paper. In other words, you do not need to write your idea on one sheet of paper and your goals on another sheet.
3. Some of what you write will be full sentences and some will be lists or clauses. Use whatever seems most appropriate.

How to Do the Project

1. Come up with an idea for your show and write it down in 25 words or fewer. Do research, if needed, to solidify your idea.
2. Decide on at least three goals or objectives for the program and write those underneath your show idea.
3. Write one sentence describing the target audience for your show.
4. Jot down the style of your show.
5. List the people, by job title, that you think you will need to help complete your show. Don't forget to list all the people who will be on the program. After each person, list one skill or personality trait that you think that person must definitely have.
6. List the elements you will need for your program—voices, music, and sound effects, as appropriate. Be as specific as you can; for example, don't just list "opening music." Indicate with a few adjectives the type of music you want.
7. Write a rundown sheet for your program. Be sure to include timing information.
8. Make a theoretical schedule for recording and editing your program.
9. List several elements of voice that will be important to your show and tell why they will be important.
10. If you think there are any legal or ethical issues that may arise given the content of your program, list these.
11. Give your paper with all this information to your instructor to receive credit for this project.

PROJECT 2

Assess your skills.

Purpose

To help you decide whether you have the skills needed for audio work and to indicate ways you might improve on your skills.

Notes

1. Answer this honestly. Perhaps your instructor will not collect this assignment, so that all students can be honest with themselves without worrying about their grades.
2. If you think of other traits that are important, feel free to add them.
3. Reread Section 1.6 if you need to refresh your mind on what each trait involves.

How to Do the Project

1. Make a chart and write at the top of it "Trait" and "Score."
2. Under "Trait," list the following:
   Versatility
   Interpersonal skills
   Writing skills
   Computer skills
   Troubleshooting ability
   Curiosity
   Creativity
   Artistic ability
   Organizational skills
   Sense of humor
   Patience
   Pleasant-sounding voice
   Ability to speak extemporaneously
   Passion
3. Under "Score," give yourself a rating on each trait from 1 to 10, with 1 being the lowest and 10 being the highest.
4. Consider your two lowest scores and write a short paragraph discussing how you could improve in each of those areas. Also, look at your two highest scores and write a short paragraph about why those are your strengths.
5. Give the paper with your scores and paragraphs to your teacher, if instructed to do so, to complete this project.

2

THE STUDIO ENVIRONMENT

2.1 INTRODUCTION

As you'll learn as you read through this book, audio production can be accomplished in a variety of situations and in a number of different environments. Recording "in the field" may simply entail using a single microphone and portable MP3 recorder or perhaps a setup consisting of a laptop, audio software, and a microphone. On the other hand, recording in the studio may entail multiple microphones and other sound sources feeding into a large audio console or just a basic editing suite. This chapter takes a look at the more formal setting of the traditional audio studio.

The room that houses the equipment necessary for audio production work and in which the finished product is assembled is known as the **production studio**. What may initially appear to be merely a roomful of electronic equipment will become a comfortable environment once you've become familiar with the space and components that make up the production facility. If your facility has several studios, they may be labeled "Production 1" or "Prod. B" or identified with another abbreviation for the production studio, "PDX." Today, a streamlined digital "studio" may merely be a workstation desk set up in the corner of a room with a mix of computer and audio equipment, as shown in Figure 2.1.

Most radio stations utilize at least two studios. One is usually delegated as the **on-air studio** and is used for live, day-to-day broadcasting. The others are audio production studios, used for putting together programming material that is recorded for playback at a later time. This includes such items as commercials (often referred to as "spots"), features, **public service announcements (PSAs)**, and station promotional or image pieces (**promos**). Regardless of the actual physical size or shape, the production facility is the creative center for a radio station or audio production house. Often the production studio mirrors the on-air studio with the same or very similar equipment configuration

and serves as a backup for the on-air room. Some facilities also have a studio that is considered a **performance studio** or **announce booth**. It usually is smaller than the other studios and houses nothing more than microphones, headphones, copy stands, a table, and chairs. The output is normally sent to a production studio to be recorded, although sometimes it's sent directly to the on-air studio for live broadcast. A performance studio can be used for voice-over work, for taping interviews, for discussions involving several guests, or for recording a small musical group or production sound effects.

Two of the biggest concerns for studio design looked at in this chapter are acoustics and ergonomics. **Acoustics** refers to how sound "behaves" within an enclosed space, and **ergonomics** refers to design considerations that help reduce operator fatigue and discomfort. Acoustical considerations have become increasingly important because of the high-quality recordings that can be obtained within a digital environment. Although you may never build or

FIGURE 2.1 An audio production environment can be a small editing suite that combines computer and audio equipment. *(Image courtesy of Motor Racing Network Radio.)*

remodel an audio studio, an understanding of the characteristics of the production room can help you assess your facility and suggest ways in which you can improve the surroundings that you'll be working in.

2.2 THE AUDIO CHAIN

Figure 2.2 shows a simplified "map" of a typical audio production studio and how a series of audio equipment is interconnected. Starting with various sound sources, such as a microphone or an audio recorder, it shows the route or signal path that sound takes to ultimately be broadcast or recorded. This path is often called an **audio chain**, because the various pieces of equipment are literally linked together. The trip can be complicated, because the sound can go through several changes along the way. For example, it can be dubbed, or copied, from CD to a recorder; or it can be **equalized**, which is a form of signal processing. The solid lines show sound being sent to the audio console from an audio source. Then it goes through **signal processing** equipment and finally to the transmitting system, which would be normal for an on-air studio. The broken line shows the sound being sent back to an audio recorder after signal processing, which would be common for an audio production studio. In both cases, the sound can be heard in the studio through monitor speakers or headphones. You'll learn more about all of this as you work your way through this text, but for now the diagram in Figure 2.2 provides a look at where you are headed.

The equipment shown is also representative of what could be found in the typical audio production studio. A **microphone** transforms the talent's voice into an audio signal. It is not uncommon for a production facility to have one or more auxiliary microphones for production work that requires several voices. Most production rooms also have two **CD players**, enabling different CDs to be played back-to-back or simultaneously. Besides CD players, many production studios utilize various digital record and playback gear, such as the **MiniDisc (MD)** recorder (as described in the Appendix), computers with audio software programs, **compact disc recorder (CD-R)**, or **digital audio workstations (DAWs)**. How many recorders or players are found in the production room depends on the complexity of the studio and the budget of the facility. All of this equipment feeds into the **audio console**, which allows the operator to manipulate the sound sources in various ways. Signal processing equipment, such as an **equalizer**, **noise-reduction** system, or **reverb** unit, is usually put into the audio chain between the audio console and the transmitting or recording equipment; however, most signal processing is now done in postproduction using specialized audio software. Monitoring the sound during production work is accomplished with studio **speakers** or **headphones**. All of these features are further discussed throughout the text. (Compare Figure 2.2 with Figures 2.5 and 2.10 to see how the audio chain translates into the actual production studio.)

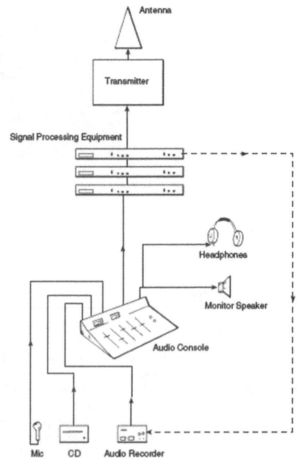

FIGURE 2.2 The audio chain shows how equipment is interconnected and how sound moves through that equipment in an audio studio.

2.3 THE STUDIO LAYOUT

Many audio production studios use a U-shaped layout (see Figure 2.3) or some variation of it, because this allows the operator to reach all the equipment control surfaces, and puts the operator immediately in front of the audio console. With the use of remote start/stop switches for any equipment that's out of the operator's reach, all equipment manipulation can occur at the audio console once everything has been set up and cued. Today most radio work is done **combo**; that is, the announcer is also the equipment operator (see Figure 2.4). Because of this setup, the equipment and operator are in a single studio, be it a production or on-air room. This type of studio layout facilitates the working combo. (If you watch re-runs of the television sitcom *WKRP in Cincinnati*, this method is also employed.) In earlier radio days, the announcer was often located in a separate room or announce booth adjacent to the studio that housed the actual broadcast equipment. Visual contact and communication were maintained via a window between the two rooms. An engineer was required to manipulate the equipment, and the announcer merely provided the voice. Many larger-market radio stations still use a similar announcer/engineer arrangement, popularly known as **"engineer-assist"** broadcasting. (The television sit-com *Frasier* used this method.)

One of the basic ergonomic considerations in putting the studio together is whether it should be a sit-down or stand-up design. Sit-down studios would have countertops at desk height and would include a chair or stool for the announcer. As the name implies, stand-up operations have counter height set for the announcer to be standing while doing production work. If you're designing a stand-up studio, make sure that even the shortest person can reach the equipment, especially equipment in turrets or countertop modules. Stand-up allows more movement and tends to provide more energy in delivery. One approach isn't really better than the other, so it is a personal choice of the individual facility.

2.4 PRODUCTION STUDIO FURNITURE

Studio furniture provides the foundation for the production studio, because all the equipment in the room sets on it, mounts in it, or is wired through it. Studio equipment is often installed on and in custom-built cabinets and counters. Although the cost can be high, such furniture can be built to the exact dimensions of the studio and for the exact equipment that will be housed in that studio. A less expensive but equally functional approach is to lay out the studio using modular stock components (review Figure 2.3). Audio studio furniture has been designed expressly for recorders, audio consoles, and other pieces of studio equipment. Using modular furniture and racks often makes it easier to reconfigure the studio or add additional equipment if the studio needs to expand.

Today's studio furniture systems also include space and cabinet modules for computer monitors and other computer equipment that's being integrated with traditional equipment in the audio studio, as shown in Figures 2.5 and 2.10. A computer monitor should be about 2 feet away from the announcer and the topmost screen line should fall slightly below eye level. Monitors that are placed too high, such

FIGURE 2.4 On-air combo operation. *(Image courtesy of Michael Parks, iHeartMedia-Harrisburg, PA.)*

as on top of a studio module, can cause neck strain. Some monitors can be kept off the studio furniture by using a special wall-mounted or ceiling-mounted TV boom. Flat-screen monitors offer more mounting options, take up less space, are aesthetically pleasing in the audio studio, and have become the standard in most audio studios. If possible, the computer keyboard should be placed in line with the monitor rather than off to the side. Sometimes the keyboard can be placed on an under-counter drawer to accomplish this, but you have to watch the operator's knee space in such a setup. If possible, avoid putting a keyboard near hard counter edges that can cause a painful problem if the operator's wrist strikes the counter edge, and make sure that the computer mouse can be reached without stretching the operator's arm.

Most studio furniture is manufactured of plywood or particleboard with a laminate surface; however, a few modern counters are employing a solid-surface countertop of Corian or similar kitchen-counter type material, as shown in Figure 2.5. Both custom-built and modular cabinets and counters are also designed to provide easy access to the myriad cables necessary to wire all the studio equipment together yet maintain an attractive image for the look of the studio. Digital equipment offers the advantage of better cable management, as linking equipment via digital inputs/outputs requires less cable than analog wiring. Other cabinets or storage modules are also available for CDs and other material that may be kept in the production studio. Furniture housing equipment may require cooling, but most digital equipment will operate fine with a passive air flow provided by back panel vents in the furniture. If a forced-air fan is required, be aware of the noise problem it could present.

Does stylish furniture make a studio sound better? Although that notion would be hard to quantify, a positive studio image

FIGURE 2.3 Many audio studios use modular furniture components arranged in a U-shaped design that allows the operator to see and reach all the equipment easily. *(Image courtesy of Graham Studios.)*

FIGURE 2.5 A solid-surface countertop on studio furniture provides the audio studio with a sleek, modern look. *(Image courtesy of Mager Systems, Inc.)*

does imply a commitment to high-quality professional production, and this often translates into more creativity, more productivity, and a better "sound" produced from that studio.

2.5 STUDIO SOUND CONSIDERATIONS

The audio production studio is a unique space, in that the physical room will have an impact on the sound produced in it. Because of this, several characteristics of sound should be considered in designing the studio, including sound isolation, noise and vibration control, and room acoustics. When sound strikes a surface (such as a studio wall), some of that sound is reflected, while some is absorbed within or transmitted through the material of the surface. Most of the sound that hits a hard, flat surface will be reflected. However, if the surface is irregular, it will break up the sound wave and disperse the reflections—a phenomenon known as **diffusion**. Sound that's absorbed into the surface is dissipated within it, but **penetration** occurs when sound goes through a surface and is transmitted into the space on the other side. Figure 2.6 illustrates that penetration, absorption, reflection, and diffusion are all characteristics that help determine the sound that is both produced and reproduced in the studio.

When a sound (such as a talent's voice) is produced, the **direct sound** is the main sound that you hear. In a production situation, it is sound that goes from the sound source straight to the microphone. On the other hand, **indirect** or **reflected sound** reaches the microphone fractions of a second after the direct sound does because it has traveled a circuitous route. Reflected sound consists of **echo** and **reverberation (reverb)**. This indirect sound has bounced off or been reflected from one surface (echo) or two or more surfaces (reverb) before reaching the microphone (see Figure 2.7). Because it's an early reflection, echo provides a distinct repetition of the sound, such as "hello—hello—hello." On the other hand, a reverb's repeated later reflections provide a continuous decay

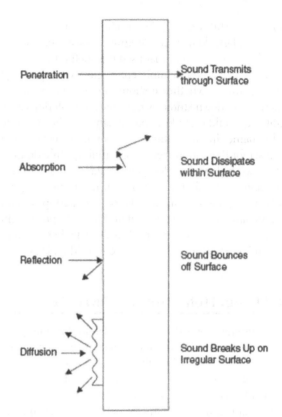

FIGURE 2.6 Sound striking an audio studio wall will reflect off, penetrate through, be diffused by, or be absorbed by that surface.

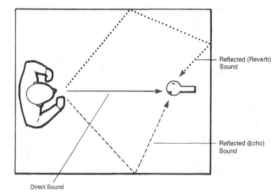

FIGURE 2.7 Direct sound takes a straight path from the talent to the microphone, but reflected sound is also produced in the audio studio.

of the sound, such as "hello—oo—oo." The components of direct and indirect sound make up what is commonly called the sound's **life cycle**.

In designing the audio studio, the goal is to manipulate these sound characteristics to create a proper sound environment for production work. When considering reflected sound, we think in terms of reverb ring and reverb route, with the same concepts applying for echo but to a lesser extent. **Reverb ring** (or **reverb time**) is the time that it takes for a sound to die out or go from full volume to silence. **Reverb route** is the path that sound takes from its source to a reflective surface and back to the original source (or a microphone, if recording). Excessive reflected sound tends to accent high and midrange frequencies, which produces a "harsh" sound; to blur the stereo image, which produces a "muddy" sound; or to cause standing waves (see Section 2.7), which produces an "uneven" sound. Reflected sound can also be **reinforced sound**, which causes objects or surfaces within the studio to vibrate at the same frequencies as the original sound in a sympathetic fashion.

Both **absorption** and diffusion are used to control reflected sound. Part of the reflected sound can be absorbed within the carpeting, curtains, and walls of the studio. Absorption soaks up sound and shortens reverb time to prevent excessive reflection. Absorption provides a **dead studio**, which has a very short reverb ring (sound dies out quickly) and a long reverb route that produces a softer sound. Excessive absorption produces a totally dead studio, which provides a "dry" sound that is unlike any normal acoustic space and isn't really desirable. In contrast, a **live studio** has a longer reverb ring and a shorter reverb route that produces a harder or more "brilliant" sound. Diffusion uses irregular room surfaces to break up sound reflections. This decreases the intensity of the reflections, making them less noticeable, but doesn't deaden the sound as much, because the sound reflections are redirected rather than soaked up. Most studio designs control reflections by a combination of absorption and diffusion techniques.

One common audio studio design is a **live end/dead end (LEDE)** approach. The front of the studio (where the operator and equipment are located) is designed to absorb and diffuse sounds. This dead end quiets some of the equipment operation noise, picks up the direct sound of the talent's voice, and absorbs the excess reflections that pass by the microphone from the live end. The live end, or back, of the studio adds a desirable sharpness to the sound by providing some reflected sound so the studio isn't totally dry. Other acoustic designs include early sound scattering (ESS), which uses a great deal of diffusion, and reflection-free zone (RFZ), which uses a great deal of absorption to control unwanted reflected sound in the studio.

2.6 STUDIO CONSTRUCTION MATERIALS

Another design consideration involves the actual construction materials used for the studio. Ideally, you want to keep penetration to a minimum by keeping outside (unwanted) sound from entering the studio and inside sound from escaping from the studio, except via the audio console. Audio studios utilize **soundproofing** to accomplish this sound isolation. Doors are heavy-duty and tightly sealed; windows are often double-glassed with the interior pane slanted downward to minimize reflected sounds; and walls, ceiling, and flooring are covered with special sound-treatment materials. For example, studio walls may be covered with acoustically treated and designed panels that both absorb and trap reflected sounds (as shown in Figure 2.8(A)). Some stations use carpeting on the studio walls, but this type of soundproofing doesn't absorb low frequencies very well. In the past, some production studios used

FIGURE 2.8 Acoustic panels and tiles help control reflected sound through both absorption (by the foam material) and diffusion (by the irregular surfaces). *(Images courtesy of Michael Parks, iHeartMedia-Harrisburg, PA; Erikk D. Lee, Auralex Acoustics, Inc.)*

egg cartons on the walls as a sound treatment. If you compare the design of an egg carton with the design of the acoustic panel shown in Figure 2.8(B), you'll see why the inexpensive egg carton route worked—to a degree.

All materials absorb sound to some degree, but each material will have a different **absorption coefficient**, which is the proportion of sound that it can absorb. A coefficient value of 1.00 indicates that all sound is absorbed in the material. On the other hand, a coefficient value of 0.00 means that no absorption occurs and that all the sound is reflected back. Hard, smooth surfaces like plaster or panel walls and hardwood floors have low absorption coefficients. Heavy, plush carpets, drape-covered windows, and specially designed acoustic tiles have higher coefficients. For example, using a 1,000 Hz tone as the sound source, the absorption coefficient of a sheet rock wall is 0.04, and that of a 2-inch Sonex foam tile is 0.81; the absorption coefficient of a glass window is 0.12, and that of a window curtain is 0.75; the absorption coefficient of a painted concrete block wall is 0.07, and that of a carpeted concrete wall is 0.37. The purpose of any soundproofing material is to help give the studio a dead sound. Soundproofing absorbs and controls excess reverb and echo and helps produce a softer sound.

In order to prevent unwanted sounds from entering a recording studio, quite often a type of **sound lock** is incorporated in the studio design. A sound lock is a small area

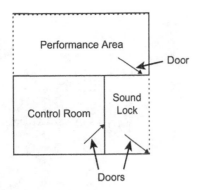

View is from above.

FIGURE 2.9 Studio sound lock. *(Figure courtesy of Sauls, S. J. (2007). Basic Audio Production: Sound Applications for Radio, Television, and Film, 2nd ed., Thomson Custom Solutions, publisher.)*

FIGURE 2.10 Any audio studio should be a comfortable and functional environment. *(Image courtesy of Arrakis Systems, Inc. www.Arrakis-Systems.com.)*

located outside both the control room and performance area that captures sound and will not allow it to pass through. A simplistic design of a sound lock is presented in Figure 2.9. (See Sauls, S. J. (2007). *Basic Audio Production: Sound Applications for Radio, Television, and Film*, 2nd ed., Thomson Custom Solutions, publisher, p. 1–13.)

2.7 STUDIO SIZE AND SHAPE

The size and shape of a production studio can also determine how reflective the studio is. As noted, the audio production studio shouldn't be overly reflective, because sound produced or recorded would be too bright and even harsh. Unfortunately, standard room construction often counters good studio design. For example, studios with parallel walls (the normal box-shaped room) produce more reflected sound than irregularly shaped studios. Sound waves that are reflected back and forth within a limited area, such as between studio walls that are parallel, can produce standing waves. In basic terms, a **standing wave** is a combination of a sound wave going in one direction and its reflected wave going in the opposite direction. If the distance between the walls is the same as the wave length (or a multiple of it), the waves interact and produce an undesirable combined sound that tends to be uneven, as previously mentioned. To help prevent standing waves, adjacent studio walls can be splayed (joined at an angle of more than 90 degrees) to help break up reflected sound and control excessive reverb and echo. The wall must lean so that the wave will bounce off and be absorbed into the carpet. Concave walls tend to collect sound, while convex walls tend to push away sound. Many people are surprised how "dead" it sounds in a recording studio (due to the lack of standing waves). Areas with many standing waves are known as "live" rooms, which are undesirable.

The actual size of the production facility is partially determined by the equipment that must be housed in it. However, in constructing the audio production room, consideration should be given to the fact that when rooms are built with height, width, and length dimensions that are equal or exact multiples of each other, certain sound frequencies tend to be boosted, and other sound frequencies tend to be canceled. As this "peaks and valleys" sound is not desirable in the audio production room, cubic construction should be avoided when possible.

2.8 STUDIO AESTHETICS

There are some studio design considerations that can be categorized as the "aesthetics" of the production room. In general, any studio should be pleasant to work in; after all, the operator may be confined to a rather small room for long stretches of time. For example, fluorescent lighting should be avoided when possible. Not only does it tend to introduce "hum" into the audio chain, but it's also harsher and more glaring than incandescent light. If possible, the studio lights should be on dimmers so that an appropriate level of light can be set for each individual operator. If the lighting causes glare on a computer monitor, use an antiglare shield or screen to diminish this problem or use track lighting that can be directed away from the screen.

Stools or chairs used in the audio studio should be comfortable and functional. Userfriendly adjustments should allow the user to set the seat height so that the entire sole of the foot rests on either the floor or a footrest. Chairs must move easily, because even though most of the equipment is situated close to the operator, they may have to move around to cue CDs or speak directly into the microphone. The production stool must also be well constructed so that it doesn't squeak if the operator moves slightly while the microphone is open. This may not be a factor for production studios designed for a stand-up operation in which there is no stool, and with counters at a height appropriate for the operator to be standing while announcing. As mentioned earlier, the stand-up operation allows the operator to be more animated in his or her vocal delivery and actually provides a better posture for speaking than a sitting position.

Many radio production rooms are decorated with music posters or radio station bumper stickers and paraphernalia. Not only does this feature keep the studio from being a cold, stark room, but it also gives the studio a radio atmosphere. Figure 2.10 shows the interior of a typical radio production studio. An audio production facility may decorate with paintings or art posters to give the facility a more business-like, professional feel.

PRODUCTION TIP 2A
Static Electricity

Static electricity can be a problem in production studios, because of the heavy use of carpeting. Most people don't enjoy getting shocked every time they touch the metal control surface of an audio console. Also, some modern audio equipment and computer systems have electronic circuits that can be disrupted by static discharges. There have been instances where audio recorders have switched into "play" mode when an operator just "sparked" the faceplate of the machine. If design factors can't keep the studio free from static, commercial sprays can be put on the carpeting or spray fabric softener can be used to provide an antistatic treatment at a modest cost. Dilute the fabric softener a bit or you'll build up a dangerous, slippery gloss on your carpets. A static touch pad can also be used in the studio—the operator merely places a finger on the pad to harmlessly discharge static buildup. Some studios have even been built with conductive laminate countertops connected to the studio's ground system to help minimize static.

2.9 ON-AIR/RECORDING LIGHTS

On-air lights (see Figure 2.11) are usually located outside the audio production room or studio. Normally they are wired so that whenever the microphone in that studio

FIGURE 2.11 When lit, the on-air or recording light indicates that a microphone is "live" in that audio studio.

is turned on, the on-air light comes on. A light outside an audio production studio will often read "recording" instead of "on air." In either case, a lit light indicates a live microphone.

Good production practice dictates that when an on-air light is on, you never enter that studio, and if you're in the vicinity of the studio, you are quiet. Inside the studio, another alert light may come on when the microphone is turned on and the announcer often says "Standby" to alert others in the studio that he or she is preparing to turn on the microphone.

2.10 HAND SIGNALS

Hand signals don't play a major role in modern audio production; however, there are situations when vocal communication isn't possible and hand signals are necessary. For example, if a voice-over talent and engineer are recording from adjacent studios with a window between them, as mentioned earlier in this chapter, they must be able to communicate with each other. There are also times when two announcers must communicate in a studio, but an open or live microphone prevents them from doing so verbally. Because of situations like these, hand signals have evolved over the years to communicate some basic production information. Figure 2.12 shows some of the basic hand signals.

Often hand signals concern getting a program started or stopped. A **standby** signal is given just prior to going on air and is immediately followed by the **cue talent** signal. To convey "You're on," this cue is given by pointing your index finger (using the same hand that gave the standby signal) at the person who's supposed to go on air. The common signal for stopping a program is the hand across the throat gesture or **cut** signal. This signal terminates whatever is happening at the moment and usually "kills" all live microphones and stops all recorders. Some hand signals are used to give directions to the talent regarding the microphone. To get the talent to **give mic level**, for example, hold one hand in front of you and open and close the thumb and fingers in a "chattering" motion to indicate that he or she should talk into the microphone so that levels can be checked.

Other hand signals are often used during a production to let the talent know how things are going or to convey some necessary information. Timing cues are given with the fingers with each finger indicating 1 minute. Using both index fingers to form a cross in front of you means there are 30 seconds left. Timing cues always indicate how much time remains in the program because there is nothing you can do about the time that has already gone by. When everything is going fine, the radio hand signal is the traditional "thumbs-up" given with clenched fist and extended thumb or the "OK" of the circle and three fingers. There is no universal set of hand signals, so you may find that your facility uses some that are different, uses

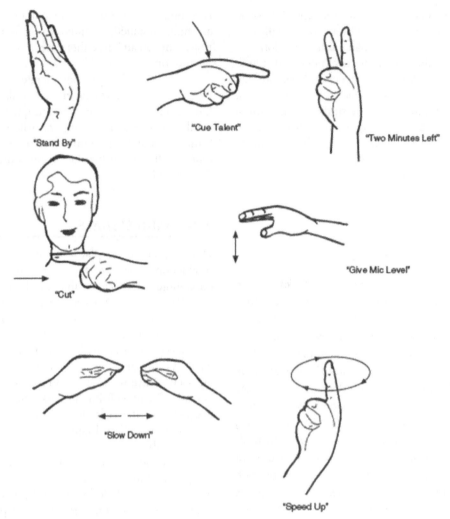

"Stand By"

"Cue Talent"

"Two Minutes Left"

"Cut"

"Give Mic Level"

"Slow Down"

"Speed Up"

FIGURE 2.12   Hand signals allow information to be conveyed in the audio studio when a "live" microphone prevents verbal communication.

some not presented here, or doesn't use any at all. In any case, an understanding of hand signals should prove helpful in certain production situations.

2.11 NOISE AND DISTORTION

Noise is inherent in any of the electronic equipment housed in audio production studios. The term **noise** refers to any unwanted sound element introduced in the production process that was not present in the original sound signal. For example, a microphone that employs an extremely long cable might add noise to the audio signal. Recorders can introduce noise from mechanical gears or just through the electronic circuits used in amplifying or recording the signal. In all audio production, the noise level should be kept as low as possible. Most audio equipment is designed to produce a **signal-to-noise ratio (S/N)** of at least 60 to 1. S/N is an audio measurement, usually in decibels, that specifies the amount by which a signal at a standard test level exceeds the level of noise produced by an electronic

component. The higher the signal-to-noise level, the better. For most analog equipment, an S/N of around 60 dB is considered good quality; modern digital equipment can show an S/N ratio of 98 dB. The S/N gives an indication of the equipment's ability to reproduce sound cleanly.

Distortion is an unwanted change in the audio signal due to inaccurate reproduction of the sound. One example is loudness distortion, which can occur when a signal is recorded at a level that is too loud for the equipment to handle. An overdriven or too loud signal sounds "muddy," and the reproduced signal does not have the same clarity or sharpness as the original. You should be aware of noise and distortion when working with audio equipment, especially analog equipment. Digital equipment frequently reduces the chance of introducing noise into your production work, but sometimes your work will be accomplished using a combination of analog and digital equipment.

Analog signals of 100 percent would equate to digital signals at −12 dB. This provides "headroom" from −12 dB to 0 in the digital domain. (Actually, there is no agreed upon standard for the relationship between 0 VU and its

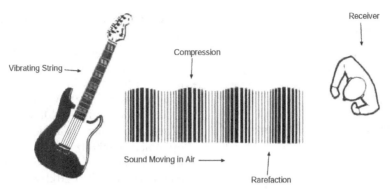

FIGURE 2.13 The production of sound requires vibrations, which are transmitted through a medium to a receiver.

equivalent in the digital world. Panasonic, for instance, states –18 dB for their professional machines, while Sony suggests –12 dB.) If the signal goes over 0 in digital, it will "clip" and will be distorted. Because of its "discrete" processing, there is basically no signal toleration above 100 percent saturation in the digital realm. As there is no headroom in digital recording, audio producers must allow some into their recordings. On the other hand, analog recording is more forgiving and allows for total saturation with little distortion if the signal peaks at just above 100 percent, as discussed in Chapter 5. This, along with the natural sound, is one of the main reasons some audio engineers still prefer to master (originate) in analog recording.

2.12 IS IT A SOUND SIGNAL OR AN AUDIO SIGNAL?

We've already mentioned how sound acts in the audio studio. It is worth your while to continue to consider sound, as this will help you understand many aspects of the production process discussed in the rest of the text. Much of what happens in the production studio has to do with manipulating sound, whether it involves a sound signal or an audio signal. When sound is naturally produced (for example, when talent speaks into a microphone), we think of that sound (his or her voice) as a **sound signal**.

In audio production, when that sound signal is then manipulated electronically (such as recorded into a digital recorder), it is called an **audio signal**. Obviously, most audio production must start at some point with a sound signal, but during the actual production process, we are often recording and manipulating an audio signal. To further complicate things, these terms are sometimes interchanged when people talk about various production processes.

2.13 SOUND DEFINED

When something vibrates, sound is generated. For example, plucking a single guitar string causes a mechanical vibration to occur, which we can easily see by looking at the string. Of course, we can also hear it. The vibrating string forces air molecules near it to come together, slightly raising the air pressure and pushing those molecules into motion, which in turn sets neighboring air molecules in motion, and on and on. Thus sound develops waves (like a stone dropped into water), which vibrate up and down and set the air molecules in a push (**compression**) and pull (**rarefaction**) motion, causing an area of higher pressure to move through the air. So in addition to the vibration noted previously, we need a medium for the sound to travel through. Of course, the medium we're usually concerned with is the atmosphere, or air. Sound can also travel through other materials, such as water or wood, but will often be distorted by the medium. Sound vibrations can't travel in a vacuum. Finally, for sound to exist technically, we need a receiver. Someone (a person) or something (a microphone) must receive it and perceive it as sound. The high-pressure area reaches receptors in our ear and we hear the vibrations as sound, or the pressure waves strike the diaphragm of a microphone, beginning the process of converting a sound signal into an audio signal.

Figure 2.13 shows a representation of sound being produced. We can't actually see sound waves, but they act very much like water waves as we've noted. The sine wave (shown in Figure 2.14C) is used to represent a sound waveform, because it can readily show the wave compression (the higher pressure portion of the wave above the center line) and the wave rarefaction (the lower pressure portion of the wave below the center line).

2.14 KEY CHARACTERISTICS OF SOUND WAVES

There are four key characteristics of sound that help determine why one sound is different from another: amplitude, frequency, timbre, and the sound envelope. A sound wave's **amplitude** relates to its strength or intensity, which we hear as volume, or loudness. The loudness of a sound can be thought of as the height of the sound wave. The louder the sound is, the higher the amplitude as shown in Figure 2.14A. As a sound gets louder, greater

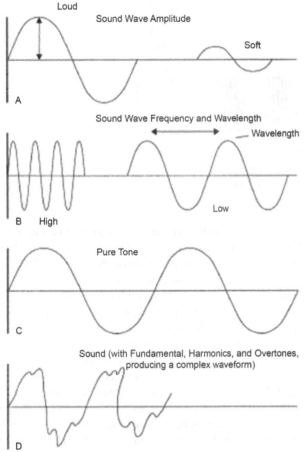

FIGURE 2.14 Characteristics of sound waves include volume, pitch, and tone, here visualized as sine waves.

compression and rarefaction of air molecules take place, and the crest of the wave will be higher while the trough of the wave will be deeper. A sound wave's actual amplitude is readily measured; however, loudness is a subjective concept. What is loud to one person isn't necessarily loud to another person. Sound amplitude is measured in **decibels** (abbreviated **dB**). The human ear is very sensitive and can hear a tremendous range of sound amplitudes, so the decibel scale is logarithmic. Near total silence is noted as 0 dB, a sound 10 times louder than this is 10 dB, a sound 100 times more powerful is 20 dB, and so on. Decibels represent the ratio of different audio levels and measure the relative intensity of sound. Sounds in the range of 0 Db (the threshold of hearing) to 120 dB (the threshold of pain) are detected by the human ear, but those sounds near and exceeding the high end can be painful and can damage your hearing. Any sound above 85 dB can cause hearing loss, but it depends on how close the listener is to the sound and how long he or she is exposed to it. The sound at many rock concerts has been measured around the 120 dB range, which explains why your ears often ring for a day or two after the show.

Frequency relates to the number of times per second that a sound wave vibrates (goes in an up-and-down cycle),

which we hear as **pitch** (see Figure 2.14B). The faster something vibrates, or the more cycles it goes through per second, the higher the pitch of the sound. Like amplitude, frequency can be objectively measured, but like volume, pitch is subjective. In audio jargon, cycles per second are known as **hertz (Hz)**. A sound wave that vibrates at 2,000 cycles per second is said to have a frequency of 2,000 Hz. When the number of cycles per second gets higher—for example, 20,000 Hz—the term **kilohertz (kHz)** is often used. It denotes 1,000 cycles per second, so 20,000 Hz could also be called 20 kHz. (You'll learn more about frequency in the next section.)

A sound's **wavelength** is the distance between two compressions (crests) or two rarefactions (troughs). Sound wavelength can vary from around 3/4 inch for a treble sound near 16 kHz to around 36 feet for a bass sound near 30 Hz. There is an inverse relationship between wavelength and pitch, so higher-pitched sounds have a shorter wavelength.

A sound's **timbre** (which is pronounced "TAM-bur"), or **tone**, relates to the **waveform** of the sound. It's the characteristic of sound that distinguishes one voice from another, even though both may be saying the same thing at the same volume and pitch. A graphic representation of a pure tone is shown as the shape of a sine wave, as in Figure 2.14C. Each sound has one basic tone that is its **fundamental**; most sound, however, is a combination of many tones with different strengths at different frequencies, so the waveform is much more complex, as shown in Figure 2.14D. These other pitches are either exact frequency multiples of the fundamental (known as **harmonics**) or pitches that are not exact multiples of the fundamental (known as **overtones**). For example, striking an A note (above middle C) on a piano would produce a fundamental tone of 440 Hz. In other words, the piano string is vibrating 440 times per second. The harmonics of this note will occur at exact (or whole number) multiples of the fundamental tone, such as 880 Hz (twice the fundamental) or 2,200 Hz (five times the fundamental). The interaction of the fundamental, harmonics, and overtones creates the timbre of any particular sound.

When sound waves combine, they will be either **in phase** or **out of phase**. If the peaks and troughs of two waves line up, they will be in phase and combine into one wave with twice the amplitude of the original waves. If the peaks of one sound wave line up with the troughs of another, they will be 180 degrees out of phase and will essentially cancel each other out, producing no sound or greatly diminished sound. Most sound (such as voice or music) is made up of a combination of sound waves that are out of phase, but less than 180 degrees off, thus producing the complex waveform mentioned previously.

A sound's wave **envelope** relates to its **duration**, or the change in volume of a sound over a period of time, as shown in Figure 2.15. Normally, a sound's wave envelope develops through four specific stages: **attack**, the time it takes an initial sound to build to maximum volume;

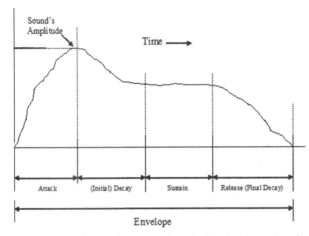

FIGURE 2.15 The sound wave envelope depicts the change in volume of a sound over a period of time. *(Image courtesy of Sauls, S. J. (2007). Basic Audio Production: Sound Applications for Radio, Television, and Film, 2nd ed., Thomson Custom Solutions, publisher.)*

decay, the time it takes the sound to go from peak volume to a sustained level; **sustain**, the time the sound holds its sustain volume; and **release**, the time it takes a sound to diminish from sustain volume to silence. In essence, decay, sustain, and release refer to the time it takes a sound to die out. Some sounds, like a percussive drum beat, have a very fast attack; other sounds, like a piano chord, have a long decay-sustain-release. Audio equipment must be able to accurately reproduce any sound wave envelope.

The envelope of sound is, in essence, its "fingerprint." For the most part, no two sound sources have the same envelope. Along with timbre, the sound envelope is another reason why a saxophone, pipe organ, and flute, all playing the same note, sound different. In addition, quite often in postproduction we manipulate the specific parts of the sound envelope to achieve a desired sound, whereby the sound engineer alters the original make-up of the sound that was initially recorded.

2.15 FREQUENCY RESPONSE

In audio production, we often mention the frequency response of equipment or, for that matter, the frequency response range of human hearing. In very general terms, the human ear is able to hear frequencies within the range of 20 to 20,000 cycles per second. For most of us, it's not quite that low or that high. In any case, production equipment, such as an audio console or monitor speaker, should be able to reproduce an audio signal in that range, and most modern equipment is measured by how well it does so. For example, a monitor speaker may have a frequency response of 40 Hz to 18 kHz, meaning that the speaker can accurately reproduce all frequencies within that range. An inexpensive broadcast microphone may have a frequency response of only 80 Hz to 13 kHz. It would not be able to pick up any of the higher frequencies—those above

13,000 Hz. This may not be a problem if the microphone were used primarily to record speech because the human voice usually falls in a frequency range of 200 to 3,000 Hz. Obviously, if you wanted to record a musical group (which often produces sounds in the full range of frequencies), you would want to use a microphone with a wider frequency response.

A frequency response curve is often used to indicate the level of frequency response because some equipment may not pick up or reproduce certain frequencies as well as others. Broadcast equipment is designed to pick up all frequencies equally well, so its response curve is considered to be flat, although few components have a truly flat **frequency response** curve.

Although there are no standard figures, the audio frequency spectrum is often divided into three regions: bass, midrange, and treble. The low frequencies (bass) are those between 20 and 250 Hz and provide the power, or bottom, to sound. Too little bass gives a thin sound, and too much bass gives a boomy sound. The midrange frequencies fall between 250 and 4,500 Hz. These frequencies provide a lot of sound's substance and intelligibility. Too little midrange gives a lack of presence, but too much midrange gives a harsh sound. High frequencies (treble) are those from 4,500 Hz to 20,000 Hz. The treble frequencies provide the sound's brilliance and sharpness. Too little treble gives a dull sound, and too much treble gives excess sparkle as well as increasing the likelihood of hearing noise or hiss in the sound.

As frequencies change, we think in terms of the musical interval of the **octave**, or a change in pitch caused by doubling or halving the original frequency. For example, a sound going from bass to midrange to treble frequencies by octave intervals would go from 110 Hz to 220 Hz to 440 Hz to 880 Hz to 1,760 Hz to 3,520 Hz to 7,040 Hz, and so on. As humans, we are subject to an awkwardly named **equal loudness principle**. That is, sounds that are equally loud will not be perceived as being equally loud if their pitch is different—we hear midrange frequencies better than either high or low frequencies. In audio production (and other forms of sound manipulation), we often compensate for this by equalization of the signal.

2.16 CONCLUSION

Unless you're building an audio production facility from the ground up, you will probably have little control over the construction of the studio; sound treatment is an important consideration, however, and methods of improving the sound environment can be put into practice in almost any situation. Although the audio studio may seem overwhelming at first, it is an environment you will become very comfortable in as you do production work. Completion of this chapter should have you in the audio production studio and ready to learn the procedures and techniques for operating all the equipment you see in front of you.

Self-Study

1. What does the radio expression "to work combo" mean?

   a) The announcer has an engineer to operate the studio equipment.
   b) The announcer operates the studio equipment and also announces.
   c) The announcer works at two different radio stations.
   d) The announcer is announcing in both the on-air and the production studio.

2. Which type of studio is least likely to contain an audio console?

   a) on-air studio
   b) production studio
   c) PDX studio
   d) performance studio

3. Which term describes sound produced in the audio studio that causes objects or surfaces within the studio to vibrate sympathetically?

   a) absorbed sound
   b) reflected sound
   c) reinforced sound
   d) diffused sound

4. In the production studio, sound that has bounced off one surface before reaching the microphone is called what?

   a) echo
   b) reverberation
   c) direct sound
   d) indirect sound

5. What does "reverb ring" in the production studio refer to?

   a) the circular route reflected sound takes before it reaches the microphone
   b) the time it takes reflected sound to go from full volume to silence
   c) just another common name for echo
   d) a sound that has bounced off two or more surfaces

6. What is the use of carpeting on the walls of some audio production facilities an example of?

   a) an inexpensive way of decorating the studio
   b) producing reverb in the studio
   c) producing a live sound in the studio
   d) soundproofing the studio

7. Studios with parallel walls produce less reflected sound than irregularly shaped studios.

   a) true
   b) false

8. Why do most production studios use a U-shaped layout?

   a) This design places equipment within easy reach of the operator.
   b) This design uses incandescent lights rather than fluorescent lights.
   c) This design necessitates custom-built cabinets.
   d) This design uses the least amount of wire to connect the equipment.

9. Static electricity is not a problem in the modern production studio, because state-of-the-art audio equipment is impervious to static.

   a) true
   b) false

10. Which hand signal almost always comes immediately after the standby hand signal?

    a) 2 minutes to go
    b) thumbs up
    c) cue talent
    d) give mic level

11. If you hold up the index, second, and third fingers of one hand in front of you, what are you telling the announcer?

    a) There are 3 minutes left in the program.
    b) There are 30 seconds left in the program.
    c) He or she should move three steps closer to the microphone.
    d) Three minutes have gone by since the beginning of the program.

12. What does one call the linking of a CD player to an audio console, the console to an equalizer, and the equalizer to an audio recorder?

    a) audio road map
    b) audio linking
    c) audio processor
    d) audio chain

13. What do we call the uneven sound that is produced when sound waves are reflected between parallel walls in such a manner that a wave reflected in one direction is combined with an identical wave going in the opposite direction?

    a) a diffused wave
    b) a standing wave
    c) an absorbed wave
    d) a sympathetic wave

14. When a "recording" light is on outside a production studio, it means a microphone is "live" in that studio.

    a) true
    b) false

15. If an audio studio has a live end/dead end design, the front of the studio is the "live end."

    a) true
    b) false

16. When sound produced in the production studio strikes a hard, flat surface, which of the following does not happen?

    a) reflection
    b) absorption
    c) penetration
    d) diffusion

17. A production studio wall that has an absorption coefficient of 0.50 will absorb half the sound striking it and reflect back half the sound.

    a) true
    b) false

18. Posters and other radio station paraphernalia should not be put up in a production studio as they will distract the announcer from doing good production work.

    a) true
    b) false

19. Which term describes what happens when the irregular surfaces of acoustic tiles break up sound reflections?

    a) absorption
    b) reflection
    c) penetration
    d) diffusion

20. What is an unwanted change in the audio signal due to inaccurate reproduction of the sound called?

    a) reverb
    b) noise
    c) distortion
    d) diffusion

21. Which statement about sound is not true?

    a) Sound is generated when something vibrates.
    b) Sound, to technically exist, must be heard.
    c) Sound vibrations develop waves by setting adjacent air molecules in motion.
    d) Sound vibrations travel faster in a vacuum than in air.

22. Which of the following is not part of a sound wave's envelope?

    a) attack
    b) decay
    c) sustain
    d) rarefaction

23. The number of times a sound wave vibrates (goes in an up-and-down cycle) per second determines which characteristic of the sound?

    a) frequency
    b) amplitude
    c) wavelength
    d) wave envelope

24. What is the standard unit of measure to gauge the relative intensity of sound?

    a) signal-to-noise ratio
    b) hertz
    c) absorption coefficient
    d) decibel

25. Sound (such as a talent's voice) that has been manipulated electronically (such as recorded on a digital recorder) is called a sound signal.

    a) true
    b) false

ANSWERS

11b. No. Crossed index fingers indicate 30 seconds left. (Reread 2.10.)

12b. No. This is not correct. (Reread 2.2.)

13b. Correct. This is the right answer.

14b. No. This is exactly what it means. (Reread 2.9.)

15b. Yes. This is a false statement. The "live end" would be the back of the studio designed to add some reflected sound giving the studio sound a desirable sharpness.

16b. This isn't a bad choice, but some sound will be absorbed and dissipated even with hard surfaces. (Reread 2.5.)

17b. No. A coefficient value of 1.00 would mean total absorption and a coefficient value of 0.00 would mean no absorption. (Reread 2.6.)

18b. Yes. This is the correct answer.

19b. No. Reflection would be sound that has bounced off a surface. (Reread 2.5)

20b. No. You're close because noise is an unwanted element introduced into the audio signal that was not present in the original sound, but there's a better response. (Reread 2.11.)

21b. No. This is a true statement. (Reread 2.13.)

22b. No. Decay is the time it takes sound to go from peak volume to a sustain level. (Reread 2.14.)

23b. No. Amplitude relates to volume and the height of a sound wave. (Reread 2.14.)

24b. No. Hertz is another term for cycles per second and is a measure of frequency. (Reread 2.14.)

25b. Correct. This is a false statement because once natural sound has been manipulated electronically, it is correctly called an audio signal; however, be aware that sometimes the terms are interchanged.

If you answered C to any of the questions:

1c. No. This is not working combo. (Reread 2.3.)

2c. No. This is just another term for production studio. (Reread 2.1 and 2.2.)

3c. Right. This is the correct response.

4c. No. Direct sound doesn't bounce off any surface before reaching the microphone. (Reread 2.5.)

5c. No. Echo and reverb are both reflected sound but distinctly different. (Reread 2.5.)

6c. No. Just the opposite would happen. Soundproofing with carpeting would help produce a dead sound in the studio. (Reread 2.5 and 2.6.)

8c. No. In a cost-conscious facility, this could be a negative. (Reread 2.3 and 2.4.)

10c. Correct. "Standby" and "cue talent" hand signals are always given one after the other.

11c. No. There is another hand signal to move the announcer closer to the microphone, and exact steps are never indicated. (Reread 2.10.)

12c. No. You are way off base with this answer. An audio processor is used to alter the sound characteristics of an audio signal. (Reread 2.2.)

13c. No. Absorbed waves would be sound reflections that have been soaked up. (Reread 2.5 and 2.7.)

16c. No. Some sound will penetrate a hard surface and be transmitted to the adjoining space. (Reread 2.5.)

19c. No. Penetration would be sound that has been transmitted through a surface. (Reread 2.5.)

20c. Yes. You're correct.

21c. No. This is a true statement. (Reread 2.13.)

22c. No. Sustain is the time a sound holds its volume. (Reread 2.14.)

23c. No. Wavelength refers to the distance between two wave compressions or rarefactions. (Reread 2.14.)

24c. No. Absorption coefficient measures the degree to which materials can absorb sound. (Reread 2.14.)

If you answered D to any of the questions:

1d. No. This seems improbable and is not working combo. (Reread 2.3.)

2d. Correct. A performance studio usually only has microphones that are fed to an audio console in either a production studio or an on-air studio.

3d. No. (Reread 2.5.)

4d. You're partly right, but echo and reverb are both indirect sound, and one is a better response to this question. (Reread 2.5.)

5d. No. While this describes reverb, there is a better response. (Reread 2.5.)

6d. Correct. Carpeting walls helps to soundproof, as would use of acoustic tiles designed for the production studio.

8d. While this may be true, it really is not the best reason. (Reread 2.3.)

10d. No. The "give mic level" signal, if used, would have been given before a "standby" signal. (Reread 2.10.)

11d. No. Time signals are usually given only to show how much time remains in a program. (Reread 2.10.)

12d. Correct. The term "audio chain" describes how broadcast equipment is connected together.

13d. No. Sympathetic waves would be sound reflections that have been reinforced. (Reread 2.5 and 2.7.)

16d. Correct. Sound is diffused when it strikes an irregular surface.

19d. Correct. This is diffusion.

20d. No. Diffusion is sound that has been broken up by an irregular surface. (Reread 2.5 and 2.11.)

21d. Correct. Sound vibrations can't travel in a vacuum.

22d. Correct. Rarefaction is not part of a sound's wave envelope.

23d. No. Wave envelope refers to a sound's duration. (Reread 2.14.)

24d. Right. The decibel is the standard unit or ratio used to measure a sound's volume.

Projects

PROJECT 1

Tour an audio facility and write a report describing it.

Purpose

To enable you to see a commercial production facility firsthand.

Notes

1. This project can be completed by touring a radio station, recording studio, or other audio production facility, but don't push a facility that seems reluctant to have you come. Although some smaller studios are happy to have you, others just aren't equipped to handle visitors.
2. Make sure that before you go you have some ideas about what you want to find out so that you can make the most of your tour.
3. Keep your appointment. Once you make it, don't change it.

How to Do the Project

1. Select a studio or station that you would like to tour. (If the instructor has arranged a tour for the whole class, skip to Step 4.)
2. Call and ask them if you could tour the facility so that you can write a report for an audio production class.
3. If they're agreeable, set a date; if they're not, try a different facility.
4. Think of some things you want to find out for your report. For example:
   a. How many production studios do they have?
   b. What types of equipment (CD player, audio recorder, and so on) do they have?
   c. What manufacturers (brand names) have they bought equipment from?
   d. How is the production studio soundproofed?
   e. How is the on-air studio different from the production studio(s)?
   f. Do they ever use hand signals during a production?
   g. Are their studios designed for stand-up operation?
   h. What is the physical layout of the studios and the station?
   i. What software programs are they using (for example, Adobe Audition)?
5. Go to the facility. Tour to the extent that they'll let you, and ask as many questions as you can.
6. Jot down notes so that you'll remember main points.
7. Write your report in an organized fashion, including a complete description of the production studio and the other points you consider most pertinent. It should be two or three typed pages. Write your name and "Audio Production Facility Tour" on a title page.
8. Give the report to your instructor to receive credit for this project.

PROJECT 2

Redesign your production studio.

Purpose

To suggest improvements to your production facility, utilizing some of the concepts mentioned in this chapter.

Notes

1. Although you may initially feel your production studio is perfect just the way it is, almost every studio can be reconfigured with improvements.
2. You won't be judged on artistic ability, but make your drawings as clear as possible.
3. You may find it useful to complete Project 1 before attempting this project.

How to Do the Project

1. Draw a rough sketch of your production studio, showing approximate dimensions, door and window location, equipment placement, and so forth.
2. Draw another sketch of the studio, suggesting changes or improvements to it. For example, if there currently is a CD player on the left side of the audio console and another on the right side, you may suggest moving them both to one side. If you notice a paneled or painted sheet rock wall in the studio, you may suggest putting acoustic tiles on that area. You may want to employ an idea for your studio that you noticed when you did Project 1. Just be creative, and try to design the best possible production studio.
3. On a separate sheet of paper, provide a reason for each change you suggest.
4. Write your name and "Studio Design" on a title page, and put your two sketches and reasons together.
5. Turn in this packet to the instructor to receive credit for this project.

PROJECT 3

Draw an audio chain flowchart for your production studio.

Purpose

To help you understand that the audio chain maps the route an electronic audio signal takes as it goes from one place to another in the production studio.

Notes

1. It may be helpful to review Figure 2.2 before beginning this project.
2. Use simple shapes to represent equipment and arrowed lines to represent the sound signal.
3. You won't be judged on artistic ability, but make your drawings as clear as possible.

How to Do the Project

1. Pick a single sound source in your production studio, such as a CD player.
2. Draw a figure to represent the CD player toward the left side of a sheet of paper, and label it appropriately.
3. Determine where the sound goes as it leaves the output of the CD player. (Most likely, it goes to the audio console.)
4. Draw a figure to the right of the CD player to represent the audio console, and label it.
5. Draw an arrowed line going from the CD player to the audio console to represent the signal flow.
6. Now determine where the sound goes next. (It could go to a signal processor or maybe directly to an audio recorder.)
7. Continue in this manner until you've drawn all the possible signal paths that the CD player sound could take. (Don't forget to include the signal to the studio monitors.)
8. Pick another sound source, such as the studio microphone, and repeat the above steps. Do the same for all the other sound sources in your studio—audio recorders, and so forth.
9. Write your name and "Audio Chain" on your sketch, and turn it in to the instructor to receive credit for this project.

3

DIGITAL AUDIO PRODUCTION

3.1 INTRODUCTION

Digital technologies have revolutionized how a person can record, edit, and otherwise manipulate an audio sound signal. From creating a music bed to adding sound effects and editing out vocal mistakes, editing is a day-to-day part of audio production work, and it's one of the most important skills a production person needs to know.

This chapter introduces concepts and practices of digital audio production. First, some basic information on how analog sound is converted to digital audio is presented. The focus then shifts to exploring some of the digital audio equipment and basic production techniques used in the modern audio studio. Finally, the chapter looks at digital audio editing and multitrack production.

3.2 THE ANALOG ROOTS OF DIGITAL PRODUCTION

Before the development and use of digital technology, audio recording relied on an analog process. **Analog** is short for "analogous," meaning "similar to" or "a replica of." An analog recording is similar to, or a replica of, an original sound. An analog signal is a continuously variable electrical signal whose shape is defined by the shape of the original sound wave (see Figure 3.1A). In the analog recording process, an electromagnetic re-creation that is similar to the original sound wave of the original sound source can be stored on vinyl recordings or magnetic tape. For example, a microphone converts sound pressure changes to changes in voltage that are sent through the microphone cable and recorded onto audio tape as changes in magnetic strength.

One disadvantage of the analog process is that each time an analog signal is recorded or processed in some fashion, it is subject to degradation because the shape of the signal changes slightly. This is more commonly referred to as "generation loss" and leads to a condition where a copy will not sound as clear as its original recording. In addition, because analog tape recording relies on magnetic pulses stored on a tape, any defect or decrease in the magnetic properties of the tape means a loss of signal quality.

3.3 THE DIGITAL PROCESS

In today's production studio, computers and other equipment use a digital recording process. By digital technology, we mean the process of converting original audio waveform information into an electrical signal composed of a series of binary numbers. All computers handle information in this manner by associating a **binary** number with letters of the alphabet and numbers, and then manipulating this data. Digital encoding is accomplished in a discrete fashion, similar to looking at individual numbers in a statistical analysis and writing them down in a set order. The audio signal starts out as analog and is converted into digital by going through four basic stages: **filtering**, **sampling**, **quantizing**, and **coding**.

First, the original analog sound signal is sent through a low-pass filter that strips off frequencies that are above and below the range of human hearing. Although originally inaudible, these frequencies can be shifted, or **aliased**, into an audible range during recording and playback, so this process is also referred to as **anti-aliasing**.

The filtered analog signal is next divided many times a second in a process known as sampling (see Figure 3.1B). Each sample of the filtered analog signal represents the amplitude of the audio signal at the moment the sample is taken. The more samples that are taken and converted to binary data, the better a digital recording system is able to re-create the sound based on its original form. To do this successfully, most digital audio equipment utilizes **sampling rates** of 32, 44.1, or 48 thousand samples per second. These rates are often labeled in terms of kilohertz

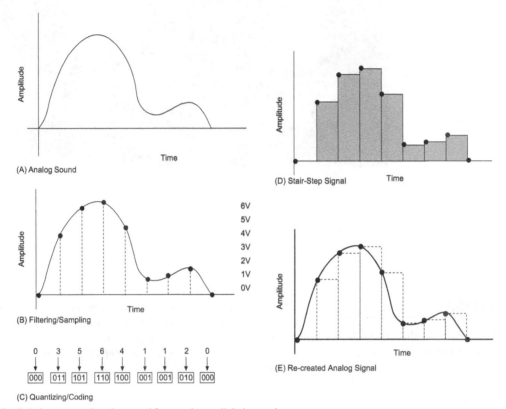

FIGURE 3.1 The digital process—changing sound from analog to digital to analog.

(32 kHz, 44.1 kHz, and 48 kHz). Again, the more samples that can be taken by a processor in digital audio equipment, the more accurate the original audio signals can be re-created, which leads to better sound quality. As a rule of thumb, the sampling rate must be at least twice the highest frequency in the audio signal to enable high-quality and accurate encoding and decoding. Because we can't hear sounds above 20 kHz and in order to insure that all frequencies are digitized, the sampling rate should be slightly greater than twice 20 kHz, which is one reason why the most common digital sampling rate is 44.1 kHz.

Quantizing and coding (the third and fourth steps in the process) are the stages that assign a numerical value to each sample. The samples taken of the **amplitude** of the audio signal can occur at any point within the range of amplitudes, from absolutely silent to very loud. Between these two extremes—or indeed between any two points, as geometry tells us—there are an infinite number of other points. Quantizing breaks up these infinite points into a more manageable number, rounding samples up or down to the nearest value. Quantization is measured by the number of bits used by a computing system, and the term "**bit depth**" refers to the quantizing levels of a sound: the more bits used by a system, the more levels there are, and the more accurate information there is about the signal. For example, a 1-bit system would signify only two quantizing levels—either no amplitude or maximum amplitude—which doesn't give much information about the sound.

Each additional bit doubles the number of levels—2 bits give four levels, 3 bits give eight levels, and so on. The standard bit rate for most digital recording is still 16 bits, although some CD, DVD, and computer recordings are being done with 20, 24, and even 32-bit technology. Sixteen bits can represent as many as 65,536 values, which is more than adequate for reconstructing high-quality sound in a computer. Keep in mind that higher bit depth also equals lower noise and better fidelity of the digital recording.

Coding refers to the process that involves assigning binary 0s and 1s in precise order corresponding to the values measured during the quantizing process. This binary or digital "word" (see Figure 3.1C) represents each individual sample's quantized (rounded up or down) voltage level at that particular moment. The analog-to-digital (A/D) converter is the electronic circuit in a digital audio system that accomplishes this task. Remember, it's the binary data that is actually recorded, not an analog representation of the signal. Because this is binary data, numerous copies can be made with no measurable loss of quality. Along with improved **frequency response**, wide **dynamic range**, and drastically reduced noise and distortion, the ability to rerecord with no decrease in quality contributed greatly to the acceptance of digital technology in audio production.

To transform the digital signal back to analog, the binary data are sent to a digital-to-analog (D/A) converter where each sample is read and decoded. A signal of the corresponding analog amplitude is produced and "held" for the individual samples, producing a stair-step signal (see Figure 3.1D) that represents the analog signal. Finally, a filter employs an exact mathematical attribute for each sample to create a correct reproduction of the original analog signal, as shown in Figure 3.1E.

3.4 REASONS FOR EDITING

The reasons for editing audio should be obvious. Rarely will you or anyone else produce the vocal track for a production exactly the way you want it on the first try. While recording the script over and over until it is perfect is possible, editing gives you the ability to eliminate mistakes without rerecording. With editing, it's more likely you'll get part of a take sounding great on one recording, part of it sounding great on another, and so on. The best final product will probably be bits and pieces from various takes that are edited to take out bad segments and keep only the exact words and phrases desired. Other production work may require editing out excessive pauses or "uhs" from a piece of news tape or language not allowed by the FCC.

In addition to eliminating mistakes, editing can help decrease the length of production work. Audio productions usually require exact times for commercials, news stories, and other programs, and editing can help your work stay true to specific time lengths or restrictions. Excessive pauses in a vocal track can be manually edited out, or in some instances, editing and playback software can be used to automatically time compress a segment to the exact length needed. Many half-hour interview programs are really edited down from an actual interview that went on much longer. Entire questions that didn't get a good response can be edited out, and long, rambling responses can be cut down to a more concise reply.

Audio editing also gives the freedom of non-linear recording, or recording out of sequence. For example, a producer might be putting together a production that uses the testimonials of several customers. It's possible that the one used first in the commercial may not have been recorded first. Editing allows you to easily rearrange the order or, again, just use a portion of what you originally recorded for your final production.

3.5 DESKTOP AUDIO PRODUCTION: THE DIGITAL AUDIO EDITOR

One of the simplest ways to get into digital production is to put together an inexpensive digital audio editor that utilizes a standard off-the-shelf computer system, plus some specialized equipment. This type of production system has the potential of replacing almost an entire production studio.

There are two main types of audio editors: 2-track and multitrack. If you're going to just edit phone calls or interviews, a simple 2-track system might be the answer.

Multitrack systems may be built for more advanced commercial or music production. Fortunately, many audio editing software programs incorporate both types of editors.

Almost any basic laptop or desktop computer system either will be (or can be) configured adequately to run audio editing software. In addition to the computer however, the necessary specialized equipment includes a **DSP (digital signal processing) audio card**, usually a PCI-bus design for desktop computers, as shown in Figure 3.2. A standard computer audio card could be used, but professional audio hardware usually offers higher sampling rate capability, greater bit resolution, and broadcast-style connectors. The DSP audio card functions as both the A/D and D/A converter, and the interface or I/O (input/output) device that moves the audio signal from its source to the editing system. In most production studios, an output from the audio console is fed into the audio card so that any equipment that runs through the board can be recorded into the desktop system.

The final component of a digital audio editing system is a software program to perform the actual recording and editing of the audio. There are several software programs that can turn a desktop or laptop computer into a powerful digital audio editor—Avid Pro Tools, Bitwig Studio, Cubase Pro, PreSonus Studio One, and Cakewalk SONAR, and Adobe® Audition®, are just a few. Most professional programs used in the audio studio fall within a range of $300 to $900, which is a relatively inexpensive investment. Some providers (including Avid and Adobe®) charge a monthly fee for cloud access to their software, which might help if your budget is thin. In addition, some very good, basic bare-bones audio editing software can be found for less than $100 or even for free on the Internet. With a modest budget, anyone should be able to build a system and begin doing digital audio production work.

PRODUCTION TIP 3A
Audacity

Audacity offers a free audio editor that can be downloaded for Windows, Mac, and some Linux operating systems at web.audacityteam.org. Many of the key features found in professional programs and mentioned in this chapter are available in the Audacity program. Users can record audio in 16-bit, 24-bit, or 32-bit resolution at a sampling rate of up to 192 kHz. Basic audio editing commands include cut, copy, paste, and delete with undo and redo functions. The program also facilitates audio effects like pitch change, noise reduction, equalization, echo, phasing, and more, and can import and export several common sound file formats such as MP3, WAV, and AIFF. Audacity is released under the terms of a GNU General Public License, where people can use the program, as well as study, copy, modify, and redistribute it under terms of the license.

FIGURE 3.2   A DSP audio card is an important component of any desktop computer-based audio editor. *(Image courtesy of Digital Audio Labs.)*

FIGURE 3.3   Some digital audio workstations resemble desktop audio editors, but employ specialized audio cards, customized boxes, and proprietary software. *(Image courtesy of SADiE Inc.)*

3.6 DIGITAL AUDIO WORKSTATIONS AND OTHER DIGITAL EDITING SOLUTIONS

The **DAW** or **digital audio workstation** is a stand-alone, hard-disk–based system that incorporates proprietary computer technology but is quite similar to desktop audio systems. Because the workstation has been developed as a dedicated audio recorder and editor, it often has built-in specialized functions and features. The basic components of the DAW are packaged as a single unit similar to the one shown in Figure 3.3. A frame houses the computer chassis, power supply, motherboard, and an internal hard disk drive that stores and manipulates the digital audio data.

A **user interface**, such as a keyboard, touch screen, or mouse, is necessary to operate the workstation. However,

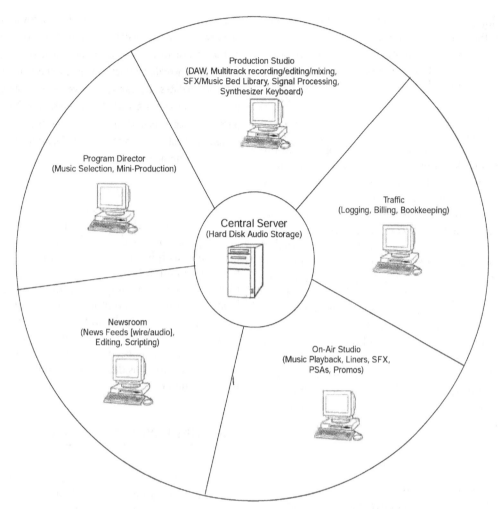

FIGURE 3.4 An all-digital facility integrates digital audio production with other aspects of the facility's operation.

many workstations have tried to become more user-friendly by making the interface include typical audio production elements, such as **faders**, **cue wheels**, and **transport buttons**. Sometimes these interfaces are shown in graphic form on a computer screen, but other systems are part of a small audio mixer. A built-in audio card provides the connection between the workstation and other audio equipment. Through the use of various ports, many workstations have the ability to network with other workstations in house or online, or interface with modems, and other components. As such, a DAW-based system can become part of an all-digital facility (see Figure 3.4) that can produce an audio spot, store it or several variations of it, play the production on air by sending the signal to a digitized transmitter, share it with other stations or production facilities, and even send logging and bookkeeping information about the airing to the appropriate personnel.

Other multitrack recorders and editors are designed to be rack-mounted in the studio or are part of a system that contains the recorder, audio mixer, and signal processing effects within an individual unit.

3.7 STRONG POINTS AND WEAK POINTS OF DIGITAL PRODUCTION

We know that digital equipment offers superior audio quality and that it doesn't suffer from various forms of audio distortion or build up any additional noise during recording or copying. But there are other advantages to digital recording that have helped endear the technology to the audio production person.

One operational advantage of the DAW is fast random access to all the material stored in the system. This ability to immediately cue up to any point leads to less time spent on basic production functions and potentially more time spent on the creative aspect of the production process.

If you record something, but then need to make it slightly shorter or longer, you can do so easily with digital equipment. For instance, if your 15-second commercial turns out to be 17 seconds long, you can remove short bits of silence throughout the commercial to get it to 15 seconds. In fact, many digital audio editing programs or workstations can do this time compression (or expansion) for you. You simply indicate how long you want something to be, and the program adjusts the material accordingly.

IDs and other types of label information can be encoded within digital media. A table of contents can provide names, times, and other data through a front-panel window on the equipment, a separate file on a CD, or through the video monitor screen of a DAW.

As there are fewer internal operating parts, repair of digital equipment is often a simple substitution of one circuit board or component for another. On the other hand, when something goes wrong with a computer hard drive, it's often not just the current project that is damaged or lost, but possibly everything else contained on the hard drive. For this reason, the need for frequent backing up of hard drives can't be overemphasized, especially if the material is sensitive or difficult to re-create. Computer programs have a tendency to crash at exactly the wrong moment, erasing hours of work in an instant if the audio files haven't been backed up. Most computer audio-editing software has an option to perform periodic automatic backups of files currently being worked on, either at timed intervals or after a specified number of edits. However, the automatic backup feature is sometimes not turned on as a default after installation of a program, so it is worth looking in the index of the program's documentation for this feature so that it can be enabled before work starts.

PRODUCTION TIP 3B
Maintaining Digital Equipment

There are few mechanical elements in digital equipment, but computer knowledge has always been important for what maintenance there is. To keep your studio computers running in their most efficient manner, you should reboot them every 3 to 5 days. Most computer programs, including editing software, will cause computer RAM to become fragmented as the program runs. You may not notice any problems with typical business programs (like word processing); however, studio computers are often running 24 hours a day, every day of the week. RAM can become so fragmented that the computer will crash, and that's a condition that you don't want to happen for a production or on-air computer. Set up a regular maintenance reboot during some station quiet time for fewer computer problems.

In addition, memory management should become a regular part of regular computer maintenance. As production work is accomplished, sound file fragments and completed productions may be left on the hard drive. In a short period of time, hours of hard disk storage space will dwindle down to minutes unless old files are deleted from the system or archived onto a removable storage medium for later use. Running disk utility programs should also be a regular maintenance routine.

Finally, be sure to check for editing software updates as well as operating system updates for your computers. Quite often, one of the easiest fixes for a computer that is running slowly or having trouble working effectively is to check for (and download) software updates. Be aware, however, that some editing software updates may conflict with the operating software of the computer, or vice versa. If you ever have any doubts or concerns, it may be best to contact your IT department or other support groups for information.

There is also a potential noise problem with digital equipment—not audio noise, but noise from cooling fans in computers or disk drives. If equipment has to be housed in a studio with live microphones, this could be a problem. Some production facilities solve this by putting computer CPUs in a separate equipment room, leaving only the keyboard, mouse, and monitor in the actual studio. If computer equipment must be in the studio, try to place it as far from microphones as possible, and build it into studio furniture to minimize any noise. Laptop computers may also be an option to consider, as their cooling components are usually much quieter.

3.8 AUDIO SYNCHRONIZATION

Digital audio signals are streams of digits broken up into digital words. If two digital signals are out of sync, then one may just be beginning a new digital word when the other is in the middle of a word. Switching from one to the other would result in an audible tick or pop. Synchronized audio signals start both new digital words at precisely the same time. Some digital production equipment is self-synchronizing, but digital audio consoles that accept many different types of digital inputs need to synchronize audio to a common clock.

Most digital audio workstations and many other pieces of digital broadcast equipment also have the ability to incorporate MIDI and SMPTE synchronization. **MIDI (musical instrument digital interface)** is an interface system that allows electronic equipment, mainly musical instruments like synthesizers, drum machines, and samplers, to "talk" to each other through an electronic language. **SMPTE (Society for Motion Picture and Television Engineers)** time code is an electronic language developed for videotape editing that identifies each video frame with an individual address. The time code numbers consist of hour, minute, second, and frame. The frame digits correspond to the 30 video frames in each second. Both MIDI and SMPTE signals can be used to reference various individual pieces of equipment and accurately start, combine, stop, or synchronize them.

3.9 LATENCY ISSUES

Latency is the short amount of time required to convert analog audio into digital audio, or to add a digital effect to audio, or to move audio from one location to another. All digital equipment and computers dealing with audio will exhibit some latency. In most individual pieces of equipment, latency is usually not an audible issue, because it is often a delay of only a few milliseconds. However, latency can be cumulative as sound goes through several normal audio processes. For example, if you add equalization (EQ) to an audio track, mix that track with several other audio tracks, and then monitor everything through an audio interface, your audio will have gone through several layers of software processing, each adding some latency effect.

In broadcast situations, latency issues arise when a live broadcast is combined with audio from a satellite or telephone feed or similar link. The linking equipment often has enough latency to produce a noticeable time delay between the live broadcast and the other audio.

Computers with faster CPU processing have helped with latency issues, as has the development of audio drivers that bypass the Windows or Mac operating system, allowing audio signals to connect directly with the sound card. Still, latency can be an issue that the audio production person must be aware of and willing to work around if it becomes noticeable in his or her production work.

3.10 DIGITAL AUDIO EDITING

Editing normally begins after you've recorded and saved audio into the computer system. Previously recorded audio files can also be imported, or directly ripped from a CD into a file. Although there are many different systems available, to gain an understanding of digital audio editing, this chapter looks at one specific system, Adobe® Audition® Creative Cloud (CC). Adobe® Audition® CC has six main work areas (Files, Media Browser, History, Editor, Levels, and Selection/View). For the purposes of this chapter, we will only be looking at four of these areas (Editor, Files, Levels, and Selection/View). The Editor screen (shown in Figure 3.5) puts the user in a single-waveform editor that is used to record, process, and edit mono and stereo audio segments. It also shows the audio sound file and is utilized to edit or otherwise process the sound. The transport buttons (located at the bottom of the Editor screen) control recording and playback functions of the audio. To the right of the transport buttons are several magnifying glass icons that allow you to zoom in and out of the waveform

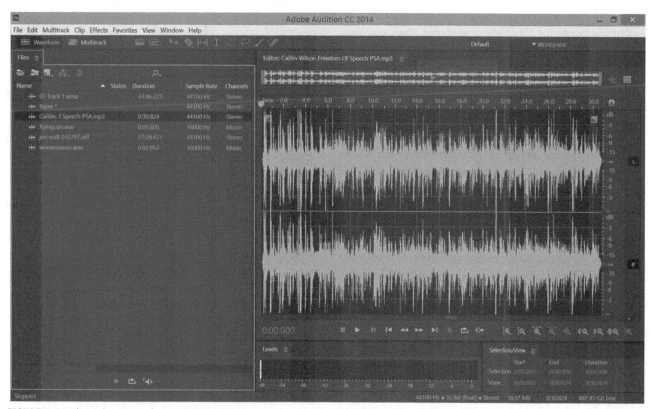

FIGURE 3.5   The main screen of an audio editing program shows various navigation and functional controls for manipulating the audio file. *(Adobe product screenshot reprinted with permission from Adobe Systems Incorporated.)*

on both horizontal and vertical axes, and a "timeline display" at the top of the screen shows timing information. The Files window lists the name(s) of any file(s) being worked on, and displays pertinent information such as format type, run time, sample size, and number of channels. The Levels window acts like a standard VU (volume unit) meter, in that it displays the audio levels on both playback and record. Finally, the Selection/View window allows the producer to see time information in more specific numeric form. There are also a variety of other functions and features on these screens that can be found via drop-down menus and shortcut buttons. As you gain more experience using this (or other) editing software, you will find these basic tools to be incredibly helpful. Becoming familiar with them as soon as possible is the best thing for you to do.

PRODUCTION TIP 3C
What's up with Adobe® Creative Cloud®?

Long-time users of Adobe® editing and production software might have been surprised in 2013 when the company replaced its CS6® version with Creative Cloud (CC)®. The main reason for any surprise was because with version CC®, Adobe® stopped distributing hard copies of its software and went exclusively with an online subscription method. In an effort to cut down on piracy (and distribution costs) Adobe® now sells and distributes its software online for a monthly or yearly fee. Users now must purchase specific month-by-month or annual plans to download software for use on their computers. And Adobe® is not alone in using this method. Users of Avid Pro Tools 12 for instance, must also subscribe on a monthly or yearly basis to use their software.

Online distribution of editing software has several advantages. For starters, users can select specific software programs to download and use, as opposed to purchasing an entire suite of software, which could cost more and often resulted in paying for software that was seldom used. Another advantage is that software updates are constantly available and many users even discover their software automatically updated when they open a program. Finally, this distribution method allows users of some software to install copies on several computers with only one account. Previously, software could only be installed using an installation key, which restricted its use to only one computer.

The Adobe® CC® suite works on Windows and Apple operating systems, along with a mobile version that can be used on tablets and smartphones for basic editing and file sharing.

To illustrate the basic audio edit, let's first record three words—*dog, cat, banana*—into our system. Basic editing programs allow you to set the sampling rate and bit depth using the sound card and software. To begin recording, click on File from the top menu bar, select New, select Audio File, and set the appropriate recording specifications (file name, 44,100 sample rate, Stereo, 16 bit depth) and then click on OK. We'll assume that you have a microphone plugged directly into your sound card or the output of your audio console feeds into the sound card. Click on the RECORD button (red circle) on the transport controls and speak the three words into the microphone, pausing slightly between each word. You can check recording levels by watching the levels window at the bottom of the screen. Finally, click on the STOP button (gray square) on the transport controls when you are finished recording.

The audio you recorded will appear as a green waveform (see Figure 3.6) on the black background in the Waveform Display area of the screen. Any silence (the pauses) will be shown as a flat green line. (These are the default screen colors, so you may see something different, depending on the settings for your system.) The audio can now be edited, processed, or stored on the hard disk as a **sound file**. These large audio files consist of an information header (noting sample rate, bit depth, and so on) and the actual data (a long series of numbers—one for each sample). CD-quality audio at 44,100 samples per second times 16 bits of data adds up to around 10 megabytes (Mb) for each minute of audio recorded.

Editing uses the mouse to define a **region** in a fashion similar to any other type of computer editing (word processing or video editing, for example). Defining a region is the same as selecting a range (start point and end point) for each edit. In this case, the edit marks are made on the computer screen, which shows a "picture" of the audio (usually the actual waveform; see Figure 3.6) and can be easily moved, or **trimmed**, to the exact position desired.

If you wanted to edit out the word *cat*, you would click an edit point just to the left of the waveform for the word *cat* and then drag the mouse to the right, stopping just before the waveform for the word *banana*. Including the pause after the word maintains the natural pacing of the audio. Selecting edit points just before and immediately after the word *cat* would increase the pause between *dog* and *banana*. In a similar manner, selecting edit points just after the word *dog* and just before the word *banana* would eliminate any pause between those words, thereby creating an unnatural flow. The area you have selected will become highlighted in white (refer to Figure 3.6 again). You can hear the audio segment that you've highlighted by clicking on the PLAY button (single green forward triangle).

If you need to adjust either edit point, click on either arrow in the space above the waveform and move the edit point left or right as required. Most editor programs have a function to zoom in on the segment of audio you're working on to fine-tune any adjustments and get the mark exactly where you want it. To complete the edit, click on Edit on the top menu bar and select Cut or click on the appropriate toolbar icon (or press CTRL + X for the keyboard shortcut). The selected audio will be cut to a clipboard and the remaining audio will automatically be joined together, leaving the words *dog* and *banana*. Keep

FIGURE 3.6   A sound file is shown on the computer screen as the waveform of the audio. The vertical axis indicates amplitude and the horizontal axis indicates time. Editing is accomplished with a simple mouse-drag operation, with the area to be edited out highlighted. *(Adobe product screenshot reprinted with permission of Adobe Systems Incorporated.)*

in mind this takes a lot longer to explain than it does to accomplish it—as you'll quickly learn!

Suppose you now want to paste the word *cat*, which we just edited out, in front of the word *dog*. Click a single edit point just to the left of the waveform of the word *dog*, then click on Edit on the top menu bar and select Paste (or press CTRL + V for the keyboard shortcut). The audio from the clipboard will be edited back in at that point so the word order is now *cat*, *dog*, and *banana*. Should you decide you don't really want to do this, you can click on Edit and select Undo (CTRL + Z), which will cancel or undo your last action. Normally, Undo can be repeated to back out of several actions. One of the biggest advantages of digital editing is the **nondestructive** nature of the process. The original audio isn't actually altered; edit information (where to start and stop playing the audio) is simply recorded in an **edit decision list** that is held in the computer program. The same audio can be used in several places within a project.

To copy a word, you'd follow a similar process. Click an edit point just after the word *dog* and drag to the end of the word *banana*. The word *banana* and the pause before it should now be highlighted. Once again go to Edit and select Copy (CTRL + C on your keyboard). Click an edit point right at the end of the word *banana* and go to Edit and select Paste. The word order should now be *cat*, *dog*, *banana*, *banana*.

In most cases, you'll want to export and save your project when you're done editing, so that it can be saved to a CD or external disc to play back later. Most programs give you the option to export in various file formats, but the most popular are WAV (which gives you high quality, but uses more hard drive space) and MP3 (which saves space, but can sacrifice sound quality). How the audio will ultimately be used may well determine what format you want to employ if you have that option.

Saving the project only creates a work file or session that your audio editing software can open. You would reopen this file if you weren't finished with your project when you saved it last and wanted to work on it more, or if you wanted to re-edit the material later. Unlike the *exported* file, which exports a single mono or stereo file, the *saved* file keeps copies of multiple work sessions intact, making it possible to go back to earlier editing points if necessary. This is especially helpful if, for example, you have created a project with dialogue or narration and an underlying music bed. Perhaps when creating it, you were using headphones and the mix between dialogue and music sounded balanced, so you exported it to play on the air. Later, however, when playing it on monitor speakers, you found that the music was too loud or not loud enough and distracted from the voice. If you had only the exported audio, but not the saved file or session, all you would have would be the finished production with the audio elements out of kilter, irretrievably mixed together. With the saved file, however, you still have the specific session where the music rests on one track and the voice on another. You can then adjust the voice or music and export the corrected audio again. Keep in mind that with many editing programs, the "save" command does not necessarily mean you are saving the entire edited audio—you are simply saving the work session and nothing more.

Regardless of what audio editor you're using, it's capable of a lot more than these few simple operations. The best way to learn what an editor can do is to try different things and read the manual or help screens. You should also check the software manufacturer's website and other online sources, as many offer tutorials or hints on how to get the most out of these programs. Even though some procedures and terminology may be different, many of the basic principles presented here would apply to any equipment that's capable of digital audio editing.

3.11 MULTITRACK EDITING TECHNIQUES

Most editing programs also allow the recording of several tracks. Figure 3.7 shows the Multitrack View screen from Adobe® Audition® CC. In this view, the first few tracks of the system are shown with some of the controls associated with those tracks. Multitrack recorders and editors give the production person a tremendous amount of creative flexibility. Each production element, such as a vocal track or music bed, can be recorded onto a separate track. Once recorded, they can be individually manipulated (such as changing volume or adding some signal processing effect), but also played back simultaneously to hear the mix of the entire production. Some editors also allow users to edit and synchronize audio for video projects.

Other software programs will have similar screens and functions. To start recording, select MULTITRACK next to the WAVEFORM button at the top of the screen and set the recording parameters as you did before. Like the editing portion of the program, pull-down menus, or right-clicking on a track will provide access to all the multitrack functions; toolbar icons also can be used to select the most commonly used functions. The audio data in each track is graphically displayed as a waveform block. In the multitrack mode, audio can easily be mixed, moved, copied, or deleted, and information can be displayed on the screen in different sizes with a "zoom" control. Each track has various virtual faders and pan controls that allow for level setting and balance control during the mix of the production. To move a waveform block with Adobe® Audition® CC, simply mouse click on it and continue to hold it down. You can now drag and drop that audio from one track to another, from one part of a track to another part, or, using right-click, you can mix, copy, divide, or delete the audio.

One of the most basic techniques used in multitrack work is **overdubbing**, or the process of adding new tracks

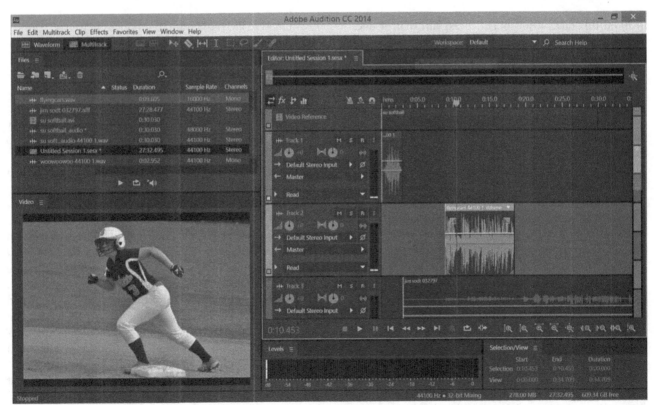

FIGURE 3.7 Most digital audio editing programs incorporate a virtual multitrack recorder. *(Adobe product screenshot reprinted with permission from Adobe Systems Incorporated.)*

to existing tracks. Refer to Figure 3.7 as you follow the steps of a basic overdub using the multitrack function of Adobe® Audition® CC. To begin recording click on File, select New Session and Multitrack, and choose the sample rate and bit resolution you want. Now click on the small R button (to enable recording) for Track 1. Then click on the master RECORD button on the transport controls to record, for example, an announcer's voice onto this track. Click the transport STOP button when you're done recording. To accomplish the overdub, click on the R button for Track 1 again, which will put this track into playback mode. Click on the R button for Track 2. This time, when you click on the master RECORD button, you will record another voice on Track 2. Click on the STOP button when you're done recording. Remember to click on the Track 2 R button when you've finished recording to disable recording on this track.

With a few other mouse clicks, you can adjust the volume levels between the two tracks. You could also import a musical WAV or MP3 file into Track 3, providing a music bed for your production. With Adobe® Audition® CC, you can rip music into the system by clicking on File and then selecting Extract Audio from CD. There are a number of parameters you can choose, but the main options will be to select the track you want to import into the system or select the time to import just a portion of a track. Once you've set the various components and clicked on OK, the audio will be ripped into the system. The exact steps described are unique to this software, but most systems will employ a similar type of process.

Overdubbing allows you to build a production in layers, because you don't have to record everything at the same time. If you decided later to add a sound effect to the production, you could just record it onto another track, move it to the spot where you want it to occur, and set the proper volume. Playing back all the tracks at the same time would let you hear how all the elements fit together.

If you've recorded a vocal on one track, but a portion of the vocal contains a mistake, another multitrack technique known as the **punch in** allows you to record over just the part that contains the mistake, and leave the rest of the track undisturbed. To punch in using Adobe® Audition® CC in the multitrack mode, double-click on the track you want to punch in on. This takes you to the waveform mode and allows you to edit that specific track. You can record several different "takes," and the program will allow you to select the one you feel is best. It's important not to change volume levels or microphone position of the original setup so that the new recording is consistent with the original recording.

Most multitrack recorders can be put in record mode by pressing a RECORD button for the individual track you want to record on or putting that track into a record ready mode, then pressing a master RECORD button on the multitrack. The technique is used for a punch in, because you play back the track to the point where you want to rerecord, punch in the RECORD button, and record over the portion of the track that you wish to correct. You punch out of record mode at an appropriate point after you've fixed the mistake. With a punch in, essentially you're doing an edit on the fly. You might need to practice a few times so you know exactly where you want to punch in and punch out. Try to choose logical spots, such as the end of a sentence or at a pause between words, or the punch in may be noticeable.

Bouncing or **ping-ponging** tracks is the process of combining two or more tracks on a multitrack recorder and rerecording and transferring them to another vacant track of the same recorder. If you were building a music bed, for example, record drums on Track 1 and a bass on Track 2. Then bounce those tracks to Track 3. Now you have Track 1 and Track 2 free, and you can record stereo strings on those two tracks, and, if you want, bounce all three tracks to Track 4. Bouncing is a technique used to increase the number of tracks that you can utilize, however there are both advantages and disadvantages to bouncing tracks. For example, once several tracks are combined, they can't be uncombined, so you can't change the balance of the mixed tracks, and if you equalize or otherwise process the track, it affects all elements on the track.

On the other hand, bouncing tracks together can reduce the hard disk space and processing time needed to manipulate audio. If you have a large number of tracks, making a "submix" of some of them will speed up CPU processing. Again, you can just highlight the tracks you want to combine and choose a "mix down" option for selected files. **Live bouncing** is a similar technique; at the same time several tracks are being mixed onto a vacant track, a live track is added to the mix. Again, it's a technique to get a maximum number of sources onto a minimum number of tracks.

3.12 MULTITRACK VOICE EFFECTS

Multitrack techniques open the door to many special effects that can turn a basic production into a creative masterpiece. Voice doubling, chorusing, and stacking are three forms of overdubbing that can give your next production a unique sound. They are often used to add a "thickness" to the vocal by making it seem like several voices speaking at the same time. Voice doubling is exactly what the name implies. Record your voice on Track 1. Now, while monitoring Track 1, record your voice again on Track 2. Even though you read the same script, it's impossible to record both tracks exactly alike. The effect, also known as "**voice dubbing**," will be closer to two people reading the same script at the same time.

Chorusing is taking voice doubling one step further. Record at least two additional tracks in sync with the original track, creating a "chorus" effect. The more additional tracks you record, the larger your chorus will sound. It's a way for one announcer to become a group of announcers. **Stacking** is a similar multitrack technique in which an announcer "sings harmony" to a previously recorded track. As it is actually a form of delay, chorus effects are often included on audio editing software.

One announcer can appear to be two *different* announcers trading lines using a technique referred to as **dovetailing**. On Track 1, the announcer records the *odd* lines of the script, while reading the even lines silently. This leaves space between the odd lines on Track 1. Now record the *even* lines of the script on Track 2 while mentally reading the odd lines to keep the timing correct. When both tracks are played back, you get the effect of two announcers reading a dialogue script. Obviously, you must slightly change one of the voices so that it sounds like two different announcers, and you'll probably need to practice a few times or do some editing to get the timing correct. Remember, you usually "mentally" read lines slightly faster than you do when you actually speak them. It can take a little time to perfect this technique, but it can be well worth the effort.

Of course, the same effect can be produced in another fashion. Just record the lines for each announcer on the appropriate track of the multitrack, changing one voice slightly. Now cut, move, and paste the lines to create the dialogue. You can create the natural pacing of two announcers by slightly overlapping the end of one line with the beginning of the next.

To produce a **slapback echo** with a multitrack recorder, record your voice on Track 1. Now, dub Track 1 to another track, but without synchronization. Play back both tracks, and you'll get a unique echo effect. As this is another form of delay, look for this special effect as a menu item on your editing software's special effects section.

3.13 TRACK SHEETS

As you can tell by reading about these various techniques, multitrack recording can become complicated. Good production practice dictates that you keep notes of what material is recorded on which track. Figure 3.8 shows

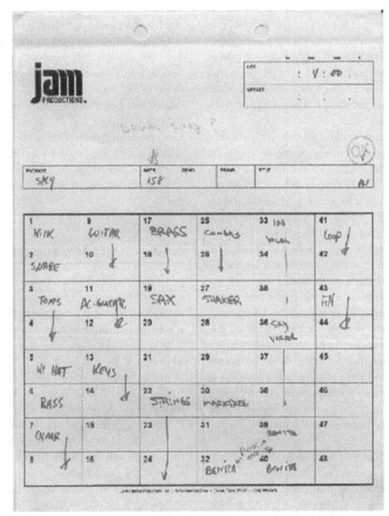

FIGURE 3.8 A tracking sheet showing full instrumentation, plus three solos with a pencil-note to "pick one" for the mixdown. The star and OK mean it's been mixed, and multitrack production master (for later syndication) created. *(Courtesy: JAM Creative Productions, Inc.)*

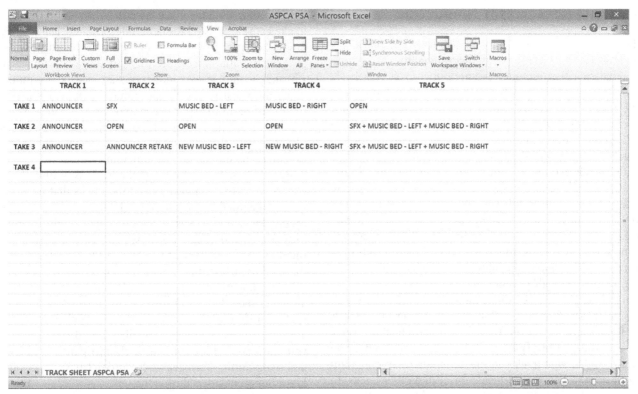

FIGURE 3.9   A track sheet created using Microsoft Excel shows what material is recorded on which track during a multitrack recording session.

an example of a track sheet used at a professional audio production facility. Although there is no universal form, Figure 3.9 shows an example of a **track sheet** created using Microsoft Excel, to be used for keeping notes regarding a multitrack production. The top of the page indicates the number of tracks the recorder has; the edge of the page indicates the various takes, and the boxes show what was put on each track during each take. In this example, during the first take, an announcer vocal was put on Track 1, a sound effect was put on Track 2, a music bed was put on Tracks 3 and 4, and Track 5 was left open. This doesn't mean that these elements were all recorded at the same time, but these were the first elements assigned to those tracks.

During the second take, the announcer vocal was not changed, but the sound effect on Track 2 and the music bed on Tracks 3 and 4 were all bounced to Track 5. On the third take, Tracks 1 and 5 did not change; however, a new music bed was put on Tracks 3 and 4, and the announcer vocal was rerecorded on Track 2. You could also put any signal processing settings, music bed cuts used, or special notes regarding the production on the track sheet.

Most editing programs allow you to name each track on-screen, as you edit and there may be no need for a hard copy. However it is still good practice to keep some form of a paper track sheet during production in the event of computer failure or file loss.

3.14 THE MIX DOWN

Regardless of how many tracks you work with in a multitrack production, you will probably mix the production down into a stereo mix with just a left and right channel. For many programs, including Adobe® Audition® CC, there is a menu item selection to "mix down" all files.

Certain tracks from the multitrack master may be panned to the right channel, some tracks may be panned to the left, and some may be balanced to both left and right. This is where the production person needs a good monitoring system and some good judgment to create just the right mix.

Most likely you're working in stereo and hearing the final mix in stereo, however you also need to hear the same mix in mono, because while sometimes a mix sounds great in stereo, it may not in mono. Often by listening in cue (assuming that the cue signal is in mono), you can tell if a stereo mix is not going to be compatible with mono. It's possible that portions of the audio signal will be out of phase when the left and right channels are combined for the mono signal, causing a cancellation and diminishment of the sound at different spots. A fine-tuning of some track equalization, a slight volume change, or minor panning adjustment will usually correct the mono signal, but it's always important to make sure the stereo and mono signals are compatible.

3.15 MULTITRACK SPOT PRODUCTION

There is no "standard" way to record using multitrack techniques, but let's look at the production of a commercial spot that includes two announcer voices, a background music bed, and two sound effects. In most cases, the music bed would begin first at full volume for a few seconds, and then fade under the vocals and be held at a background level. The first announcer begins his or her voice-over on top of the music bed, probably trading lines with the other announcer until the end of the copy. Sound effects are added at the appropriate points, and the music bed is brought up to full volume at the end of the voice-over. The music bed ultimately ends cold or may fade out. Of course, all of this takes place in a 15-, 30-, or 60-second time frame. Assuming a production studio setup with multitrack audio editing capability, here is what is happening during this "typical" production.

Each sound source (microphone, CD, etc.) should be assigned to a stereo track. Put one announcer on Track 1, the other announcer on Track 2, a sound effect on Track 3, a music bed on Track 4, and another sound effect on Track 5. The first track you record is extremely important because it's often the reference for all the other tracks. In audio production, the music bed or vocal track is often put down first. In this case, the music bed is a preproduced and timed music bed from a CD production library. We just have to dub it or rip it from the CD into the computer, either by importing it or setting the multitrack recorder to record on Track 4 only (the music bed). In either case, you want to have the entire music bed at full volume, even though you'll be fading it under the announcers. The various levels can be balanced during the final mix.

Next, put Track 4 into a "safe" or "sync" mode. This allows you to overdub on the other tracks at the same time that you monitor the music bed track. Set Tracks 1 and 2 into the record position. Since the announcers trade lines throughout the commercial, they should be recorded at the same time, which will provide a more natural feeling of interplay and continuity than if they were recorded separately. While monitoring Tracks 4, 1, and 2, record the vocals on Tracks 1 and 2. Music should be faded at the appropriate place, so the announcers get the proper feel of reading the spot, but remember the actual blend of all

the elements will take place later. At this point, if the vocal tracks came out okay, the announcers could leave, and a production person would finish the spot. That makes sense to do if the announcers are high-priced talent, and that's why many multitrack productions are started by putting down the vocal tracks. Suppose one announcer misspoke the final line in the spot. Depending on the complexity of the spot, it would be possible to set up Tracks 4 and 1 to play, and have the announcer rerecord a portion of Track 2 to correct the mistake with a punch in.

By additional overdubbing, the various sound effects would be added by playing back the tracks that were previously recorded and recording only those tracks that were assigned to sound effects, doing one effect at a time. If you miscue a sound effect, you don't have to reset every element and start all over again. You would merely rerecord the track that had the mistake on it or click and drag the misplaced element to the desired location. Once you have all tracks recorded, you can begin to "mix down" the final spot. You should play back all tracks several times first to adjust the levels. If you're using digital multitrack software, you can often adjust volume levels and pans so they will automate during playback and you don't have to manipulate a bunch of faders during the mix.

3.16 CONCLUSION

This chapter is significant in your development as a production person. Audio editing is a very important concept as more emphasis is being placed on sound bites and ear-catching productions. The modern production person must also be skilled in working in the multitrack environment to be successful. Although this chapter is only designed to give you some basic familiarity with the equipment and techniques, it should serve as a good starting point. Like almost everything about audio production, the real learning comes when you go into the studio and try out different concepts. You'll only become really skilled in multitrack production work or audio editing if you spend time in the studio to understand the creative possibilities it offers. The type of equipment or particular software that is used is not nearly as important as *how* this equipment is used to convey information in a meaningful way.

Self-Study

1. In which area is digital technology superior to analog technology?

   a) signal-to-noise ratio
   b) frequency response
   c) dynamic range
   d) all of the above

2. Which of the following is a reason to edit audio?

   a) to eliminate mistakes
   b) to record out of sequence
   c) to cut to exact length
   d) all of the above

3. A standard off-the-shelf desktop or laptop computer can be converted into a basic digital audio editor by adding appropriate software and what other component?

   a) an ESP audio card
   b) an ESPN audio card
   c) a DSP audio card
   d) an ASP audio card

4. An audio sound file that has been edited with a non-destructive system has been permanently altered so the original audio can't be restored.

   a) true
   b) false

5. Musical instruments (such as synthesizers or samplers) can interface with digital audio equipment through an electronic communications language known as what?

   a) DSP
   b) Audacity
   c) MINI
   d) MIDI

6. Where would you place the edit marks to edit out the word "two" from the phrase "one … two … three"?

   a) just after letter e of *one* and just before letter *t* of *three*
   b) just before letter *t* of *two* and just before letter *t* of *three*
   c) just before letter *t* of *two* and just after letter *o* of *two*
   d) no correct way to mark this edit is shown here

7. Which stage of the digital recording process breaks down the analog signal into discrete values or levels?

   a) filtering
   b) sampling
   c) quantizing
   d) coding

8. Which term describes the process when two or more tracks of a multitrack recording are combined and rerecorded on another vacant track?

   a) overdubbing
   b) bouncing
   c) punching in
   d) combo-cording

9. In multitrack recording, what is the process of adding new tracks to existing tracks called?

   a) bouncing tracks
   b) overdubbing tracks
   c) punching in tracks
   d) ping-ponging tracks

10. Why can digital recordings be copied over and over with no measurable loss of sound quality?

   a) The sound signal is a continuously variable electrical signal, the shape of which is defined by the shape of the sound wave produced.
   b) The original sound source is sampled over 44,000 times per second.
   c) It is binary data that is recorded, which can be accurately copied.
   d) Because they sound better when you're wearing headphones, as opposed to monitor speakers.

11. One announcer can appear to be two different announcers reading a dialogue script using which multitrack recording technique?

   a) voice doubling
   b) chorusing
   c) dovetailing
   d) stacking

12. Most digital equipment promises a longer interval between breakdowns than comparable analog equipment; however, when there is a technical problem, repair time for the digital equipment will probably also be longer.

   a) true
   b) false

13. The main reason announcer voices are often recorded first on a multitrack recording is because vocal sources are always assigned to Track 1 and Track 2.

   a) true
   b) false

14. Most audio editing software can save files in a number of different audio formats. Which are the two most common?

   a) .WAV and .WMA
   b) MP3 and AIFF
   c) VOC and PCM
   d) WAV and MP3

15. Which digital sampling rate is used most frequently?

   a) 32 kHz
   b) 44.1 kHz
   c) 48 kHz
   d) 88.2 kHz

16. A track sheet is a manufacturer's specification sheet that lists the number of tracks that a multitrack recorder has.

   a) true
   b) false

17. Multitrack productions are usually mixed down to a 2-track stereo master. If the stereo master is slightly out of phase when combined into a mono signal, it can be corrected by each of the following adjustments to one of the stereo tracks EXCEPT.

   a) change track equalization
   b) move track to a vacant track
   c) change track volume
   d) change track panning

18. During multitrack recording, which technique allows you to rerecord just a portion of a track to correct a mistake, while leaving the rest of the track undisturbed?

   a) punching in
   b) overdubbing
   c) voice doubling
   d) bouncing

19. What is another term for "bouncing tracks" on a multitrack recorder?

   a) chorusing
   b) overdubbing
   c) ping-ponging
   d) dovetailing

20. The "drag-and-drop" function of a multitrack audio editor would allow you to move audio segments from one part of a track to another, but not from one track to a completely different track.

   a) true
   b) false

ANSWERS

If you answered A to any of the questions:

1a. You're partially correct, but this is not the best response. (Reread 3.3 and 3.7.)
2a. Although this is one reason for editing audio, there is a better answer. (Reread 3.4.)
3a. No. You didn't "sense" the correct answer. (Reread 3.5.)
4a. You're not quite right with this answer. Although audio edited with a destructive system is ultimately physically changed, most non-destructive editing systems provide an Undo function that can restore the original audio. (Reread 3.10.)
5a. No. DSP is a piece of computer hardware, not a communication language (Reread 3.5 and 3.8).
6a. No. This would leave no space or pause between the words one and three. (Reread 3.10.)
7a. No. Filtering cuts off unwanted high frequencies before sampling begins. (Reread 3.3.)
8a. Wrong. Overdubbing is the normal technique for multitrack recording. (Reread 3.11.)
9a. No. This is another multitrack recording technique. (Reread 3.11.)
10a. No. This refers to an analog signal. (Reread 3.2 and 3.3.)
11a. Wrong. Though this is a "voice doubling" technique, it's another multitrack trick. (Reread 3.12.)

12a. No. Only half of this statement is true. Although digital equipment will probably break down less often than analog equipment, repair time for digital equipment also promises to be shorter in many cases. (Reread 3.7 and Production Tip 3B.)

13a. No. It really doesn't matter what sources are assigned to what tracks. (Reread 3.15.)

14a. No. Although audio editors might use both of these formats to save files, only one of them is very common. (Reread 3.10.)

15a. No. This is another sampling rate that some digital equipment is capable of utilizing. (Reread 3.3.)

16a. No. A track sheet is a way of keeping notes during a multitrack production. It usually lists track numbers and what is recorded on each track. (Reread 3.13.)

17a. Wrong. This will often correct the mono signal. (Reread 3.14.)

18a. Yes. The punch in is also known as an insert edit because you can just rerecord a portion of a track.

19a. No. This is a different multitrack technique. (Reread 3.11 and 3.12.)

20a. Wrong. You would be able to move the audio from track to track as well as anywhere on an individual track. (Reread 3.11.)

If you answered B to any of the questions:

1b. You're partially correct, but this is not the best response. (Reread 3.3 and 3.7.)

2b. Although this is one reason for editing audio, there is a better answer. (Reread 3.4.)

3b. No. You might have just taken a "sporting" guess to get this answer. (Reread 3.5.)

4b. Yes, because most non-destructive editing systems have an Undo function that enables edited audio to be restored.

5b. No. Audacity is an audio editing software program, not a communication language. (Reread 3.8 and Production Tip 3A.)

6b. Yes. This would edit out the word *two*, but it would maintain the natural pause between *one* and *three*.

7b. No. Sampling is when the signal is "sliced" into thousands of individual samples (voltages). (Reread 3.3.)

8b. Correct. Bouncing or ping-ponging tracks is the multitrack technique described here.

9b. Yes. This is a primary advantage of multitrack recording and one of the most basic techniques used in multitrack production.

10b. No. But you're heading in the right direction. Sampling rate is important for exact reproduction of the original sound, but it isn't really the reason why digital recordings can be copied over and over. (Reread 3.2 and 3.3.)

11b. Wrong. Although this is a technique for adding voices, it's another multitrack technique. (Reread 3.12.)

12b. Yes. Only part of this statement is true. Digital equipment repair time should be shorter than analog.

13b. While it doesn't really matter what sources are assigned to what tracks, announcers are often recorded first because it's easier to lay down other tracks to the vocals, and high-priced talent can be finished with their part of a production once their tracks are recorded, even if the entire spot isn't completed.

14b. No. Although audio editors might use both of these formats to save files, only one of them is very common. (Reread 3.10.)

15b. Correct. This is the most common digital sampling rate.

16b. Correct. A track sheet is a way of keeping notes of what is recorded on each track of a multitrack production.

17b. Correct. Just moving a track's location won't help.

18b. No. This technique allows you to add additional tracks to a production. (Reread 3.11.)

19b. No. This is a different multitrack technique. (Reread 3.11 and 3.12.)

20b. Correct. You can move audio from track to track as well as anywhere on an individual track.

If you answered C to any of the questions:

1c. You're partially correct, but this is not the best response. (Reread 3.3 and 3.7.)

2c. Although this is one reason for editing audio, there is a better answer. (Reread 3.4.)

3c. Correct. A DSP audio card is a standard audio interface for the PC platform.

5c. No. You're probably confusing this with something small. (Reread 3.8.)

6c. No. This would lengthen the pause between *one* and *three* and sound unnatural. (Reread 3.10.)

7c. Yes. This is correct.

8c. Wrong. Punching in is a technique that allows you to record a segment of a track without affecting the material before or after that segment. (Reread 3.11.)

9c. No. This is another multitrack recording technique. (Reread 3.11.)

10c. Correct. Binary data can be accurately recorded over and over, making digital copies sound exactly like the original.

11c. Correct. By reading even lines of a script on one track and then odd lines on another track, one announcer can sound like he or she is talking to another person with this technique.

14c. No. Although audio editors might use both of these formats to save files, neither of them is very common. (Reread 3.10.)

15c. No. This is another sampling rate that some digital equipment is capable of utilizing. (Reread 3.3.)

17c. Wrong. This will often correct the mono signal. (Reread 3.14.)

18c. No. This technique is a form of overdubbing that allows you to "voice double." (Reread 3.11.)

19c. Yes. This is another term for bouncing tracks.

If you answered D to any of the questions:

1d. Correct. Digital technology offers all these technical improvements over analog.

2d. Right. All of these are good reasons for editing audio.

3d. No. You got "bitten" on this answer. (Reread 3.5.)

5d. Yes. MIDI (musical instrument digital interface) is the correct response.

6d. No. One of the other answers is the correct way to mark this edit. (Reread 3.10.)

7d. No. Coding is when a series of binary digits are assigned to each individual sample. (Reread 3.3.)

8d. Wrong. There is no such thing. (Reread 3.11.)

9d. No. This is another multitrack recording technique. (Reread 3.11.)

10d. No. This has nothing to do with digital audio's reproduction or ability. (Reread 3.2 and 3.3.)

11d. Wrong. Although this is a technique for adding voices, it's another multitrack technique. (Reread 3.12.)

14d. Yes. These are the two most common formats that audio editors might use to save files.

15d. No. This is not a common sampling rate. (Reread 3.3.)

17d. Wrong. This will often correct the mono signal. (Reread 3.14.)

18d. No. This technique allows you to combine tracks and transfer them to a vacant track. (Reread 3.11.)

19d. No. This is a different multitrack technique. (Reread 3.11 and 3.12.)

Projects

PROJECT 1

Undertake digital audio editing.

Purpose

To gain practice editing a vocal sound file using a digital editing system.

Notes

1. If you're not sure of what you're doing, ask the instructor for assistance.
2. Remember, you're to do several edits, not just one.
3. You'll be judged on how clean your edits are, so be sure you make your edit marks accurately.

How to Do the Project

1. Familiarize yourself with the operation of the digital editing system in your production studio. If you have questions, ask your instructor.
2. You may use the material on the website associated with the text, or you may record your own. If you can record your own, select some news copy or a weather forecast from a news wire service, or write something similar and record it into your editing system. Label this sound file as required by your system.
3. Do your edits as follows:
   a. Press the PLAY button, and listen to what is recorded.
   b. Select something you wish to edit. Write down on a piece of paper the part you plan to edit, with a few words before and after it. Put parentheses around what you plan to take out. For example: "Today's weather calls for (sunny skies and) a temperature of 70 degrees."
   c. Stop playback so that it's at the exact place you wish to edit—in our example, just in front of "sunny."
   d. Make your beginning edit mark.
   e. Continue playing the audio until you get to the end of your edit—in our example, just before "a."
   f. Make your end edit mark.
   g. Preview the edit. You can probably adjust either edit mark so that it is accurately positioned. If necessary, do so.
   h. Perform the actual edit, and listen to it. If you've made some type of error, "undo" the edit, and start again.
4. Repeat the above steps for another edit. Do a few more similar edits.
5. When you've done several edits, record and save them all.
6. Label your recording with your name and "Digital Audio Editing."
7. Turn in the recording to your instructor to receive credit for the project.

PROJECT 2

Using a digital audio editor, build a short music bed and record a "voice and music bed" spot.

Purpose

To continue to give you experience with digitally based recording and editing.

Notes

1. This project incorporates a single announcer voice and a music bed. There are many ways to accomplish it, so don't feel like you must follow the production directions exactly.

2. You will use music from the website to build your music bed. If you don't have access to the textbook's website, ask your instructor for music or provide your own from a CD or other source.
3. You'll be laying the same few seconds of music down several times, so select some music that will lend itself to this. Find a piece of music that won't sound like it has been cut off abruptly.
4. The project is written to make a music bed that is 60 seconds long.
5. You will also need to write a script that can be read in about 50 seconds. Write the copy about a desktop item, such as a pen, stapler, or paperclip holder.

How to Do the Project

1. Activate your computer editing program.
2. Select and import music from the textbook's website or another source, and record a sound file of about 20 seconds.
3. Select a 10- to 15-second region of music from the sound file, and lay it down on one track made available through your computer editing program.
4. Try to loop the music piece so that it creates a continuous loop on playback. You may need to trim from, or add to, the music to make this work.
5. Once you have your loop, copy and paste it to create a 60-second music bed.
6. Listen to the segment and improve upon it so that it could be used as a music bed under someone talking about a commercial product. You may want to change volume or edit to tighten it a bit. What you can do and exactly how you do it will depend on the program that you're using, and the effect you are trying to achieve.
7. Now record the script. Rerecord or edit to correct vocal mistakes, to get the timing correct, or just to make it sound the way you want.
8. Set playback levels for both the vocal track and the music bed. Both should start at full volume.
9. Record the spot. Start the music bed at full volume and then start the vocal track as you simultaneously fade the music slightly so that the vocal is dominant.
10. As the vocal ends, bring the music bed back to full volume and then fade it out at 60 seconds.
11. Make sure you have a good balance between the music bed and announcer voice. If you need to do it over, just cue up, make necessary adjustments, and try again.
12. Label your final project with your name and "Voice and Edited Music Bed Spot."
13. Give the project to your instructor to receive credit for this project.

PROJECT 3

Write and record a 60-second "concert commercial."

Purpose

To develop your skill in creating a spot for a concert by a recording artist, a type of spot frequently heard on radio.

Notes

1. This project assumes that you have some type of digital audio editor in your studio; however, it can also be accomplished with other equipment.
2. The production incorporates two announcer voices, a music bed, and several "clips" from the artist's songs.
3. You can write the actual script as you want it, but follow the basic concept given below.
4. Exact production directions aren't given in this project, because there are many ways to accomplish the recording and editing.

How to Do the Project

1. Import or record four or five segments of songs from the artist's biggest hits. Each clip should be about 10 seconds long, but some may be longer and others shorter.
2. You should pick segments of the songs that are readily identifiable—usually the chorus or "hook" of the song. Try to end the clip at the end of a phrase so that it isn't just cut off in the middle of a word. Make sure the last clip you use doesn't fade out, as you want to finish the spot with a strong, natural ending.
3. Start recording each segment a bit before the point you actually want and record a bit beyond the actual end. Mark the true start point and end point using your audio editor and then label and save each segment.
4. Record the announcer lines with both announcers trading lines. However, the announcers should read the last line together. If your editor has effects capability, add some reverb or a flanging effect to the voices.
5. Your script should use lines similar to this:

Announcer 1:	Appearing live at County Stadium, (Artist Name)!
Announcer 2:	Join (artist's name) on June 25 for (his/her/their) first-ever concert in (your city)!
Announcer 1:	Here's the summer concert you've been waiting for!
Announcer 2:	Tickets are just $50 and can be purchased at the County Stadium Box Office or at www.getticketshere.com!
Announcer 1 & 2:	Don't miss this chance to see (artist's name) in concert at County Stadium on June 25!

6. Using your audio editor, assemble the spot in a sequence that alternates between announcer tracks and song clips, beginning with Announcer 1's first line.
7. Add a background music bed that will pull the whole spot together. You might edit an instrumental segment from one of the artist's songs to do this, or you might find a music bed that fits the style of the other music used in the spot.
8. When you mix the music bed in, you should start with the bed at full volume for a couple of seconds and then fade it under the announcer segments. When the music clips are playing, the music bed should be faded out entirely.
9. Listen to the finished commercial. If it's good enough, prepare it for the instructor.
10. Label the finished project with your name and "Concert Commercial" and give it to your instructor to receive credit for this project.

PROJECT 4

Write and record a report that compares various recording/editing software programs.

Purpose

To allow you to further investigate some of the computer software that is available for digital production systems.

Notes

1. Prepare a list of questions that you want answered before you begin your research.
2. This project requires obtaining material through the Internet from various manufacturers.
3. The manufacturers shown at the end of this project are not the only ones who produce audio editing software. You could also complete the project by looking at other programs.

How to Do the Project

1. Choose five audio editing software providers from the list below. Your final report will discuss your findings of only the five you choose.

2. Develop some questions that you want answered about audio editing software. For example:
   a. What are the computer system requirements for each software program?
   b. What exactly can each do in the area of audio, radio, and video production?
   c. How much do they cost?
   d. Are there any problems associated with these programs?
   e. What sample rate and bit depth does each program use?
3. Gather information by going to the website of the manufacturers.
4. If your facility has this software or you know of a station or facility in your area that does, try to talk with someone who uses it on a regular basis to further research the programs.
5. Organize your material into an informative report that compares the various software programs. Write your name and "Editing Software Programs Comparison" on the title page of your report.
6. Record your report using the recording software of your choice from the list below. Use a music bed at the beginning and end of your recorded report. Be sure to save your final recorded report as one MP3 file.
7. Give the written and recorded report to your instructor to receive credit for this project.
   Adobe® Audition® CC
   creative.adobe.com/products/audition
   Audacity
   sourceforge.net/projects/audacity
   Avid Pro Tools
   www.avid.com
   Cubase
   www.steinberg.net
   SAW Studio
   www.sawstudio.com
   Sonar
   www.cakewalk.com
   Sound Forge
   www.sonycreativesoftware.com/soundforgesoftware
   TwistedWave
   www.twistedwave.com

PROJECT 5

Record a public service announcement that uses a sound effect.

Purpose

To develop your skill in creating a radio spot that incorporates a sound effect to provide an effect or transition.

Notes

1. This project assumes that you have enough familiarity with your studio equipment to accomplish basic recording and production techniques.
2. The production incorporates a single announcer voice, two distinct music beds, and a sound effect to provide a transition between the music beds.
3. You will need to write a simple script that can be read in about 20 seconds. Write the copy about an environmental concern, such as water pollution, littering, or forest fire prevention. The spot should follow a "problem–solution" format.
4. There are many ways to accomplish this project, so don't feel like you must follow the production directions exactly.

How to Do the Project

1. Select music beds that are appropriate for the style of the spot. You should use a more somber-sounding bed for the "problem" and a more upbeat sound for the "solution."

2. Import and record the music bed and script (voice) onto the computer. First record the "problem" part of the spot, using a serious tone of voice and mixing the appropriate music bed with it. Then record the "solution" with a lighter vocal delivery, slightly faster pace, and more up-tempo music.

3. Begin recording the music bed at full volume. Start the vocal track and simultaneously fade the music bed slightly so that the vocal track is dominant while you record the rest of the script.

4. Once you are finished recording, save your work.

5. Now, find an appropriate sound effect. It will be used to separate the two music beds and provide a transition between the "problem" and "solution" parts of the spot.

6. Import or record the sound effect into the project.

7. Set correct playback levels for the vocal/music bed tracks and the sound effect. Then cue both to the beginning sound.

8. It may take you several attempts to get the spot to come out correctly. If you need to do it over, just cue everything and try again.

9. Label the finished PSA, with your name and "Sound Effect Radio Spot," and turn it in to your instructor to receive credit for this project.

4

MICROPHONES

4.1 INTRODUCTION

In audio production, microphones are used to record voice, music, or sound effects in both the studio and the field. The studio microphone (see Figure 4.1) takes on an important role, because it is usually the first element in the audio chain. It's the piece of equipment that changes the announcer's voice into an electrical signal that can then be mixed with other sound sources and either recorded, streamed, or broadcast. Because the purpose of the microphone is to change sound energy into electrical energy, it is known as a **transducer**, which is a device that converts energy from one form into another.

While digital microphones are becoming more common, many microphones are still analog-based. Putting an analog-to-digital converter in an external microphone preamplifier or immediately after the microphone in the audio chain, or within the microphone housing itself could quickly transform any microphone's electrical output to a digital signal. For example, Audio-Technica's AT2020 USB microphone, shown in Figure 4.2, has an analog-to-digital converter built into the housing, with a USB output for direct connection to a computer or digital recorder. Manufacturers including National Semiconductor and Akustia have developed integrated circuits that convert a microphone sound signal directly into a digital bit stream

FIGURE 4.1 The studio microphone changes the announcer's voice into an electrical signal. *(Image courtesy of Neumann/USA.)*

FIGURE 4.2 With a USB digital output, this microphone is ideal for computer-based recording. *(Image courtesy of Audio-Technica U.S. Inc.)*

output signal, making for a truly digital microphone. It's likely that we'll continue to see more digital microphone developments in the future.

This chapter looks at different types of microphones and some of their key characteristics. Selecting the wrong microphone for a particular situation can result in a poor recording that can't be fixed with any postproduction processing. The astute audio production person takes advantage of a microphone's qualities to achieve the best possible sound quality for either live broadcast or digital recording.

4.2 CLASSIFYING MICROPHONES

There is no one universal, standard microphone that will work in all recording situations. For example, a microphone that is perfect for voice-over work in the studio may not work well for recording a sound effect in the field. There are, however, specific types of microphones that work better than others in certain situations. Microphones are usually classified by two key characteristics—the way in which they convert sound into electrical current, and their pickup pattern. Generally speaking, there are two ways that microphones convert sound into electricity, and there are two types of pickup patterns. In both categories, there are some variations.

Some radio stations use a standard microphone throughout their facilities; however, the general audio production facility tends to take another approach, which is to have multiple types of microphones available. That makes sense, because it allows the facility to be flexible when handling a number of different recording situations.

4.3 DYNAMIC MICROPHONES

The **dynamic microphone** refers to one way in which sound is converted into electricity. Commonly known as the **moving-coil microphone** or occasionally the **pressure microphone**, its sound-generating element is constructed of a thin, plastic diaphragm, a magnet assembly, and a voice coil, as shown in Figure 4.3. The flexible diaphragm, located near the head of the microphone, responds to the pressure of sound waves. The diaphragm is positioned so that the changes in pressure cause an attached small coil of wire to vibrate. This coil rests within the field of a permanent magnet, and movements of the diaphragm result in a disturbance of the magnetic field. This induces a small electric current into the coil of wire, which is the audio output signal.

The dynamic microphone is a good general-purpose microphone commonly used in radio and audio production studios. Because of its relatively simple construction, it is modestly priced, and produces very low noise with excellent **frequency response** (the accurate reproduction of both high and low frequencies). Another reason for its popularity is its sturdy design. The dynamic microphone can

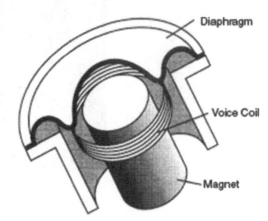

FIGURE 4.3 The internal components of a dynamic microphone include a diaphragm, magnet, and voice coil. *(Image courtesy of Shure Incorporated.)*

withstand a moderate amount of abuse, which often occurs in broadcast settings and in field production. Dynamic microphones can handle extremely high sound levels, which make it almost impossible to overload them, and they are not usually affected by temperature or humidity extremes. This style of microphone is also fairly insensitive to wind, and this feature, along with its ruggedness, makes it an excellent remote or field recording microphone.

The main disadvantage of the dynamic microphone is that it has difficulty reproducing certain vocal characteristics. The microphone has a tendency to exaggerate plosives (popping on *p*) and sibilance (hissing on *s* or *c*). Dynamic microphones also can lose some light, delicate sounds, because the mass of the diaphragm requires a fairly high level of sound to move it, and even though the dynamic microphone is fairly rugged, all microphones are fragile to some extent and should be handled with care like any other piece of audio production equipment.

4.4 CONDENSER MICROPHONES

The other way a microphone changes sound into electricity is to use an electronic component (a **capacitor**) for transduction. This type of microphone is commonly referred to as a **condenser microphone**. The sound-generating element consists of a charged conductive diaphragm and an oppositely charged metallic backplate separated by an insulating material that creates an air space between them to form a sound-sensitive capacitor (see Figure 4.4). The thin metal or metal-coated plastic diaphragm responds to sound waves, changing the distance between the diaphragm and the backplate; this alteration changes the **capacitance** (the resistance to electrical voltage buildup) and generates a small electrical signal that is further amplified within the electronics of the microphone. This fluctuation of electrical current is the audio output signal.

A condenser microphone requires a power supply, sometimes a battery, to charge the backplate, the diaphragm,

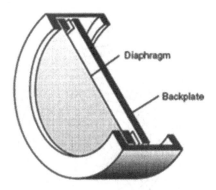

Condenser Microphone

FIGURE 4.4 The internal components of a condenser microphone include two oppositely charged plates—a moveable diaphragm and a fixed backplate. *(Image courtesy of Shure Incorporated.)*

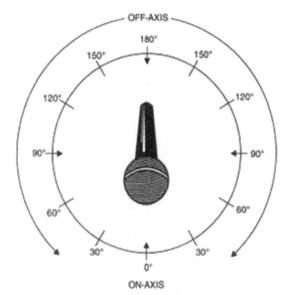

FIGURE 4.5 On-axis sound, picked up directly from the front of the microphone, fully impacts the microphone's diaphragm.

and related electronic components. Some condenser microphones utilize small internal power supplies or **phantom power** rather than a battery. Phantom power usually comes from a recorder or an audio console through the microphone cable and back to the microphone.

Because all condenser microphones are powered, the electronics of the microphone can produce a little noise, and there is a limit to the sound level that the microphone can handle. Still, the condenser microphone is an excellent choice, because it's fairly rugged and produces excellent sound quality and wide frequency response. This microphone also has excellent transient response (the measure of how well a microphone's diaphragm responds to changes in sound waves), but can be distorted at high recording volumes. Although the dynamic microphone is the most commonly used radio production microphone, the condenser microphone is also frequently found in the modern audio production facility. Built-in microphones on many portable audio recorders are often condenser microphones that provide fairly good quality on both consumer and professional models.

4.5 MICROPHONE PICKUP PATTERNS

In addition to categorizing microphones by the way they convert sound into electricity, they can also be classified by the direction in which they are most sensitive to sound, also known as their **pickup patterns**. Microphones are constructed so that they have different directional characteristics (Figure 4.5). Sound picked up at the front of the microphone, or at 0 degrees, is said to be **on-axis**. Sound picked up from the microphone's side is 90 degrees **off-axis**, and sound picked up from the rear of the microphone is 180 degrees off-axis. A microphone's housing, through the use of small openings and ports, can be designed so that unwanted sound (often off-axis sound) is canceled out or weakened as it enters the microphone.

The on-axis sound received directly from the front of the microphone fully impacts the microphone's diaphragm. An

understanding of the microphone pickup patterns helps the user place the microphone in the best location relative to the sound source, in order to maximize the pickup of desired sound and minimize the pickup of unwanted background noise.

The most common pickup patterns are omnidirectional and cardioid, but there are variations. Although most microphones will have one fixed pickup pattern, **multidirectional microphones** have switchable internal elements that allow the microphone to employ more than one pickup pattern.

4.6 THE OMNIDIRECTIONAL PICKUP PATTERN

The **omnidirectional** microphone is also known as a **nondirectional** microphone. These two terms may seem to contradict each other since *omni* means "all" and *non* means "no." Both terms are correct, however, because this microphone picks up sound in all directions, but that also means that it has no particular pickup pattern. Think of a basketball with a microphone in the middle. No matter where the sound comes from around the basketball, the microphone responds to it equally well. Figure 4.6 illustrates the pickup pattern for a typical omnidirectional microphone.

Omnidirectional microphones are used whenever it is desirable to pick up sound evenly from all sides of the microphone, including above and below it. Omnidirectional microphones are commonly used outside the studio when the ambience of the location needs to be picked up along with a person's voice. Of course, the fact that these microphones pick up sound equally well from all directions can also be a disadvantage. You may pick up unwanted background noise (such as traffic noise, air conditioner noise, or crowd noise) in addition to the voice you

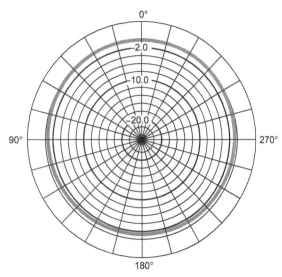

FIGURE 4.6 The omnidirectional pickup pattern shows that the microphone picks up sound equally well from all directions.

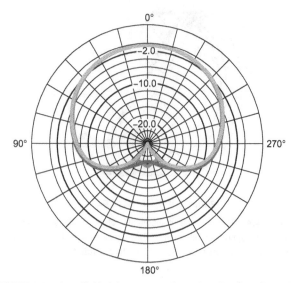

FIGURE 4.7 A cardioid pickup pattern shows that the microphone picks up sound mainly from the front and side, but not very well from the back.

want to record. Omnidirectional microphones used in a highly reflective room may also produce a "hollow" sound, because they tend to pick up more reverb from the room than other types of microphones.

4.7 THE CARDIOID PICKUP PATTERN

The **cardioid** microphone picks up sound from mainly one direction—the front of the microphone. Its pattern is actually heart-shaped, hence the name cardioid (see Figure 4.7). Another way to visualize this pickup pattern is to think of an upside-down apple, where the stem represents the microphone and the rest of the apple approximates the cardioid pickup pattern. Although the microphone picks up sound from the front and sides, the level of pickup from the sides is only about half that of the front. The microphone doesn't pick up sound well from the rear at all—less than one-tenth of the sound it can pick up from the front.

Variations on the basic cardioid design include **super-cardioid**, **hyper-cardioid**, and **ultra-cardioid** microphones. They continue to offer great rejection of sound from the sides, but each microphone also picks up a more narrow scope of on-axis sound. These microphones are even more directional in nature and are sometimes referred to as **unidirectional**. Cardioid microphones are very popular because they reject unwanted sounds (excessive reverb, feedback, background noise), but the talent must be careful to stay "on mic" and not move too far off-axis, especially when using super-, hyper-, or ultra-cardioid microphones.

Another microphone pickup pattern is the **bidirectional** microphone, which picks up sound from the front and the rear of the microphone, or on-axis and off-axis (see Figure 4.8). Its pickup pattern can be visualized as a

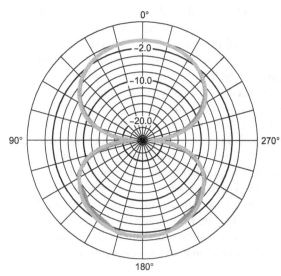

FIGURE 4.8 The bidirectional pickup pattern shows that the microphone picks up sound from the front and back of the microphone.

figure 8, with the microphone located at the intersection of the two circles. It was often used for radio dramas so that actors could face each other, but it is not a common pickup pattern for most of today's microphone use. It is, however, a good microphone for the basic two-person in-studio interview.

4.8 POLAR RESPONSE PATTERNS

Keep in mind that pickup patterns refer to the three-dimensional area around the microphone in which it best "hears" or picks up the sound. Figures 4.6, 4.7, and 4.8 are actually visual representations of a microphone's **polar pattern**. Polar patterns are two-dimensional drawings or

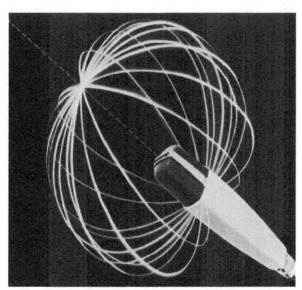
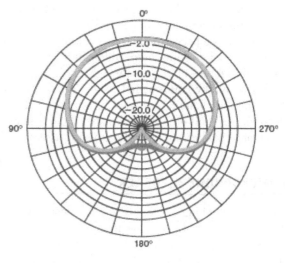

FIGURE 4.9 A three-dimensional view of the cardioid microphone pickup pattern alongside its two-dimensional polar pattern representation. *(Three-dimensional image courtesy of Sennheiser Electronic Corporation.)*

representations of a microphone's pickup patterns, and are used mainly due to the difficulty of displaying three dimensions on a printed page. Compare the cardioid pickup pattern and polar pattern shown in Figure 4.9. This distinction is brought up here because although pickup patterns and polar patterns are technically different, in practice, the terms are often used interchangeably.

A polar pattern is drawn around a full circle, or 360 degrees. As noted earlier, the line from 0 to 180 degrees represents the microphone's axis, and sound entering the microphone from the front (0 degrees) is on-axis. The concentric circles of a polar pattern show decreasing dB levels and allow us to see how the sound will be attenuated as it is picked up off-axis. For example, Figure 4.7 shows that sound picked up directly in front of the microphone will experience no attenuation or weakening. On the other hand, sound picked up from either side (90 degrees off-axis) will be attenuated by about 5 dB, and sound coming directly from the rear of the microphone (180 degrees) will be attenuated by almost 20 dB.

4.9 IMPEDANCE OF MICROPHONES

Another factor sometimes used to categorize microphones is **impedance**, a characteristic that is similar to resistance and common to audio equipment. Impedance is expressed in ohms, and microphones can be either high-impedance (10,000 ohms or higher) or low-impedance (600 ohms or less). Microphones that fall between the 600–10,000 ohm range are known as medium-impedance microphones. Most professional microphones are low-impedance, as they provide the best frequency response, and most professional audio equipment is designed to accept this type of microphone. High-impedance microphones are also quite

limited in the length of microphone cable that can be used with them before hum and severe signal loss occurs. High-impedance microphones should not be plugged into audio recorders or other equipment designed for low-impedance; similarly, low-impedance microphones should not be used with high-impedance equipment. If impedance is mismatched, sound will be distorted. There are impedance converters that can convert one type of impedance to the other and some microphones can automatically switch from one impedance level to another.

4.10 SENSITIVITY OF MICROPHONES

Sensitivity refers to a microphone's efficiency or ability to create an output level. For the same sound source (say, a particular announcer's voice), a highly sensitive microphone produces a better output signal than a less sensitive microphone. To compensate for this, the gain control (volume) must be increased for the less-sensitive microphone; this increased gain also produces more noise however. Although different sensitivity-rating systems can be employed, condenser microphones generally have high-sensitivity specifications, and dynamic microphones have medium sensitivity. Obviously, if you're trying to pick up a loud or close-up sound, a microphone with a lower sensitivity rating would be desirable.

4.11 PROXIMITY EFFECT AND BASS ROLL-OFF

Using a microphone sometimes produces a sound phenomenon known as the **proximity effect**. This is an exaggerated bass boost that begins as the sound source gets about 2 feet from the microphone. The effect should be most noticeable as the announcer gets about 2 or 3 inches from the microphone and is especially noticeable with microphones that have a cardioid pickup pattern. Although it could help deepen or add fullness to a normally high voice, the proximity effect is usually compensated for by a **bass roll-off switch** on the microphone. This switch, when turned on, will electronically "roll off," or block some of the bass frequencies that would be boosted by the proximity effect.

4.12 MICROPHONE FEEDBACK

Feedback is a screeching or whining sound generated when sound picked up by a microphone is amplified, produced through a speaker, picked up again by the microphone, amplified again, produced through a speaker again, and so on, creating a loop. Reducing the speaker volume or turning off the microphone usually ends the feedback. Feedback is a common microphone problem in public address situations and in production studios with audio consoles that do not have muting systems. Feedback is usually not a problem in broadcast media production, because the monitors are automatically muted when the microphone is switched on. Occasionally, announcers can produce feedback in the production studio by operating their headphones at an excessive volume or accidentally picking up a stray speaker signal from an outside source, such as another studio speaker close by.

4.13 MULTIPLE-MICROPHONE INTERFERENCE

Sometimes, when two or more microphones receiving the same sound signal are fed into the same mixer, the combined signal becomes electronically **out of phase**. This happens because the sound reaches each microphone at a slightly different time so that while the sound wave amplitude is up on one microphone, it's slightly down on the other. Under these circumstances, the resulting sound will have frequency peaks and cancellations causing very poor sound quality. This situation is also known as **multiple-microphone interference** and can be avoided by remembering a 3-to-1 ratio. If the microphones are about 1 foot from the announcer (or sound source), they should be at least 3 feet apart from each other. This way the pickup patterns won't overlap (see Figure 4.10A). Another solution to this problem is to place microphones that must be close together head-to-head, so they will receive the signal at the same time (see Figure 4.10B). Although multiple-microphone interference isn't usually a problem in the audio studio, it can occur in some video production studios or remote situations.

4.14 STEREO

We hear sound in stereo because we have two ears, and most sounds arrive at one ear before the other. For

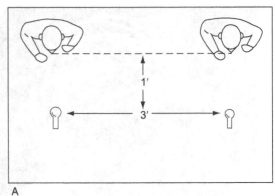

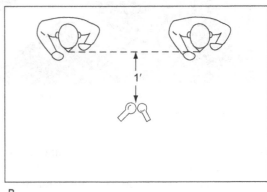

FIGURE 4.10 In a studio production with multiple announcers, you can avoid multiple-microphone interference by keeping the distance between the two microphones three times the microphone-to source distance (A). If you have limited studio space, another method is to place the microphones head-to-head (B).

instance, when hearing any sound (crowd noise, traffic, sirens) your right ear hears a slightly different perspective than your left. This occurrence allows us to locate sound by turning our head until the sound is "centered." When stereo sound is compared to monophonic (mono) sound, it's apparent that stereo adds both "depth" and "imaging" to the sound.

Imaging is the apparent placement of the sound between the left and right planes and provides location of the sound in space. **Depth** is the apparent placement of the sound between the front and back planes and provides the ambience of the space where the sound is produced. Good stereo sound allows us to record and reproduce sound as it appears in real life.

4.15 STEREO MIKING TECHNIQUES

Most audio production work is done with a mono microphone. Even in stereo studios, often the same mono microphone signal is simply sent to the left and right channels. However, there may be times when you will want to employ true stereo microphone techniques. There are a number of different techniques for stereo miking, but the

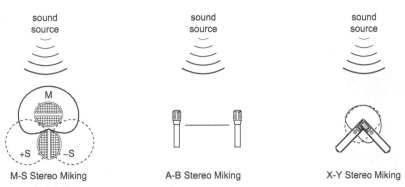

sound source   sound source   sound source

M-S Stereo Miking   A-B Stereo Miking   X-Y Stereo Miking

FIGURE 4.11   The most common techniques to microphone sound for stereo.

three most common techniques are known as A-B, X-Y, and M-S miking (see Figure 4.11).

Since the ultimate goal of stereo recording is to have separate sound signals come from the left and right speakers, one way to accomplish this is through **A-B miking**, which splits a pair of omnidirectional or cardioid microphones to the left and right of center about 3 to 10 feet apart. A good rule to follow is to separate the microphones by about half the width of the sound source. For example, if you were recording a band on a 10-foot-wide stage, you'd space your microphones about 5 feet apart. This A-B technique (also known as **spaced pair** or **split pair**) uses one microphone to feed the left channel of the stereo signal and another one to feed the right channel. As long as phase problems are accounted for and the sound source remains a relatively equal distance from the microphones, a true stereo signal will be obtained. However, when using this technique, the sound is often "spacious," with a great deal of separation, especially as you place the microphones farther and farther apart. As a result room ambience may become overbearing, with individual background sounds seeming to "wander" into the stereo image.

The **X-Y miking** (also called **cross-pair**) technique requires placing two cardioid microphones facing left and right like crossed swords forming an X- and Y-axis, as shown in Figures 4.10B and 4.11. The angle formed between the heads of the microphones is usually 90 degrees, but can vary from 60 to around 135 degrees, with the right microphone facing the left side of the sound source and the left microphone facing the right side. This angle has an impact on the stereo effect and requires some experimenting to get it right. If the microphones become too parallel, the stereo effect is minimal; if the angle is too wide, there will seem to be a "hole" in the stereo image. If the heads of the microphones are close together, this X-Y miking allows the sound signal to reach both microphones at the same time and there will be no phase problems. Although this approach solves some of the split-pair arrangement problems, it can cause a loss of "focus" on the image, because the microphones are off-axis to the center of the stereo image.

M-S miking, or **mid-side miking**, offers superior imaging by using microphones arranged in an upside-down "T" pattern. The mid microphone (often a cardioid or super-cardioid, but sometimes an omnidirectional) is aimed at the sound source, and the side microphone (usually a bidirectional) is placed parallel to the sound source and perpendicular to the mid microphone to pick up the sound to the left and right. Both microphones must be fed into a mixer designed to matrix, or form the incoming signals into a stereo signal. Mid-side miking generally provides a very accurate stereo recording regarding the localization of the sound sources.

Variations of these techniques, using different types of microphones and microphone placements, are also employed to achieve stereo miking. If you intend to do a great deal of stereo recording, you may employ a pre-built stereo microphone that can duplicate the X-Y technique using a single microphone with specially designed internal sound-generating structures (see Figure 4.15).

4.16 SURROUND SOUND

Surround sound is more difficult to record than stereo sound, so it is often put together in a postproduction environment. The difficulty occurs because of the number of channels involved. In its most common form, surround sound provides 5.1-channel sound, that is, six separate audio channels. To accomplish this, a center channel is added to the stereo left and right channels facing the front of the listener, and left and right surround channels are added to the rear. These are all full-frequency response main channels. The sixth channel, designated as ".1," is a limited-response channel for bass frequencies only, often reproduced by a powered subwoofer. Surround sound that is designated 7.1 would add a left- and right-side channel to the 5.1 configuration and 6.1 would add just a "top" channel.

It is possible to record audio to accommodate all of these channels. One way involves placing cardioid and/or super-cardioid microphones in a circle, each pointing at a different place on the circumference of the circle to

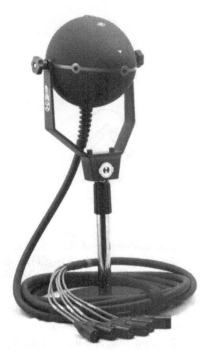

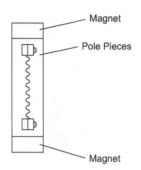

FIGURE 4.13 The internal components of the ribbon microphone center around a thin, corrugated metallic ribbon.

FIGURE 4.12 A microphone designed specifically for surround sound recording. *(Image courtesy of Holophone, a Division of Rising Sun Productions Limited.)*

approximate the various channels. A variation of the stereo spaced pair technique can also be used with multiple omnidirectional or cardioid microphones. Arrange three front microphones (left, center, right) about a foot apart and two rear microphones (left, right) about 2 feet behind the front microphones and about 2 feet apart from each other. Of course, if you're using cardioid microphones, the rear microphones should be rear facing. Generally, the.1 or bass channel is just a low-pass signal from one of the microphones or a combination of the bass signals from all the microphones.

Another (and less complicated) method is to use a surround sound microphone, such as the Holophone (see Figure 4.12). Its elliptical shape emulates the characteristics of the human head, and sound waves bend around the microphone as they would around the head, providing sound spatiality and directionality. This microphone captures sound with miniature receiving elements positioned around, and flush with, the oval surface of the microphone.

The problem with making realistic surround sound is that there are so many channels, front and rear, that the mono signal often needs to be altered so that it conveys the characteristics people expect as they hear sound coming from different directions. This is best undertaken in postproduction, where the sounds can be manipulated in less-hectic fashion.

Surround sound isn't currently used much for radio, however it is fairly standard for film and television, where dialogue is handled by the left–center–right speakers, ambient sounds and some special effects are handled by

the rear surround speakers, and big explosive sounds are handled by the subwoofer. Surround sound helps convey "movement" of sound as effects and other sounds can swirl around the listener.

4.17 SPECIAL PURPOSE AND OTHER TYPES OF MICROPHONES

There are several other variations of dynamic and condenser microphones that employ several variations of pickup patterns, and while they are less likely to be encountered in the contemporary audio production or broadcast studio, you may still see them in television work or remote field production.

For many years, the **ribbon microphone** was common in broadcasting and a few manufacturers still produce this style of microphone. The ribbon microphone is a type of dynamic mic that contains a sound-generating element consisting of a thin, corrugated metallic ribbon suspended in the field of a magnet. Sound waves vibrate the ribbon to generate an electrical output signal (see Figure 4.13). The ribbon microphone has an excellent warm, smooth sound, but it is bulky and very fragile and has been largely replaced by the condenser microphone.

A **regulated phase microphone** can be thought of as part dynamic and part ribbon microphone. A wire coil is attached to or impressed into the surface of a circular diaphragm. The diaphragm is suspended between two circular magnets that are designed to be acoustically transparent (the magnets don't impede the sound waves striking the diaphragm in any way). An electrical current is generated as sound waves vibrate the diaphragm. Because of its design, the regulated phase microphone exhibits some of the qualities of a ribbon microphone with some of the ruggedness of the dynamic microphone.

Most **tube microphones** are merely modern condenser microphones that incorporate a vacuum tube into the electronics of the microphone. Tube microphones provide superior dynamic range, outstanding clarity, and the "warmth" of classic analog sound that many find missing with solid-state and digital electronics. Tube microphones

FIGURE 4.14 Lavaliere microphones are used more in television than radio because of their unobtrusive design. *(Image courtesy of Audio-Technica U.S., Inc.)*

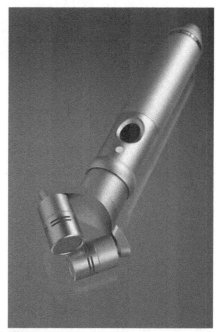

FIGURE 4.15 Stereo microphones often employ multiple sound-generating elements as part of a single microphone. *(Image courtesy of RODE Microphones.)*

are most often found in high-end audio production facilities.

There are other microphone types that are often designed for a specific use and are frequently considered to be another variety of microphone. For example, the **lavaliere microphone** (also known as **lav**) (see Figure 4.14) is a tiny microphone that can be unobtrusively clipped to an announcer's lapel or tie. There are both dynamic and condenser models, and although the lavaliere microphone is occasionally used in radio remote situations, its small size makes it more appropriate for television than radio.

A **stereo microphone** like the one shown in Figure 4.15 incorporates small, multiple sound-generating elements as part of a single microphone housing that can duplicate various stereo microphone techniques, such as the X/Y method.

Wireless microphones (also known as **RF** (radio frequency), **FM**, and **radio microphones**) are used mainly in production situations where a microphone cable might hinder a recording or production. A wireless microphone is really part of a three-part system, which includes the microphone itself, a radio transmitter, and a radio receiver (see Figure 4.16). The audio signal generated (employing either a dynamic or condenser element) is sent from the microphone by a low-power transmitter rather than a cable. This transmitter is either in the microphone housing or contained in a small battery pack worn by the talent. The transmitted signal is picked up by a receiver located nearby and converted from a radio frequency signal into an audio signal at that point. A common problem associated with wireless microphones is interference. Because they broadcast on specific FCC-assigned frequencies that are not exclusive to FM microphones, they sometimes pick up interference from other radio frequency users. Additionally, wireless microphones can be susceptible to interference from other electrical devices in use nearby, such as video monitors and fluorescent lighting.

FIGURE 4.16 Because of its wireless design, an FM microphone system can be used when a microphone cable would get in the way. *(Image courtesy of Shure Incorporated. Used by permission.)*

The **pressure zone microphone** or **PZM** (also known as a **boundary**, **plate**, or **surface-mount microphone**) is a small microphone capsule mounted next to a sound-reflecting plate, as shown in Figure 4.17. Designed to be used on a flat surface, such as a tabletop, the microphone picks up sound from all directions above the table surface. It also receives both direct and reflected sound at the same time, because the microphone is so close to the reflective

FIGURE 4.17 The PZM is designed to be used on a flat surface, such as a tabletop. *(Image courtesy of Crown Audio, Inc.)*

FIGURE 4.18 A shotgun microphone has a long tube or barrel that is "aimed" at the sound source. *(Image courtesy of Sennheiser Electronic Corporation.)*

surface. This design boosts the incoming sound signal, which improves the clarity of the sound.

Figure 4.18 displays the **shotgun microphone**. A microphone capsule at one end of a tube (or barrel) is "aimed" like a gun toward the sound source. The design of the microphone rejects sounds from the side and rear but picks up a very narrow angle of sound from the front of the microphone's barrel. The highly directional nature of the microphone makes it good at picking up sound from a considerable distance; however, sound quality is somewhat less than that achieved by standard cardioid microphones.

A **parabolic microphone** is another kind of microphone that is used primarily for field and sports production. A parabolic microphone employs either a wireless or wired cardioid or omnidirectional microphone, which can either be permanently attached or switched out as shown in Figure 4.19. What makes a parabolic microphone unique is that it uses a large, concave, plastic "bowl" to collect sound waves, much like a satellite receiving dish collects radio waves. The design of the parabolic microphone makes it an excellent choice for collecting ambient and background sound from a great distance, in order to help enhance the realism of a sporting or field event. Parabolic microphones are often used at field level for sporting events in order to capture the sounds of the game. They can also be used to pinpoint a specific sound source if necessary. Because of their size and their primary use, parabolic microphones

are not used in studio recordings. However, if you ever find yourself engineering the sound for a sporting event (on radio, television, or online) you will find that these microphones are very useful to help create a professional-sounding production.

4.18 MICROPHONE ACCESSORIES

A production person will also find one or more of the following microphone accessories necessary for proper audio production work: windscreens, pop filters, shock mounts, and stands or booms. The most common **windscreens** are ball-shaped foam accessories that can be placed over the head or front of the microphone (as shown in Figure 4.20) to help reduce plosive sounds. Announcing words that emphasize *p*, *b*, or *t* sounds naturally produce a sharp puff of air that can produce a pop or thump when hitting the microphone's diaphragm. Not only do windscreens prevent popping, but they also help keep dust out of the internal elements of the microphone and can provide some cushion if the microphone is accidentally dropped. Windscreens can also be built into the grill of a microphone.

Another windscreen design consists of a porous, film-like material that is suspended within a circular frame. The **pop filter** (or **blast filter**) apparatus is attached to the microphone stand and positioned in front of the

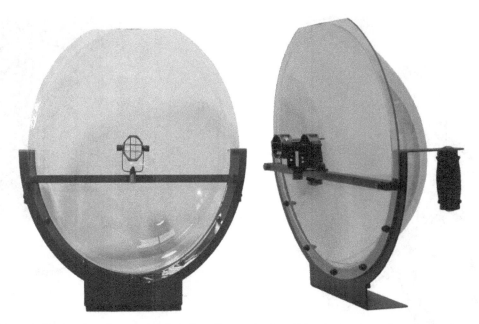

FIGURE 4.19 This parabolic microphone has a mount that allows the operator to switch out microphones depending on the recording situation. The dish is approximately 24 inches wide and the entire assembly can be mounted onto a tripod. *(Image courtesy of Jony Jib Camera Solutions, Inc., www. jonyjib.com; copyright 2012.)*

FIGURE 4.20 Microphone windscreens help reduce plosive sounds. *(Image courtesy of Shure Incorporated.)*

microphone rather than placed on the microphone itself, as shown in Figure 4.21.

Another microphone accessory, the **shock mount**, is often used to isolate the microphone from any physical vibrations (or shocks) that may be transmitted through its stand. The microphone is suspended, usually by an arrangement of elastic bands, and isolated from the stand or boom to which it is attached (see Figure 4.22). If the microphone stand is accidentally bumped, the sound from the bump will not be passed on and amplified by the microphone.

FIGURE 4.21 A pop filter is a porous, film-like material suspended within a circular frame that is positioned in front of the microphone. *(Image courtesy of Shure Incorporated. Used by permission.)*

Another microphone accessory that can help create professional-quality sound is one of sE Electronics' Reflexion Filter® systems. As shown in Figure 4.23, a microphone is placed in a concave stand lined with sE's patented

FIGURE 4.22 The microphone shock mount isolates the microphone from the microphone stand. *(Image courtesy of Neumann USA.)*

FIGURE 4.23 The X1 Studio Bundle by sE Electronics includes their Reflexion Filter® X (RF-X), a shock mount, pop filter, and microphone. The bundle helps the producer to create the acoustics of a professional studio's vocal booth in practically any location. *(Image courtesy of sE Electronics International, Inc., www.seelectronics.com; copyright 2015)*

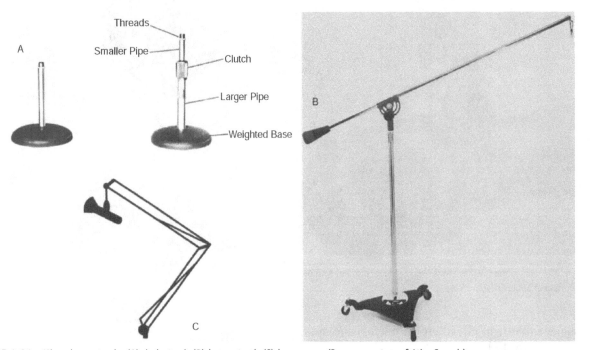

FIGURE 4.24 Microphone stands: (A) desk stand; (B) boom stand; (C) boom arm. *(Images courtesy of Atlas Sound.)*

sound absorption and diffusion materials, which help keep unwanted room reflections from reaching the microphone capsule, thus simulating the acoustics of a professional vocal booth. The use of a pop filter helps create an even better reproduction of "studio sound" in a non-studio environment.

Various microphone stands or **booms** are used to position the microphone for different recording situations. Some stands consist of two chrome-plated pipes, one of which fits inside the other. A rotating clutch at one end of the larger diameter pipe allows the smaller pipe to be adjusted to any height desired. At the other end of the larger pipe is a heavy metal base (usually circular) that supports all of the pipes in a vertical position. Other microphone stands utilize a single pipe that's at a fixed height. In either case, the microphone is attached to the top of the pipe by a standard thread and microphone-stand adapter (see Figure 4.24A). Floor stands adjust for the announcer in a standing position, and desk stands are used for the seated announcer.

A **boom stand** is a long horizontal pipe that attaches to a large floor stand (see Figure 4.24B). One end of the boom is fitted with the standard thread for the microphone, and the other end is weighted to balance the microphone. The horizontal pipe allows the boom stand to be away from the announcer while positioning the microphone in a workable relationship to him or her. The boom stand is frequently found in video production, when the microphone must be out of the picture. A **boom arm** (see Figure 4.24C) is a microphone stand especially designed for use in the radio studio. It consists of metal rods and springs designed somewhat like a human arm. The microphone attaches to one end, and the other end goes into a mounting base. The whole unit can then be attached to a countertop or other production studio furniture so that the microphone can be placed in close proximity to the audio console.

4.19 MICROPHONE USAGE

An understanding of microphone types, their pickup patterns, their accessories, and other characteristics is useful only if you can apply this knowledge to everyday audio production use. It seems reasonable to assume that the microphone in a studio will be a dynamic microphone, because that's the one most commonly used in broadcasting, but keep in mind it could also be a condenser microphone. Perhaps the more important consideration is its pickup pattern. In the studio, we mainly want to pick up voices and some studio ambience. A cardioid pickup pattern, with its pickup of front and side sounds, works best to accomplish this while not picking up unwanted sounds from the rear or from the side.

Studio interviews may require you to mic a group of people seated around a table. You could use a single omnidirectional microphone in the center of the table, or you could use a separate cardioid microphone for each individual at the table. Another option is to use the PZM.

Coverage of sporting events may require the use of microphones with several different pickup patterns. Announcers could use cardioid microphones while omnidirectional microphones would be used to pick up crowd noise. Super- or hyper-cardioid microphones or parabolic microphones could be used to pick up activity on the field or court. Sometimes the venue and its settings will determine what microphones should be used; proper microphone techniques will differ, for instance, between a basketball game played indoors and a football game played outside in the snow.

PRODUCTION TIP 4A
Microphone-to-Mouth Relationship and Setting Levels

You see it all the time in home-made online videos: announcers and hosts practically eating their microphone while they speak ... and their audio sounds terrible. "Swallowing the microphone" is not good microphone technique. When using a microphone in the studio, keep in mind two basic rules concerning distance from the microphone and position of the microphone. First, your mouth should be no closer than about 6 inches from the microphone. That's about the length of the "hang loose" hand sign, and a good way to remember mic-to-mouth distance. You may find, however, that you need to be closer to, or farther away from the mic because of the strength of your voice or the vocal effect that you're trying to achieve. Remembering to "hang loose" is a good starting point for using a microphone.

Second, position the microphone so that you are not talking *directly* into it. The mic should be positioned level with your nose and tilted down toward your mouth. Be sure to talk beneath the front of the microphone, or place the microphone below your mouth with the front of it tilted up toward your mouth, allowing you to talk across the top of it.

As if that were not enough, beginning announcers often misuse a microphone by blowing into it to see if it's live or to set a level. This is the worst way to test a microphone and can actually damage it. In fact, the higher the quality of the microphone, the more likely that it will be damaged in this manner. The best way to set a microphone level is to read several lines of your script or ad lib some material. If you just count ("Testing... 1... 2... 3...") or simply say "check" over and over, you're not getting enough variation in sound to set an accurate level. Most people don't count or say a single word over and over again in quite the same tone and volume as when they speak several sentences.

And as always, test and check your microphone levels before you begin recording, and adjust them as necessary.

4.20 CONCLUSION

Remember, there is no one universal, standard microphone to use in every audio production. When using microphones, don't be afraid to experiment. The bottom line for any production should be how good it sounds. It's good to be flexible in audio production work, because sometimes you won't have a wide variety of microphones available. But if you do have a full complement of microphones in your audio production facility, don't be afraid to try different ones. If you use what you've learned in this chapter, you should be able to obtain clear, appropriate sound under a variety of circumstances.

Self-Study

1. What is another name for the dynamic microphone?

   a) condenser
   b) pressure
   c) capacitor
   d) PZM

2. The dynamic microphone's sound-generating element is constructed of a diaphragm, a permanent magnet, and a voice coil. Into which of these is a small electrical current induced during use?

   a) diaphragm
   b) magnet
   c) coil
   d) none of the above

3. In what way does the condenser microphone differ from the dynamic microphone?

   a) The condenser microphone needs a power supply, and the dynamic microphone doesn't.
   b) The dynamic microphone has a diaphragm, and the condenser microphone doesn't.
   c) The dynamic microphone has better sound quality than the condenser microphone.
   d) The condenser microphone is bidirectional, and the dynamic microphone is omnidirectional.

4. True 5.1 surround sound is accomplished by adding what to the basic stereo setup of left and right front channels?

   a) a single rear or surround channel
   b) both left and right rear or surround channels
   c) both left and right rear or surround channels plus a center front channel
   d) both left and right rear or surround channels plus a center front channel plus a bass channel

5. Which microphone pickup pattern picks up sound on all sides?

   a) hyper-cardioid
   b) cardioid
   c) omnidirectional
   d) bidirectional

6. Which microphone would probably be most appropriate for conducting a radio interview on the sidelines at a football game?

   a) unidirectional
   b) omnidirectional
   c) bidirectional
   d) parabolic

7. Which microphone would probably be most appropriate for a sportscaster in an open press box at a baseball game?

   a) cardioid
   b) nondirectional
   c) bidirectional
   d) omnidirectional

8. A microphone is a transducer, which means it can pick up sound equally well in all directions.

   a) True
   b) False

9. Which of the following is most likely to exaggerate the bass sounds of a person's voice?

   a) feedback
   b) proximity effect
   c) multiple-microphone interference
   d) bass roll-off

10. What is the purpose of a shock mount?

   a) to reduce plosive sounds
   b) to keep the announcer's head at least 12 inches away from the microphone
   c) to isolate the microphone from mechanical vibrations
   d) to prevent static electricity discharges

11. Which type of microphone stand can be farthest away from a person and still allow the person to be close to the microphone?

   a) boom stand
   b) floor stand
   c) desk stand
   d) shock mount

12. Which stereo miking technique uses two microphones crossed (like swords) at a 90-degree angle to each other?

   a) split-pair miking
   b) mid-side miking
   c) M-S miking
   d) X-Y miking

13. Which type of microphone uses small, multiple sound-generating elements within a single microphone housing?

   a) ribbon mic
   b) wireless mic
   c) stereo mic
   d) boundary mic

14. A microphone's pickup pattern is exactly the same thing as a microphone's polar response pattern.

   a) true
   b) false

15. A pop filter attaches directly to the head of a microphone.

   a) True
   b) False

16. Which microphone would be best at picking up sound when the sound source is a considerable distance from the microphone?

   a) lavaliere microphone
   b) PZM
   c) RF microphone
   d) parabolic microphone

17. Sound picked up from the rear of a microphone is 90 degrees off-axis.

    a) true
    b) false

18. When testing a microphone's level, what should you do?

    a) Say the word "testing" over and over again.
    b) Read the first several lines of your copy.
    c) Count from 1 to 50.
    d) Blow into the microphone as hard as you can.

19. Microphone impedance refers to a microphone's ability to create an output signal.

    a) true
    b) false

20. Which term describes a "screech" that occurs when sound is picked up by a microphone, amplified, fed back through a speaker, and picked up again, over and over?

    a) proximity effect
    b) multiple-microphone interference
    c) feedback
    d) bass roll-off

21. Which of the following is *not* a way that microphones are commonly classified?

    a) a microphone's sound-generating element
    b) a microphone's size
    c) a microphone's pickup pattern
    d) a microphone's impedance

22. Which stereo miking technique splits a pair of omnidirectional or cardioid microphones to the left and right of center about 3 to 10 feet apart?

    a) Left/Right miking
    b) M-S miking
    c) X-Y miking
    d) A-B miking

23. Which microphone has a wire spiral embedded in a circular diaphragm as part of its sound-generating element?

    a) dynamic microphone
    b) ribbon microphone
    c) moving coil microphone
    d) regulated phase microphone

24. Multiple-microphone interference can be avoided if the microphones employed are at least three times as far from each other as they are from the sound source.

    a) true
    b) false

25. Which of the following is *not* another term for a wireless microphone?

    a) FM microphone
    b) RF microphone
    c) radio microphone
    d) PZM

If you answered A to any of the questions:

1a. No. This is a different type of microphone. (Reread 4.3 and 4.4.)

2a. Wrong. The diaphragm feels the pressure. (Reread 4.3.)

3a. Right. The condenser microphone power supply is needed to charge the backplate and diaphragm.

4a. No. This would not provide surround sound. (Reread 4.16.)

5a. No. Hyper-cardioid microphones pick up sound mainly from the front. (Reread 4.7.)

6a. Wrong. Although a unidirectional microphone can be used in the field, it picks up sound from mainly one direction, so it might not pick up enough of the crowd ambience. (Reread 4.7.)

7a. Yes. It would pick up the sportscaster without much of the background noise.

8a. No. Transduction has nothing to do with a microphone's pickup pattern (Reread 4.1)

9a. No. Feedback is a screeching noise caused by having open microphones near speakers. (Reread 4.11 to 4.12.)

10a. No. A pop filter or windscreen is used to reduce plosive sounds. (Reread 4.18.)

11a. Right. This is the best selection.

12a. No. This is a different stereo miking technique. (Reread 4.15.)

13a. No. As the name implies, a ribbon microphone employs a thin metallic ribbon as part of a single sound-generating element. (Reread 4.17.)

14a. Wrong. A pickup pattern is the three-dimensional shape of the area around the microphone in which it hears the sound best; a polar response pattern is the two-dimensional representation of this. (Reread 4.8.)

15a. No. (Reread 4.18.)

16a. Wrong. Lavaliere microphones are designed to be attached to the announcer's clothing. (Reread 4.17.)

17a. No. Sound that is 90 degrees off-axis would be coming from the side of a microphone. (Reread 4.5.)

18a. Wrong. Doing this will not give a true indication of the variations of your speech (Reread Production Tip Box A)

19a. No. Impedance is an electrical characteristic similar to resistance. (Reread 4.9 and 4.10.)

20a. No. Proximity effect is a characteristic of microphones that accents the bass response. (Reread 4.11 to 4.12.)

21a. No. Microphones are categorized by their sound-generating elements. (Reread 4.2, 4.3, and 4.4.)

22a. No. There is no such term. (Reread 4.15.)

23a. Wrong. Dynamic microphones do have a wire coil as part of their sound-generating element, but it's not configured like this. (Reread 4.3 and 4.17.)

24a. Yes. This is a true statement.

25a. No. FM microphone is another term for a wireless microphone. (Reread 4.17.)

If you answered B to any of the questions:

1b. Correct. The dynamic microphone is also called a pressure or moving-coil microphone.

2b. Wrong. The magnet sets up the field. (Reread 4.3.)

3b. No. Both microphones have a diaphragm. (Reread 4.3 and 4.4.)

4b. No. This would not provide surround sound. (Reread 4.16.)

5b. Wrong. The cardioid picks up sound on all but one side—usually the side right behind the microphone. (Reread 4.7.)

6b. Correct. An omnidirectional microphone would pick up from all sides, thus easily miking both the interview and some crowd noise.

7b. No. The crowd noise would tend to drown out the announcer. (Reread 4.7.)

8b. Yes, this is the correct answer.

9b. Right. When an announcer gets too close to the microphone, the bass may be exaggerated.

10b. No. For one thing, the announcer's head should be about 6 inches away, not 12. (Reread 4.18 and Production Tip 4A.)

11b. No. A person must stand right beside a floor stand. (Reread 4.18.)

12b. No. This is a different stereo miking technique. (Reread 4.15.)

13b. No. Wireless microphones employ a transmitter (often part of the microphone) and a receiver, but this is not a description of this system. (Reread 4.17.)

14b. Right. This is the correct response.

15b. Correct. Pop filters are not attached directly to a microphone.

16b. Wrong. PZMs are designed to be placed on a flat surface, such as a tabletop. (Reread 4.17.)

17b. Yes. Sound that is picked up from the rear of a microphone is 180 degrees off-axis.

18b. Correct. Doing this allows you to get the most accurate levels in regard to variations in your voice and reading style.

19b. Correct. Microphone sensitivity refers to a microphone's ability to create an output level.

20b. No. This is not correct. Multiple-microphone interference is a phase problem that creates peaks and cancellations in the sound. (Reread 4.11, 4.12, and 4.13.)

21b. Correct. Although microphones come in a wide variety of sizes, they are not usually categorized by size.

22b. Wrong. This technique requires the microphones to be close together. (Reread 4.15.)

23b. No. The ribbon microphone uses a thin ribbon suspended between two magnets for its sound-generating element. (Reread 4.17.)

24b. No. The statement is true. (Reread 4.13.)

25b. No. RF microphone is another term for a wireless microphone. (Reread 4.17.)

If you answered C to any of the questions:

1c. No. This is another name for the condenser microphone. (Reread 4.3 and 4.4.)

2c. Correct. The current is in the coil.

3c. No. The condenser microphone usually has slightly better sound quality than the dynamic microphone. (Reread 4.3 and 4.4.)

4c. No. This wouldn't provide true surround sound although you're getting close. (Reread 4.16.)

5c. Yes. An omnidirectional microphone picks up on all sides.

6c. No. Although this microphone could pick up the interview nicely, it wouldn't get much crowd noise. (Reread 4.7.)

7c. No. A bidirectional microphone picks up on two sides, and the sportscaster would be on only one side. (Reread 4.7.)

9c. No. This will create a distorted signal. (Reread 4.11 and 4.13.)

10c. Yes. This is a special type of microphone holder that suspends the microphone.

11c. No. A desk stand has to be right in front of the person. (Reread 4.18.)

12c. No. This is a different stereo miking technique. (Reread 4.15.)

13c. Yes. This is the correct answer.

16c. No. RF microphones are wireless, but they don't have exceptional distance pickup characteristics. (Reread 4.17.)

18c. Incorrect. All this will do is make the person taking your levels angry. (Reread Production Tip Box A.)

20c. Correct. That screeching, howling sound is feedback.

21c. No. Microphone pickup patterns are used to categorize microphones. (Reread 4.2 and 4.5.)

22c. Wrong. This miking technique requires crossed microphones at an angle that can vary from 90 to 140 degrees. (Reread 4.15.)

23c. No. The moving coil microphone is just another name for the dynamic microphone. (Reread 4.3 and 4.17.)

25c. No. Radio microphone is another term for a wireless microphone. (Reread 4.17.)

If you answered D to any of the questions:

1d. No. A PZM does not indicate a type of microphone construction. (Reread 4.3 and 4.17.)

2d. Wrong. (Reread 4.3.)

3d. No. Both dynamic and condenser microphones can have omnidirectional or bidirectional pickup patterns. (Reread 4.3, 4.4, 4.5, 4.6, and 4.7.)

4d. Yes. These components when added to the stereo left and right front channels make up 5.1 surround sound.

5d. No. A bidirectional microphone picks up sound from the front and back of the microphone. (Reread 4.7.)

6d. No. This microphone is better for picking up crowd or field noise and not for interviews (Reread 4.17.)

7d. No. This is another term for nondirectional. (Reread 4.7.)

9d. Wrong. Bass roll-off is an electronic "turn down" of bass frequencies. (Reread 4.11.)

10d. No. This is not the correct answer. (Reread 4.18.)

11d. No. Although this is often used in conjunction with a microphone stand, it won't determine the distance between announcer and microphone stand. (Reread 4.18.)

12d. Correct. This is a description of X-Y stereo miking technique.

13d. No. Boundary microphones employ a single sound-generating element in conjunction with a sound-reflecting plate. (Reread 4.17.)

16d. Right. Parabolic microphones will pick up sound from a point some distance away.

18d. Wrong. In addition to not even getting a level for your voice, all you will do is get spit on the microphone and possibly damage it. (Reread Production Tip Box 4A.)

20d. No. Bass roll-off is a switch on some microphones that electronically turns down the lower frequencies. (Reread 4.11 and 4.12.)

21d. No. Microphone impedance is used to categorize microphones. (Reread 4.2 and 4.9.)

22d. Correct. This describes the A-B miking or spaced pair technique.

23d. Yes. This is the correct response.

25d. Correct. The PZM is a type of boundary microphone that is designed to be placed on a flat surface, such as a tabletop.

Projects

PROJECT 1

Position microphones in various ways to create different effects.

Purpose

To enable you to experience proximity effect, feedback, multiple-microphone interference, and the differences in sound quality that occur when a microphone is placed at different distances and angles from an announcer.

Notes

1. You may need some help from your instructor or an engineer in setting up the equipment.
2. For the proximity effect, try to find a microphone that doesn't have a bass roll-off switch, or make sure it's switched off if it does.
3. Don't allow feedback to occur for too long. It can be damaging to all the electronic equipment—and your ears. You may have to plug the microphone into something other than the audio board if the board automatically shuts off the speakers when the microphone is turned on.
4. For multiple-microphone interference, make sure that the microphones are less than 3 feet apart. Omnidirectional microphones will demonstrate the effect the best.
5. Use a cardioid microphone for the distance and angle experiments, because it will demonstrate the points better.

How to Do the Project

1. Set up an audio board so that two microphones are fed into it and so that the sound of those two microphones can be recorded on an audio recorder.
2. Put the recorder in record mode and activate one of the microphones. Start talking about 2 feet away from the microphone, and keep talking as you move closer until you are about 2 inches from it. As you talk, say how close you are to the microphone, and mention that you're experimenting with the proximity effect. Stop or pause the recorder.
3. Position a microphone so that it's close to an activated speaker. Turn on the audio recorder and talk into the microphone. Record a short amount of the feedback and turn off or pause the recorder.
4. Position two microphones in front of you that are less than 3 feet apart. Put the audio recorder in record mode and talk into the microphones, saying that you're testing for multiple-microphone interference. Turn off or pause the recorder.
5. Position one microphone in front of you. Put the audio recorder in record mode. Position yourself 12 inches from the microphone and talk into it. Then position yourself 6 inches from the microphone and talk directly into it. Then talk across the top of the microphone. Move 6 inches to the side of the microphone, and talk into it with your mouth positioned to speak across the top of it. Get behind the microphone, either by moving behind it or by turning it around, and talk from about 6 inches away. Describe each action as you do it.
6. Listen to the recording to hear the various effects and to see that you have, indeed, recorded all the assignments. If some of them didn't turn out as well as you would have liked, redo them.
7. Write a brief observation of each effect explaining what you hear on your recording.
8. Turn in your recording and your observations to your instructor to receive credit for this project. Make sure you put your name on it and label it "Microphone Placement."

PROJECT 2

With several other students, make a recording using stereo miking techniques.

Purpose

To give you experience using standard microphones to employ various stereo miking techniques.

Notes

1. Which technique you employ will depend on the microphones that are available at your facility.
2. Because this project requires several people, your instructor may assign it as a group project.
3. The "How To" section that follows uses X-Y techniques for illustration purposes; you can adjust it for any of the other techniques.

How to Do the Project

1. At one end of your studio (or room) arrange at least three students in a left–center–right configuration. This could also be three groups of students or even a small musical group, as long as they're arranged so that specific sections can be identified as left, center, or right.
2. Set up two cardioid microphones to record onto an audio recorder.
3. Arrange the microphones in an X-Y position as described in the text, in line with the "center" of the group.
4. Begin recording and have the students or group do the following:
   a have the left say something
   b have the right say something
   c have the center say something
   d have all three sections say something different, but at the same time
   e have just the center say something
   f have the left and right say something
   g have all three sections say something
5. As you are recording the variations in Step 4, make sure to identify each segment by having one student say something like, "This is just the 'left' group," before the group says anything.
6. Play back your recording. Are you able to "hear" the stereo image correctly? Does the left group "appear" to be located left? Make some observations about the recording results and write them down.
7. Repeat the same recording, but use a single microphone; that is, record in mono.
8. Listen to the second recording. What differences do you hear between the two recordings? Write down your observations again.
9. Turn in your recording and the observation sheet labeled with your name and "Stereo Miking" to your instructor to receive credit for this project.

PROJECT 3

Compare sound from different types of microphones.

Purpose

To learn the differences between dynamic and condenser microphones and between cardioid and omnidirectional pickup patterns.

Notes

1. You may need help from your instructor or engineer in setting up the equipment.

2. Don't be concerned if you can't discern differences between the dynamic and condenser microphones. The pop and hiss effect doesn't occur for all voices.
3. If your facility has microphones with other sound-generating elements (such as ribbon or regulated phase) or other pickup patterns (such as bidirectional or ultra-cardioid), add those to the exercise.

How to Do the Project

1. Select a dynamic and a condenser microphone from those at your facility.
2. Attach the microphones to an audio recorder. You can put both microphones through an audio console and use them one at a time, or you can attach each microphone to the recorder in its turn.
3. While you're recording, say the following into the dynamic microphone: "Peter Piper picked a peck of pickled peppers," and "She sells seashells by the seashore." Turn off or pause the recorder.
4. Repeat Step 3, but use the condenser microphone.
5. Select a microphone with a cardioid pickup pattern and another microphone with an omnidirectional pickup pattern from those at your facility.
6. Attach the microphones to an audio recorder as in Step 2.
7. Position the cardioid microphone on a stand and, while recording, walk all the way around the microphone, saying where you are as you move: for example, "Now I'm right in front of the microphone; now I'm 90 degrees to the right of the front of the microphone." Turn off or pause the recorder.
8. Repeat Step 7, using the omnidirectional microphone.
9. Listen to the recordings, and write down any observations you have about the differences among the various microphones.
10. Turn in the paper and your recordings to your instructor to receive credit for this project. Be sure to include your name on both and call the assignment "Microphone Comparisons."

PROJECT 4

Diagram/Apply miking techniques to various on-campus or local sporting events

Purpose

To see how a variety of sporting events and conditions can affect microphone selection and placement.

Notes

1. You may want to consult your instructor, chief engineer, or team coach(es) about which microphones to use and where they could be placed in your campus stadiums.
2. You may want to consider and address how many production assistants you would need to properly mic the field or stadium.
3. If your school does not have a competitive team for a sport listed above, create a diagram that would show how you would mic the field for intramurals instead.

How to do the project

1. Draw diagrams of the playing fields, courts, and rinks at your school:
   - Football
   - Baseball
   - Softball
   - Basketball

- Tennis
- Soccer
- Lacrosse
- Field Hockey
- Swimming
- Track and Field

2. Consider the conditions of each location. Where are the crowd bleachers located? Where is the press box located? Are there any "special effects" or traditions you should be aware of (i.e., cannons being fired after each home touchdown)? Are there any issues with Public Address speakers or bands?

3. Consider the possible weather conditions, especially if the game is played outdoors. Consider the acoustics of sports played indoors. Use these considerations to determine not only where you will place your microphones, but also what microphones you will use.

4. Fill out your diagrams with microphones clearly located where you would place them. In addition to turning in a diagram for each sport listed above, turn in a typed/printed report that lists the types of microphones you would use, where you would place them, and why. Turn in the diagrams and your paper to your instructor to receive credit for this project. Be sure your name is on both the diagrams and your report.

5

THE AUDIO CONSOLE

5.1 INTRODUCTION

In the modern audio studio, the spotlight is often on computer editing rather than the **control board**; however, the **audio console** is still a primary piece of equipment in most production facilities. It can sometimes be more difficult to understand than other equipment, but it is important that you are familiar with its operation, because most other pieces of equipment operate in conjunction with the audio console. Therefore, unless you can operate the audio console, you can't really utilize other studio equipment, such as a CD player or audio recorder. Audio consoles can range from a simple field mixer (see Figure 5.1) to a typical on-air radio console (see Figure 5.2) or a complex control board used for music recording (see Figure 5.3). Note that a console, used in on-air studios at broadcast stations, is quite often referred to as a "board." Typically, the term "mixer" denotes a smaller console, particularly those employed in remote applications. A typical on-air board performs fewer functions than a production console and is fairly easy to understand, while a larger production

FIGURE 5.1 Even a small audio mixer often provides several inputs, equalization, and other audio processing effects. *(Image courtesy of Behringer North America, Inc.)*

console can be intimidating and difficult to understand (based solely on the expanded functions it can perform). Regardless of their complexity, all control boards have basic similarities. Even though you may run across many different audio consoles in your production work, a thorough knowledge of any one should enable you to use any control board after a brief orientation.

Most boards include some method for input selection (such as source selectors), and volume control (such as faders), as well as some method for output selection (such as program/audition switches). They should also have some method of indicating to the operator the strength of the signal (through VU meters), and a way of allowing the operator to hear the mix of sources (through monitor/cue speakers or headphones). Boards also have amplifiers at various stages so that the signal is loud enough when it eventually goes to the transmitter or an audio recorder. These amplifiers are buried inside the board and are not something the board operator can usually control. In addition, audio consoles can have many other special features to help the board operator work more efficiently and creatively. The board may look intimidating because of all the buttons, knobs, and levers, but most of these are repeats, as the board has many different inputs and outputs. These will be explained in detail as we begin to explore the operation of the audio console.

5.2 THE DIGITAL AUDIO CONSOLE

The majority of audio consoles that you'll find in various production facilities will be digital, like the one shown in Figure 5.2. Although the first generation of digital boards were quite expensive, most current digital audio consoles are just slightly more expensive than their analog counterparts and often include additional features that justify any small extra expense. Most digital boards also have the capability to handle both analog and digital inputs with

FIGURE 5.2 Some digital audio consoles feature a self-contained, modular architecture, much like older analog boards. *(Image courtesy of Radio Systems.)*

FIGURE 5.3 Audio boards found in recording studios and some production facilities are often more complex and have more features than basic audio consoles. *(Image courtesy of AMS Neve.)*

modular designs that include plug-in channels that can be easily updated from an analog to a digital format as the studio adds additional digital equipment.

There are several options for digital audio console design. Some boards emulate traditional audio consoles and are self-contained units, as shown in Figure 5.2. The various audio sources are fed directly into the stand-alone console, which follows the most common wiring practice and allows console replacement to be a simple matter of "pick up the old, drop down the new." Other digital consoles are designed so that the "board" is merely an interface or control surface for an audio "engine" that can be installed in a rack in the studio or in an equipment room down the hall (see Figure 5.4). In this instance, the audio console works like the keyboard for your computer. The audio engine is essentially a router that accepts both analog and digital inputs with a capacity that usually exceeds even the largest analog board. Generally, analog inputs are converted to digital signals, and digital signals with different sample rates are synchronized at the audio engine.

Another feature of many digital audio consoles is the inclusion of a touch screen interface (review Figure 5.4)

FIGURE 5.4 Many digital audio consoles are merely control surfaces that work in conjunction with a separate audio router that allows the electronics to be centralized in a rack or equipment room. *(Image courtesy of Wheatstone Corporation.)*

or similar type of LCD (liquid crystal display) window. Because digital boards can be so readily customized to individual users, such as assigning different sources to different channels, a visual display helps with programming the board setup and operating the console.

5.3 AUDIO CONSOLE FUNCTIONS

Any control board has three primary functions: mix, amplify, and route audio. First, the audio console enables the operator to select any one or a combination of various inputs. In other words, it must first be determined where the signal is coming from: a microphone, a CD player, or an audio recorder, for example. Audio consoles are sometimes referred to as "mixing boards," because of their ability to select and have several inputs operational at the same time. Much of the production work you do will be a mix of voice, music, and sound effects through the audio console.

The second function of the control board is to **amplify** the incoming audio signal to an appropriate level. Although all sound sources are amplified to a degree, some sound sources (such as microphones) produce such a small electrical current that they must be further amplified to be used. What is also meant by "amplify" is that the volume of an audio signal going through the console can be raised or lowered. You'll learn more about this level adjustment later in this chapter. The third function of the audio console is to enable the operator to route these inputs to a number of outputs, such as monitor speakers, the transmitter, or an audio recorder. This function allows the operator to determine where the signal is going and to provide a means for listening to the signal.

5.4 COMPUTERS AND AUDIO CONSOLES

Rather than having a physical audio board containing individual input and output modules with multiple controls, in some instances, the operator may simply manipulate a "virtual console" on a computer screen, as shown in Figure 5.5. This is common practice on digital audio workstations that include mixing functions and other audio software programs. Such virtual boards are often part of a digital audio storage and studio system, such as Broadcast Electronics' AudioVAULT, which is a complete suite of software modules. The primary sections include a control screen for live or automated announcer operation and a waveform editor, and additional sections incorporate traffic and music scheduling, among other things. Other companies also offer computerized radio automation systems.

5.5 BASIC AUDIO CONSOLE COMPONENTS

All control boards, whether digital or analog, operate in basically the same way. For the purpose of simplicity, let's consider a small audio console with just two channels. Figure 5.6 shows such a board. Each channel (M-1 and

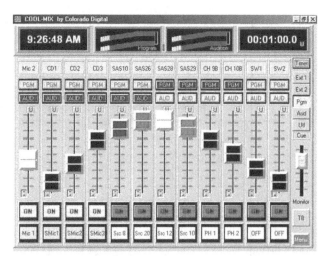

FIGURE 5.5 "Virtual" audio console controls emulate the functions of the traditional audio board. *(Image courtesy of Arrakis Systems, Inc.)*

L-1) has two inputs (labeled A and B), so that, for example, a microphone could be assigned to channel M-1 (position A), and channel L-1 could have a CD player assigned to input A and an audio recorder assigned to input B. In general terms, a channel refers to the path an audio signal follows. On an audio console, a **channel** refers to a group of switches, faders, and knobs that are usually associated with one or two sound sources (glance ahead to Figure 5.9). On the board in Figure 5.6, note the individual input selectors, output selectors, volume controls, and on/off switches associated with each individual channel. The cue, headphone, and studio monitor gain controls are associated with both channels, as are the VU meters.

5.6 INPUT SELECTORS

The **input selectors** on this particular audio console are pushbuttons that can be put in either an A or B position. This feature allows two different sound sources to be associated with each channel. Although channel M-1 and channel L-1 look identical, there is a major difference between them. Channel M-1 has been designed to accept only microphone-level inputs. Microphones generally do not have amplifiers built into them, whereas CD players and audio recorders have already put their signals through a small amount of amplification. When a signal from a sound source comes into channel M-1, it is sent through a preamplification stage that is not present for signals coming into channel L-1. In other words, a microphone-level input allows a signal to catch up to a stronger signal coming into the line position in terms of amplification. Then both signals often go through additional amplification.

The way the input selector switches in Figure 5.6 are arranged, a microphone comes into the M-1 channel (A input), a CD player comes into the L-1 channel (A input), and an audio recorder comes into the L-1 channel

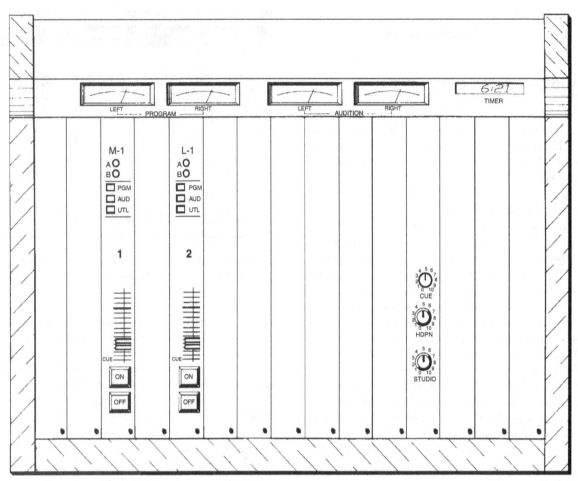

FIGURE 5.6 A simple two-channel audio console.

(B input). Nothing is assigned to the B input of channel M-1. Wiring the equipment to the board involves running a cable from the microphone, CD player, and audio recorder to the back (or bottom) of the audio console. Such wiring is usually done in a semipermanent-to-permanent way by the engineer and should not be tampered with by others.

Not all audio boards have input selector switches. Some radio production boards have only one input per channel that must be at the **microphone level**, or other inputs that can only accommodate equipment that has been preamplified and is ready for a **line level**. On boards of this type, usually only microphones can be patched into the first two inputs, and only CD players, audio recorders, and other line-level equipment can be patched into the remaining inputs.

On the other hand, some boards have input selector switches that have three or more positions for one input. For these boards, it's possible to patch a CD player at position A, another CD player at position B, and an audio recorder at position C all into the same input. The use that the facility is going to make of the various pieces of equipment has to be carefully studied, because of course no two pieces of equipment could be used at the same time on a single channel. Regardless of the configuration of an audio

board, the first two channels (from the left) are often utilized as microphone-level channels. Channel 1 is normally the main studio microphone, and channel 2 is often an auxiliary microphone.

Digital audio consoles are changing the concept of input selection somewhat. Most digital boards allow the user to assign *any* audio source to *any* channel using a type of audio router, as shown in Figure 5.7. In other words, the console configuration can be personalized for each individual user, so that channel 3 can be associated with a CD player for one person, an audio recorder for another person, and a microphone for the next person. Once the source is selected for any of the console's channels, the source name is usually displayed on an alpha display for that channel module. Digital consoles and virtual audio console controllers often refer to "input" as "sources" and "output" as "destinations."

5.7 INPUT VOLUME CONTROL

The input **volume controls** shown in Figure 5.6 are called **sliders**, or **faders**. They are merely **variable resistors**. Although they are called volume controls, or **gain controls**, they don't really vary the amount of amplification of the

FIGURE 5.7 On most digital consoles, any available audio source can be assigned to any one of the console's input channels. *(Image courtesy of Wheatstone Corporation.)*

signal. The amplifier is always on at a constant volume. Raising the fader (moving it from a south to a north position) decreases the amount of resistance to this signal. When the fader is raised and the resistance is low, a great deal of the signal gets through. The dynamic is like that of a water faucet. The water volume reaching the faucet is always the same, even when the faucet is closed. When you open the faucet (decrease the resistance), you allow the water to flow, and you can vary that flow from a trickle to a steady stream.

Some older boards have rotary knobs called **potentiometers**, or **pots**, instead of faders, although these are getting harder and harder to find. Pots provide the same function as faders. As the knob is turned clockwise, the resistance is decreased and the volume is increased. Most production people consider the fader to be easier to work with. For one thing, the fader gives a quick visual indication of which channels are on and at what level which is much harder to see with a rotary knob.

The numbers on both rotary knobs and faders may be in reverse order on some audio consoles to show their relationship to resistance. For example, if a knob is turned completely counterclockwise, or off, it may read 40; at a 12 o'clock position, it may read 25; and completely clockwise, it may read 0. These figures represent decreasing amounts of resistance and thus higher volume as the knob is turned clockwise. Modern boards with fader volume controls often use a range of numbers from 255 to 0 to 110 or 115. Although the same relationship to resistance is true (the more the fader is raised, the less resistance), these numbers actually relate to decibels and the VU meter. If the board has been set up properly, the 0 setting on the fader will produce a 0 reading on the VU meter (see Figures 5.10 and 5.11). Some boards avoid using any numbers at all and merely use equally spaced indicator lines to provide some kind of reference for various knob and fader settings.

Of course, most boards have more than the two channels of our example board. In most production studios, boards have 10 to 20 channels, but there are smaller and larger boards. Each channel has its own gain control and with two inputs per channel, more than 30 individual pieces of audio equipment can be manipulated through the console. In professional audio production facilities, consoles with more than 20 channels are not uncommon. With the digital board's ability to assign various sources to channels, the trend is for smaller boards, as you usually work with only two to four channels active at any one time.

In addition to the gain controls just mentioned, some boards have a **gain trim**, or **trim control**, that fine-tunes the volume of each input (as shown in Figure 5.9). For example, if the sound signal coming from a CD player has the left channel louder (stronger) than the right channel, a stereo trim control can decrease the left level or increase the right level until the sound signal is equal for both channels. Each input channel on an audio console usually has a gain trim feature, and often there is a similar trim adjustment for the program and audition output of the audio board. Although these trim controls (sometimes referred to as "pad") may be on the face of the audio board, they are more often an internal adjustment that is taken care of by the engineer when the control board is initially set up.

5.8 MONITORING: SPEAKERS AND HEADPHONES

Once the signal is through the input gain controls, it's amplified in a program amplifier and then sent several places (see Figure 5.8). One of these is a **monitor amplifier**. This amplifies the signal so that it can be sent into a **monitor speaker** to enable the operator to hear the signal that is going out. Boards usually contain a simple potentiometer to control the gain of the monitor speaker. For instance, our example board has a pot labeled STUDIO, which controls the volume of the monitor speakers. This control in no way affects the volume of the sound being sent out to an audio recorder (or transmitter, for example). It controls only the volume for the person listening in the control room. In a broadcast application, the on-air monitor presents the actual output of the station after it has been broadcast. In a true broadcast situation, you must monitor the actual air signal. If you are only listening to the system's program signal, you would not be aware if the signal goes off the air. In a production environment, the studio monitors are speakers used to hear what is routed through designated channels on the console. A common mistake of beginning broadcasters is to run the studio monitors quite loud and think all is well, while in reality they have the signal going through the audio board (and therefore to a recorder or on the air) at a very low level. It's important for the operator to be aware of the level of sound going out the line, which is discussed regarding VU meters in section 5.10.

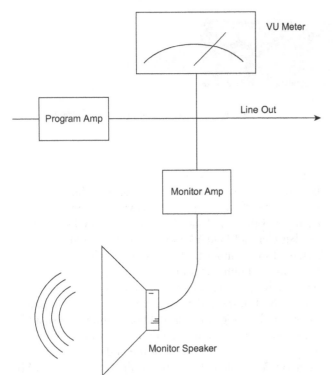

FIGURE 5.8 A block diagram of the monitor amplifier section of an audio console.

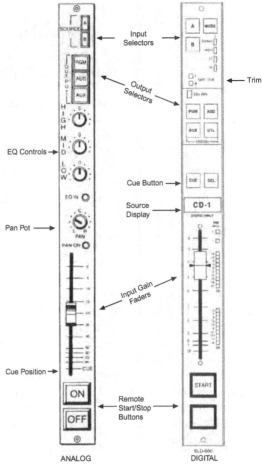

FIGURE 5.9 Whether digital or analog, a single channel of an audio console consists of a group of switches, pushbuttons, and a fade that can be used to manipulate a sound source that is assigned to that channel. *(Image courtesy of Wheatstone Corporation.)*

Most audio consoles also have a provision for listening to the output of the board through **headphones**. Because live microphones are often used in production work, the monitor speakers are muted when the microphone is on so that **feedback** doesn't occur. To be able to hear an additional sound source, such as a CD, headphones are necessary. Audio consoles often allow you to monitor any of the outputs with headphones by selecting an appropriate switch. There is usually a volume control to adjust the signal level going to the headphones. On our example board, this control is labeled HDPN. As such, headphones allow one to hear all sound sources, be it in production or on-air, when the studio or control room microphone is engaged.

5.9 CUE

Another function found on most audio boards is called **cue**, which allows you to preview an input sound source. Both rotary pots and fader controls go into a cue position, which is below the off position for that control. If you turn the rotary pot all the way counterclockwise to off, it will reach a detent, or stop. If you keep turning the knob (with a little extra pressure), it will click into the cue position. Faders are brought down, or south, until they click into cue (see Figure 5.6 again). Some faders can be put into cue with a separate push button (see Figure 5.9) that, when depressed, puts that channel into cue regardless of where the fader control is set. Cue position is usually marked on the face of the audio console.

In the cue position, the audio signal is routed to a cue amplifier and then to a small speaker either built into the control board or located just adjacent to it. As the quality of this small internal speaker is usually marginal at best, the cue signal is often sent to a small, but better quality, external speaker located near the audio console. Some audio consoles send the cue signal to the main studio monitor speakers or may be monitored by headphones. The program signal is automatically turned down, or dimmed, when a channel is put into cue and the cue signal is heard on top of the program signal. On our example audio console, the cue volume level is controlled with the pot labeled CUE.

As the name implies, this position is designed to allow the operator to cue up a sound source. For example, an audio recording can be cued to the exact beginning so that the sound will start immediately when the recorder is turned on. If an input is in cue, the signal doesn't go to any other output such as the transmitter or an audio recorder. Its only purpose is to allow off-air cueing. Many beginning announcers and production people forget to move the volume control out of the cue position after cueing up

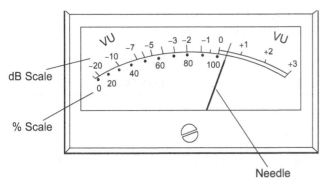

FIGURE 5.10 The standard VU meter incorporates both a percent and decibel scale.

FIGURE 5.11 VU meters with fluorescent or liquid crystal displays can indicate volume changes more quickly and accurately than electromechanical meters. *(Image courtesy of Dorrough Electronics.)*

the sound source. If the control is left in the cue position, the signal won't go out on the air or be routed to an audio recorder. It will play only through the cue speaker.

5.10 VU METERS

Another place the signal is sent after program amplification is the volume unit indicator, or **VU meter** (see Figure 5.10). This is a metering device that enables the operator to determine what level of sound is going out the line.

One common type of VU meter has a moving needle on a graduated scale. Usually the top position of the scale is calibrated in **decibels (dB)**, and the lower portion of the scale is calibrated in percentages. In audio engineering, a reading of 0 dB is 100 percent volume, or the loudest you want the signal to go. The VU meter is important for consistent audio production work. As noted in Chapter 2, "The Studio Environment," how loud something sounds is very subjective. What is loud to one announcer may not be deemed loud by another, especially if they each set the monitor speaker volume differently. The VU meter gives an electronic reading of volume that is not subjective.

The accuracy of VU meters is sometimes questioned in two areas. First, VU meters have trouble indicating transients—sudden, sharp, short increases in volume of the sound signal. Most VU meters are designed to indicate an average volume level and ignore these occasional sound bursts. Second, VU meters tend to overreact to the bass portion of the sound. In other words, if a sound signal is heavy in the bass frequencies, it will probably show a higher VU reading than it would if the total sound signal were being accurately read. In spite of these concerns, the VU meter remains the best indicator of volume levels in broadcast production.

Generally, an operator should control the signal so that it stays approximately between 80 percent and 100 percent. When the needle swings above 100 percent, we say the signal is **peaking in the red** because that portion of the VU meter scale is usually indicated with a red line. This is a warning to the operator to lower the gain with the fader or pot because the sound is **overmodulated**. Digital equipment is very unforgiving in regard to recording "in the

red." Most digital equipment will not tolerate recording at any level above 100 percent and will distort or add "pops" to any recorded signal that exceeds it. Analog equipment is more forgiving in that it simply distorts sound, but good production practice dictates recording everything below 100 percent.

Often there is a metal peg at the far end of the VU meter to prevent the needle from going off the dial. Allowing the gain to become so high that the needle reaches the upper peg is called **pegging the meter**, or **pinning the needle**, and should be avoided to prevent damage to the meter as well as distortion of the signal.

When the signal falls below 20 percent consistently, we say the signal is **undermodulated** or **in the mud**, and the operator should increase the volume. If it's necessary to adjust the level during the program, we say that the operator is **riding the gain** or **riding levels**. "Gain" is an audio engineer's term for loudness or volume. This is why a radio operator was first called a disc jockey. He would play record "discs" and "ride" the gain.

Our example audio board (Figure 5.6) has a set of VU meters that indicates the left and right volume of the program sound going out and another set of meters that shows the audition output. Most boards have multiple VU meters. For example, they might have separate meters for each individual channel. Boards that have multiple outputs also have multiple meters. For example, a board that is stereo, like our example board, will have one meter for the right channel and one for the left.

On some boards, the VU meter isn't an electromagnetic meter at all, but is rather a succession of digital lights (LEDs, or light-emitting diodes) that indicate how high the volume is. Other electronic meters (see Figure 5.11) replace the LEDs with liquid crystal or fluorescent displays; the advantage of these over mechanical meters is that they can indicate volume changes more quickly and accurately.

5.11 OUTPUT SELECTORS

The most common arrangement for the **output selectors** on an audio board is a bank of three buttons for **program,**

audition, and **auxiliary** outputs. The example board in Figure 5.6 has this type of configuration except the auxiliary output is labeled UTL for **utility**. When no button is pushed in, the output is stopped at this point. When the program button (PGM) is pushed, the signal normally goes to the transmitter if the board is in an on-air studio, or to an audio recorder if the console is in a production room. The program position would be the normal operating position when using an audio console.

If the audition button (AUD) is depressed, the signal is sent to an audition amplifier and can then be sent to the monitor speakers, an audio recorder, another studio, and so on. The audition signal will *not* normally be sent to the transmitter. The purpose of the audition switch is to allow off-air recording and previewing of the sound quality and volume levels of a particular signal. For example, you could be playing a CD on one CD player through channel 3 of the audio console (in the program position) and at the same time be previewing another CD on a different CD player through channel 4 in the audition position. Each channel of the audio console can be used either in the program or audition position.

The auxiliary (AUX) or utility (UTL) position is just like either the program or audition position and is another output for the audio board (see Figure 5.6). For instance, some studios are set up so that the auxiliary position of one control board feeds another studio and becomes an input on the audio console in the other studio. Such as the news booth sending a feed to the on-air studio. Unlike most input selectors, which allow only one input to be selected at a time, output selectors usually allow more than one output to be active. If all output buttons were depressed, you could send the same audio signal to three different locations at the same time.

The configuration of output selector buttons varies from board to board. If there are a large number of outputs—six, for example—then there may be six output buttons for each input. The input signal will be sent to whichever buttons are depressed, or selected. Sometimes there are no output selectors; every input either goes out or it doesn't, depending on the master volume control. Other boards have buttons labeled SEND that determine where the signal goes. Some boards just use a single toggle switch in place of push buttons. The functions are exactly the same, depending on which way the toggle is switched, for example, program or audition. Typically, if the toggle is in the middle, or neutral position, the signal stops at that point (just like having no button pushed in a three-button bank).

It helps to distinguish between the basic radio board from consoles with input and output channels used in professional sound studios. Audio consoles are often referred to by their number of inputs and outputs, so a 16-in/4-out console has 16 inputs and 4 outputs. These are called multi-channel consoles (as in Figure 5.3) and have two definitive sides: input and output. The basic stereo board, such as those often found in radio broadcast studios

(see Figure 5.2), are not multi-channel consoles since all inputs are sent to an established left and right stereo output.

5.12 OUTPUT VOLUME CONTROL

There is no output volume control on our example board in Figure 5.6 that can be adjusted by the operator. Some audio consoles include a **master fader** that controls the volume of the signal leaving the board. Even if an input sound source is selected, program output is selected, and fader volume is up, if the master fader is all the way down, no signal will go from the console. Many boards have more than one output gain control. Again, a stereo board requires two masters: one for the right channel and one for the left. If there are multiple volume controls, then there's usually a master gain control that overrides all the other output volume controls. In other words, if the master is down, the signal won't go anywhere, even though one or more of the output volume controls are up.

5.13 REMOTE STARTS, CLOCKS, AND TIMERS

Other bells and whistles that frequently appear on audio boards include **remote start switches** (see Figures 5.6 and 5.9). These are usually located below each individual channel fader, and if the equipment wired into that channel (such as a CD player or audio recorder) has the right interface, it can be turned on, or started, by depressing the remote start. This makes it easier for the operator to start another sound source while talking into the microphone without having to reach off to the side and possibly be pulled off microphone. For most consoles, these switches also turn that channel on and off. In other words, even if you have the fader for that channel turned up, no sound signal will go through the channel until it has been turned on. However, some consoles are set up so that they will automatically turn the channel on when the fader is moved upward.

Many control boards include built-in clocks and timers. Digital clocks conveniently show the announcer the current time (hours, minutes, and seconds), and timers (see Figure 5.6) can be reset at the start of a CD to count up the elapsed time or count down the remaining time. Many timers will begin automatically when an "on" or remote start button is pressed.

5.14 EQUALIZERS AND PAN POTS

Many sound boards include simple **equalizers**. These increase or attenuate certain frequencies, thus altering the sound of the voice or music by changing the tonal quality of the sound. In some instances, they help eliminate unwanted sound. For example, an audio recording might include an electronic hum that could be lessened by filtering out frequencies around 60 Hz. Equalizers can also be used for special effects, such as making a voice sound like it is coming over a telephone. It's important to note that when you equalize a sound, you can affect both the unwanted and wanted sound; equalization is often a compromise between eliminating a problem and keeping a high-quality, usable audio signal.

Usually the equalizers are placed somewhere above each input volume control. In Figure 5.9, the input signal of the analog channel is split into three frequency ranges: high, midrange, and low. Turning the control clockwise increases the volume of that range of frequencies; counterclockwise rotation of the control will decrease its level. The "EQ in" button on this channel is actually a bypass switch that allows the operator to hear and compare the signal with and without equalization. Equalizers and other similar equipment will be discussed in further detail in Chapter 8, "Signal Processing and Audio Processors."

Audio console channel inputs that are **stereo** often have a **pan knob**, or **pan pot** (pan meaning "panoramic"). By turning (panning) this knob to the left or right, you can control how much of the sound from that input goes to the right channel and how much goes to the left channel output. In other words, if the pan pot of a channel is turned toward the L position, the level would sound stronger, or louder, from the left speaker. Normally, the pan pot would be in the center position and the input sound would be directed equally to the left and right outputs of that channel. **Monaural** input channels may have a **balance control**, which serves a similar purpose.

5.15 OTHER FEATURES

Some boards have a built-in **tone generator**. This reference tone is usually placed on an audio recording before the actual program material begins. The tone generator sends out a tone through the board that can be set at 100 percent, using the board VU meter. The VU meter on the source to which the signal is being sent (such as an audio recorder) is simultaneously set at 100 percent. After the two are set, any other volume sent through the board will be the exact same volume when it reaches the recorder. A tone generator allows for this recording consistency; otherwise, sounds that register at 100 percent coming through the board might peak in the red and be distorted on the audio recorder. The tone on the recording is also used during playback. The audio engineer or board operator listens to the tone and sets the recorder VU meter to 100 percent. That way the recording will play back exactly as it was recorded.

Some audio consoles have a **solo switch** above each input. When this switch is on, only the sound of that particular input will be heard over the monitor. Other boards have a **mute switch** for each channel, which prevents

the signal from going through that channel when it's depressed. A mute switch acts just like an on/off switch.

A few boards have a **talk-back switch**, which is a simple intercom system consisting of a built-in microphone and a push-button control that turns the microphone on or off. The normal position of this switch is "off" so that the button must be pushed in to activate the microphone. The signal from the talk-back microphone is sent to a speaker or headphones in another studio—for example, a performance studio—which would allow the operator at the audio console to communicate with the announcer in a studio at a separate location.

PRODUCTION TIP 5A
Manipulating Faders

The type of transition that you are planning to have can affect how you are operating the faders of your board. If you know you will be doing a segue (explained shortly), you can set the fader levels for the new sound you will be bringing in ahead of time by using the audition function. Then, when it's time to take out one sound and bring in the other, you can simply hit the appropriate off and on buttons below the faders. That way you don't need to be concerned about getting the fader to the right position in order to have the new sound at the right level.

Working with the on/off buttons instead of the faders is also a good idea if you are doing a program where people are speaking. You can set the faders for each person's speaking level ahead of time and then, when the program begins, you can just push the "on" buttons and all the levels will be correct. This mode of operation is also advantageous if you are both talking and playing music. Set your microphone level ahead of time and then each time you will be talking, you can simply push the "on" button and not have to look down to make sure you have set the fader in the right position.

Of course, if you need to do a fade-in, fade-out, or cross-fade, this method won't work. For these transitions, you will want to have the "on" buttons already pushed and then bring the faders up or down to execute the fade.

5.16 SOUND TRANSITIONS AND ENDINGS

As mentioned earlier in this chapter, one of the functions of the audio console is to mix two or more sound sources together. Often this mix is really a sound transition, or the merging of one sound into another. In audio production, one basic transition is the **fade** (gradually increasing or decreasing volume), in which you mix one sound with silence. For example, to **fade-in** a CD means to slowly increase the volume from silence to the desired level (see Figure 5.12A). A **fade-out** accomplishes just the opposite, as the CD goes

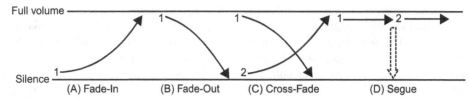

FIGURE 5.12 Sound transitions are used frequently in audio production work.

from full volume to silence (see Figure 5.12B). Most compact discs are recorded so that they fade out at the end naturally, but in production work it's sometimes necessary to end a song early, and the production person can do so with a manual fade-out. Conversely, to allow a piece of music to have a natural ending, it is sometimes necessary to fade in after the music has started. This is known as a "dead roll," since the music is started with the fader down.

The other common transitions are the cross-fade and segue. As the name implies, a **cross-fade** occurs when one sound is faded down as another sound is faded up (see Figure 5.12C). There is a point as the two sounds cross when both sounds are heard. Because of this, care should be taken in choosing music to cross-fade. Some combinations of songs can sound extremely awkward. The speed of a cross-fade is determined by the board operator and depends on the type of effect desired; however, most cross-fades are at a medium speed to give a natural, brief blending of the two sounds. A **segue** is quite different; it is the transition from one sound to the next with no overlap or gap (see Figure 5.12D). A segue can best be accomplished when the first song ends cold. Music with a **cold ending** doesn't fade out, but rather it has a natural, full-volume

end. Unlike fades of any kind, segues are accomplished with both sounds at full volume. Most disc jockey work is a mixture of cross-fades and segues, but all the sound transitions are used frequently in audio production work.

In addition to the fade-outs and cold endings, one other music ending that should be noted is the **sustain ending**. Songs with a sustain ending will hold the final chord or notes for a short period of time and then very gradually fade out. This is distinctly different from either a normal fade-out or a cold ending.

5.17 CONCLUSION

If you have followed the descriptions and explanations offered in this chapter, the audio console should be a less frightening assemblage of switches, knobs, and meters than it was when you began. You should begin to feel comfortable working with the controls of the board in your own studio. You should also be aware that most modern audio consoles will be a digital board or it may be a virtual audio console that you maneuver with a few mouse clicks or a touch screen.

Self-Study

1. It's possible to have music from an audio recorder go into a control board and then come out and be recorded on another recorder.

   a) true
   b) false

2. In Figure 5.9, according to the pan pot position on the analog channel, what is the relationship between the sound signal going to the left channel and the signal going to the right channel?

   a) Left channel volume would be less than right channel volume of the same signal.
   b) Left would be the same as right (similar change).
   c) Left would be greater than right (similar change).
   d) There would be no signal going to the right channel.

3. If the digital fader in Figure 5.9 was at 3 and you moved it to 7, what would you have accomplished?

   a) amplified the signal
   b) put the channel in cue
   c) decreased the resistance
   d) decreased the volume

4. In Figure 5.6, sound on channel L-1 would not get to program out to be broadcast or recorded unless what happened?

   a) The fader is at 0.
   b) The PGM output selector switch is on.
   c) The HDPN volume control is at 3.
   d) The A input selector switch is set to CD.

5. Audible sound comes from what in Figure 5.8?

   a) the VU meter
   b) the monitor amplifier
   c) the program amplifier
   d) the monitor speaker

6. In Figure 5.10, a reading of 50 percent on the scale is roughly equivalent to which reading on the dB scale?

   a) −6 dB
   b) −4 dB
   c) 50 dB
   d) −10 dB

7. Which expression describes a 20 percent reading on the VU meter in Figure 5.10?

   a) peaking in the red
   b) turning up the pot
   c) being in the mud
   d) riding the gain

8. Which expression describes a 3-dB reading on the VU meter in Figure 5.10?

   a) riding the gain
   b) broadcasting in stereo
   c) peaking in the red
   d) pegging the meter

9. A line-level channel of an audio console would have what type of equipment assigned to it?

   a) CD player and audio recorder
   b) microphone and CD player
   c) microphone and audio recorder
   d) only microphones

10. Which choice most accurately describes the monitor/speaker?

    a) an input
    b) a mix
    c) an output
    d) an equalizer

11. Look at Figure 5.6. What happens if the PGM/AUD/UTL output selector switch of channel M-1 is in the audition position?

    a) Sound will not reach the input A switch.
    b) Sound can be going to an audio recorder.
    c) Sound can be going to the transmitter.
    d) Sound will go to the cue speaker.

12. Which statement about the master volume control on an audio board is true?

    a) must be up for sound to leave the board
    b) is required only if the board is stereo
    c) controls only the volume of the line inputs
    d) controls only the volume of the microphone inputs

13. Which control found on an audio console might be used to help eliminate electronic hum on an audio recording?

    a) the pan knob
    b) the gain trim
    c) the solo button
    d) the equalizer

14. Which statement about the cue position on a fader is true?

    a) allows sound to go to the transmitter
    b) sometimes substitutes for the trim control
    c) sends sound to a small speaker in the audio board
    d) allows sound to fade from the left to right channel

15. Which of the following can help assure that the level that's being recorded on an audio recorder is the same as that coming from the audio board?

    a) tone generator
    b) remote switch
    c) digital timer
    d) pan pot

16. Which of the following statements about digital audio consoles is *not* true?

    a) Digital consoles are more expensive than analog consoles.
    b) Digital consoles include some type of display screen.
    c) Digital consoles do not accept analog inputs.
    d) Digital consoles offer several design architectures.

17. Which sound transition occurs when one CD is faded down at the same time as another CD is faded up?

    a) fade-in
    b) fade-out
    c) segue
    d) cross-fade

18. On an audio console, which term refers to a group of switches, faders, and knobs that are usually associated with one sound source, such as a CD player?

    a) input selector
    b) remote start switch
    c) channel
    d) output selector

19. A segue is the basic sound transition in which one sound is mixed with silence.

    a) true
    b) false

20. Most audio consoles used in radio broadcasting are identical to the consoles used in music recording.

    a) true
    b) false

21. Which feature of an audio console would allow the operator to alter the tonal quality of a sound going through the board?

    a) mute switch
    b) gain trim control
    c) equalizer control
    d) talk-back switch

22. How can a channel of the digital audio console shown in Figure 5.9 be put into cue?

    a) by moving the fader up to 0
    b) by moving the fader down to 7
    c) by depressing the START button
    d) by depressing the CUE button

23. Which feature of an audio console is a simple intercom system?

    a) talk-back switch
    b) tone generator
    c) pan knob
    d) output selector

24. On an audio console, a pan pot and a balance control both serve a similar function.

    a) true
    b) false

25. If you were attempting to segue from one song to another, how should the first song end?

   a) fade-out
   b) cold ending
   c) fade-in
   d) sustain ending

ANSWERS

If you answered A to any of the questions:

1a.   Right. A tape recorder can be both input and output and an audio console can link two recorders.

2a.   No. Check the setting of the pan pot. (Reread 5.14.)

3a.   No. The fader never actually amplifies the signal. This is done with preamp or amplifier circuits. (Reread 5.7.)

4a.   No. When a fader or pot is at 0, it is usually on. (Reread 5.7 and 5.11.)

5a.   No. This indicates level, but you hear nothing from it. (Reread 5.8 and 5.10.)

6a.   Right. You read the scale correctly.

7a.   No. The needle would be at the other end of the scale for this. (Reread 5.10.)

8a.   No. "Riding the gain" means to keep the volume at proper levels, and this is not being done. (Reread 5.10.)

9a.   Correct. Both a CD player and an audio recorder are line-level inputs.

10a. No. You're at the wrong end. (Reread 5.3, 5.5, and 5.8.)

11a. No. The input selector switch is before the output selector switch and really has nothing to do with it. (Reread 5.6 and 5.11.)

12a. Right. The purpose of the master volume is to allow all appropriate sounds to leave the audio console.

13a. No. This controls how much sound is going to the left and right channels. (Reread 5.14.)

14a. No. When a channel is in cue, sound will not go to the transmitter. (Reread 5.9.)

15a. Right. If the board VU meter and the tape recorder VU meter are both set at 100 percent tone, the levels should be the same.

16a. No. Although the difference between the price of an analog board and a digital console is becoming less and less, many digital boards are slightly more expensive than their analog counterparts. (Reread 5.2.)

17a. No. This is a different sound transition. (Reread 5.16.)

18a. No. This is one of the switches involved, but this is not the correct term. (Reread 5.5 and 5.6.)

19a. No. You've confused this with the fade. (Reread 5.16.)

20a. No. Although there are similarities between all audio consoles, those most often found in radio are more basic than those used in the music recording studio. (Reread 5.1.)

21a. No. This control turns off a console channel. (Reread 5.14 and 5.15.)

22a. No. This will turn the volume of the channel up. (Reread 5.7 and 5.9.)

23a. Correct. The talk-back feature allows a board operator to talk with an announcer in another studio via a simple intercom system.

24a. Yes. Both can be used to adjust how much of the audio signal goes to either the left or right channel output.

25a. No. A segue should happen with both songs at full volume with no overlap or silence between songs; if the first song fades out, this won't happen. (Reread 5.16.)

If you answered B to any of the questions:

1b.   Wrong. A tape recorder can be fed into an audio console and another tape recorder can be placed at the output. (Reread 5.3 and 5.5.)

2b.   No. Check the setting of the pan pot. (Reread 5.14.)

3b. No. Cue would be even below the "infinity" mark. (Reread 5.7 and 5.9.)

4b. Right. When the PGM output selector switch is in the "on" position, normally sound would be sent to the transmitter or an audio recorder.

5b. No. This amplifies so that the sound can come out, but you don't hear the sound from it. (Reread 5.8.)

6b. No. You went in the wrong direction. (Reread 5.10.)

7b. No. If anything, the pot is being turned down. (Reread 5.10.)

8b. No. This is not correct. (Reread 5.10.)

9b. No. A microphone should come into a microphone input, because it needs to be amplified to reach the line level of amplification. A CD player would be plugged into the line position. (Reread 5.3 and 5.6.)

10b. No. This is incorrect. (Reread 5.3, 5.5, and 5.8.)

11b. Yes. The purpose of the audition position is to send the sound somewhere other than the transmitter. Sound would not necessarily have to go to an audio recorder, though it could.

12b. No. A master volume control functions the same for both mono and stereo. Stereo boards, however, usually have two master volume controls, one for each channel. (Reread 5.12.)

13b. Wrong. This allows fine-tuning of input or output levels on an audio board. (Reread 5.7 and 5.14.)

14b. No. The two have nothing to do with each other. (Reread 5.7 and 5.9.)

15b. No. This will turn on the recorder remotely but will do nothing about levels. (Reread 5.13 and 5.15.)

16b. No. Whether it's a simple alpha display on the channel module or a full LCD screen, because they can be readily reconfigured, all digital boards have some type of display screen. (Reread 5.2.)

17b. No. This is a different sound transition. (Reread 5.16.)

18b. No. This is one of the switches involved, but this is not the correct term. (Reread 5.5 and 5.13.)

19b. Correct. The basic sound transition that mixes one sound with silence is the fade, not the segue.

20b. Yes. This is false, because radio consoles are usually not as complex as the boards used in music recording.

21b. Wrong. This control varies the input level of a sound source of a console channel. (Reread 5.7 and 5.14.)

22b. No. Although some consoles are put into cue by moving the fader down, it would have to go all the way down and usually "click" into a cue position. (Reread 5.7 and 5.9.)

23b. Wrong. The tone generator provides a reference level for setting correct recording levels. (Reread 5.15.)

24b. No. This is not false because both controls can be used to adjust the amount of audio signal that is sent to the left or right output. (Reread 5.14.)

25b. Yes. The first song should have a cold ending if you're going to segue into the next song.

If you answered C to any of the questions:

2c. Right. The pan pot is set so that more of the sound from the input goes to the left channel.

3c. No. Changing the level from 3 to 7 would increase resistance. (Reread 5.7.)

4c. Wrong. The HDPN volume control only controls the volume of the headphone's audio. (Reread 5.8 and 5.11.)

5c. No. This amplifies sound, but you don't hear sound from it. (Reread 5.8.)

6c. There is no such reading on a VU meter. (Reread 5.10 and review Figure 5.10.)

7c. Right. "In the mud" is the term for an extremely low reading.

8c. No. It is worse than that because the needle is all the way to the end of the scale. (Reread 5.10.)

9c. No. A microphone should come into a microphone input because it needs to be amplified to reach a usable level. An audio recorder would be plugged into the line position. (Reread 5.3 and 5.6.)

10c. Right. Sound comes out to the monitor/speaker.

11c. No. The audition position is not normally used to send the sound to the transmitter. (Reread 5.11.)

12c. No. It controls all the sound that is set to leave the board. Line and microphone positions have no bearing on it. (Reread 5.6 and 5.12.)

13c. No. This allows you to hear one input channel by itself. (Reread 5.14 and 5.15.)

14c. Right. Cueing is just for the person operating the board.

15c. No. This is simply a type of clock. It has nothing to do with levels. (Reread 5.13 and 5.15.)

16c. Yes. Digital boards have modules that accept analog signals and convert them to digital.

17c. No. This is a different sound transition. (Reread 5.16.)

18c. Yes. This is the correct term.

21c. Right. EQ controls increase or decrease certain frequencies of the sound, thus changing the tone.

22c. No. If the fader was up, this would let us hear audio, but not in cue. (Reread 5.9 and 5.13.)

23c. Wrong. Pan pots control the amount of sound that goes to the left or right output of a channel. (Reread 5.14 and 5.15.)

25c. No. You're confused here; a fade-in isn't a way a song ends. (Reread 5.16.)

If you answered D to any of the questions:

2d. No. That would not be correct. (Reread 5.14.)

3d. Right. This is an increase in resistance to the signal; not as much of the signal gets through, and it's at a lower volume.

4d. Wrong. It wouldn't matter if you assigned the A input to a CD, audio recorder, or some other piece of equipment. (Reread 5.6 and 5.11.)

5d. Right. You hear sound from the monitor speaker.

6d. You might be thinking of half of the dB scale, but that isn't correct. (Reread 5.10.)

7d. No. If anything, the operator is not properly riding the gain. (Reread 5.10.)

8d. Right. "Pegging the meter" is correct. We could also have used the term "pinning the needle."

9d. Wrong. Microphone signals must be amplified to reach line level and should only come into a microphone input. (Reread 5.3 and 5.6.)

10d. No. An equalizer affects frequencies. (Reread 5.3, 5.5, 5.8, and 5.14.)

11d. Wrong. The cue speaker will be activated when a channel is put into cue, regardless of where the PGM/AUD/UTL switch is. (Reread 5.9 and 5.11.)

12d. No. It controls all the sound that is set to leave the board. Microphone and line positions have no bearing on it. (Reread 5.6 and 5.12.)

13d. Right. This can help eliminate high frequencies where scratches reside.

14d. No. You're confusing this with a pan pot. (Reread 5.9 and 5.14.)

15d. No. This allows sound to fade from the left to right channel. (Reread 5.14 and 5.15.)

16d. Wrong. Digital boards have more than one design approach, for example, the mainframe and control-surface system. (Reread 5.2.)

17d. Yes. This is the correct response.

18d. No. This is one of the switches involved, but this is not the correct term. (Reread 5.5 and 5.11.)

21d. Wrong. This control is a form of studio intercom. (Reread 5.14 and 5.15.)

22d. Yes. On this console, any channel can be put into cue merely by pressing the cue button associated with that channel, regardless of where other switches or buttons are set.

23d. Wrong. Output selectors determine where sound goes when it leaves the audio console. (Reread 5.11 and 5.15.)

25d. No. A segue should have no overlap or silence between songs; if the first song has a sustain ending, this won't happen. (Reread 5.16.)

Projects

PROJECT 1

Learn to operate an audio console.

Purpose

To make you proficient operating some functions of the audio console, and to practice basic sound transitions.

Notes

1. Audio boards are generally one of the more complicated pieces of audio equipment. It may take you a while to master a board, but don't despair. Take it slowly, and don't be afraid to ask for help.
2. Audio boards all have the same general purpose. Sounds come into the board, are mixed together, and are sent out to somewhere else.
3. The actual exercise should be done rapidly. You won't be judged on aesthetics. In other words, when you're fading from one source to another, do it quickly. Don't wait for the proper musical beat, phrase, or pause.

How to Do the Project

1. Familiarize yourself with the operation of the audio console in your production studio. Learn the inputs, the outputs, the method for changing volume, and other special features of the board.
2. As soon as you feel you understand the board, do the following exercise while recording it on an audio recorder. Practice as much as you like first.
   a. Cue up a CD, play part of it, and fade it out.
   b. Using the studio microphone, announce your name and the current time.
   c. Begin a second CD.
   d. Cross-fade to another CD, and then fade it out.
   e. Bring in an auxiliary microphone, and ad-lib with another announcer.
   f. Fade-in a CD, segue to another CD, and fade it out.
   g. Announce something clever on the studio microphone.
3. Listen to your recording to make sure it recorded properly.
4. Label your recording "Audio Console Operation" and include your name. Hand the assignment in to your instructor to receive credit for this project.

PROJECT 2

Diagram and label an audio board.

Purpose

To familiarize you with the positioning of the various switches and controls so that you can access them quickly.

Notes

1. Some boards are very complicated and have more functions than are discussed in this chapter. Usually, this is because they're intended to be used for sound recording of music. If you have such a board, you need to label only the parts that you will be using frequently.

2. If you can't find controls for all of the functions given in this chapter, ask for help. Because there are so many different brands and types of boards, sometimes functions are combined or located in places where you can't identify them easily.
3. You don't need to label each switch and knob. If your board has ten inputs, it will obviously have ten channel-volume controls. You can circle them all and label them together, or make one label that says "Input Volume Controls" and draw arrows to all ten.
4. You will be judged on the completeness and accuracy of your drawing. You won't be graded on artistic ability, but be as clear as possible.

How to Do the Project

1. Sketch the audio console in your production studio.
2. Label all the basic parts: input selectors, channel gain controls, VU meters, output selectors, and master gain controls (if your board has them).
3. Also label any other parts of the board that you will be using frequently, such as equalizers, cue positions, and headphone connections.
4. Label your sketch with your name and the title, "Audio Console Diagram." Give your completed drawing to your instructor for credit for this project.

PROJECT 3

Record a two-voice commercial.

Purpose

To develop your skill in creating a spot while working with another announcer.

Notes

1. This project assumes that you have enough familiarity with your studio equipment to accomplish basic recording and production techniques.
2. The production incorporates two announcer voices and a music bed.
3. You will need to write a dialogue script that can be read in about 23 seconds. You might write copy about a restaurant you eat at frequently or some cosmetic or health product you are familiar with.
4. There are many ways to accomplish this project, so don't feel you must follow the production directions exactly.

How to Do the Project

1. Record the script (voices only) onto an audio recorder. Make sure both voices are close to the same volume level. Rerecord to correct mistakes or just to get it to sound the way you want.
2. Select a music bed that is appropriate for the style of the spot. You might find something on the website that accompanies this text.
3. You can play back the music bed directly from a CD, or you may have to record it on another recorder.
4. Set correct playback levels for both the vocal track and the music bed. Both will start at full volume. Then cue both to the beginning sound.
5. Record your completed spot onto a CD-R, so set your recorder to record mode. (Your instructor may advise you to record on different media.)
6. Begin recording. Start the music bed at full volume. Start the vocal track and simultaneously fade the music bed slightly so the vocal track is dominant.
7. As the vocal ends, bring the music bed back to full volume and then quickly fade it out as you approach 30 seconds.

8. It may take you several attempts to get the spot to come out correctly. If you need to do it over, just cue everything and try again.

9. Listen to the finished commercial. Make sure that you've gotten a good balance between the music and the voice and all recording levels are proper.

10. Label the assignment with your name(s) and "Two-Voice Radio Spot." Turn it in to your instructor to receive credit for this project.

6

DIGITAL AUDIO PLAYERS/RECORDERS

6.1 INTRODUCTION

Digital technology has long replaced conventional analog technology in many applications around the audio production studio. For instance, the CD player long ago replaced the turntable as a necessary piece of production equipment. However, even digital technology that has been considered a staple of audio production is being replaced. CD players have become a less important production tool lately as more audio files are recorded, edited, and played back on computers as separate files and not stored as often on one removable disk. Mobile and tablet computing, smartphones, and the Internet in general have allowed for more digital manipulation of files in regard to transferring between editors and storage. This chapter looks at some of the digital equipment that is commonly used in today's audio production environment. Although this chapter is not intended to be a comprehensive survey, most of the equipment mentioned still plays a role in digital playback and recording in many audio production studios.

6.2 THE CD PLAYER

The **compact disc (CD) player** was the first piece of digital audio equipment to be embraced in production and broadcast work, and is still a common source for playing back pre-recorded material. Because CDs are often played one right after the other, there are usually at least two CD players in each production room or on-air studio. This way, one disc can be cued while the other is playing on the air or recording.

CD players offer superior sound quality, greater **frequency response**, better **signal-to-noise ratio (S/N)**, improved **dynamic range**, and almost no **distortion**, compared to an analog component such as a tape deck or turntable. In addition, there is no physical contact between the transduction apparatus of the player and the surface of the CD, so little (if any) wear takes place.

The CD format also offers the convenience of quick access to material stored on the disc. CD players can instantly move from one track to another anywhere on the disc with just the push of a button. Another plus of the CD player, from a production viewpoint, is its ability to cue to a track. CD players allow the announcer to find the exact beginning of a track and start the CD instantly at that point. CD timing information, such as elapsed or remaining time and track length, is helpful for production work, and the ability to select and play segments of music or to automatically fade in at any point in the music can be a creative production tool.

The CD player shown in Figure 6.1 is typical of a unit designed specifically for broadcast use. Some facilities use consumer CD players, but these often require special interfaces and cabling to be compatible with other audio equipment and are not really designed for day-in and day-out use.

One of the operational features of many broadcast-quality units is a **cue wheel** that allows the operator to "rock" or "scrub" the CD back and forth. This allows the user to go to the exact start point within a track, and the player can be paused at that spot, ready for play. Other normal CD controls include a play button to start the CD, several controls to select or program specific tracks on the CD, and an open/close control to load the CD into the machine. Regardless of the exact design, the CD ends up in a **tray** or **well**, where it spins so the laser can read it.

The internal structure of the CD player centers around three major components: a drive motor, a **laser** and lens system, and a tracking mechanism. The drive motor spins the disc at a variable rate between 200 and 500 revolutions per minute. To make sure the data on the disc is read by the laser at a constant rate, the drive motor speeds up as the laser moves toward the outside of the disc. In the laser and lens system, a **laser diode** generates a laser light beam, and a **prism system** directs it toward the disc surface. Different types of lenses focus the laser beam exactly on the pits of data encoded on the disc. Reflected light is

FIGURE 6.1 The professional-quality CD player is built for rack mounting, heavy-duty use, and may have features not available on consumer units. *(Image courtesy of Denon Professional.)*

FIGURE 6.2 A CD must be put into a plastic housing before it can be played in a cart-style CD player. *(Image courtesy of Denon Professional.)*

directed back through the lens and prism to another wedge lens and a **photodiode**, which creates the data signal that will be converted to an audio signal. The tracking mechanism moves the laser assembly so that it follows the spiral track of data on the disc.

One design approach to standard CD players that is still in use at many radio stations and audio production facilities has been to build them so that they only hold CDs that are placed into a plastic housing or cartridge before they can be played in the unit (see Figure 6.2). The plastic cartridge increases the cost of this system, but it also gives the discs extra protection from dirt and damage.

Regardless of the type of CD player utilized, some CDs seem to display an overly metallic, harsh "digital" sound. This has nothing to do with the CD player. It usually occurs with some CDs because the original recording was intended for analog vinyl or tape playback, and engineers often added equalization to compensate for sonic limitations of vinyl recording. Many early CDs were mastered

with this same equalization. Material that has been more recently recorded for CD use does not display this problem, nor does older material that has been prepared or remastered for transfer to CD.

6.3 CDS AND CARE OF CDS

The CD is a small plastic disc, 12 centimeters (about 4.7 inches) in diameter. On one side of the disc, music is stored as a single spiral of microscopic pits and lands (flat areas) that contain the encoded information about the sound. In addition to music, data encoded on the disc tells the CD player where each track begins and ends, how many tracks are on the disc, and other timing and indexing information.

A CD stores music as 16-bit digital words, starting from the center hole, going to the outside rim of the disc. With the standard digital sampling rate, 44,100 of these digital "words" are required for each channel every second. About 650 MB of data can be stored on a standard 74-minute CD. A thin aluminium reflective coating (some older CDs may use a gold-coated reflective surface) makes it possible for the laser light to read the encoded data, while a layer of clear polycarbonate plastic protects the encoded information. The reverse side of the CD is made up of the label and a protective lacquer coating (see Figure 6.3). CDs were originally designed to hold 60 minutes of recorded material; however, about 74 minutes of music has become the standard CD length, with 80-minute lengths sometimes being used.

The CD has shown that it is not as indestructible as it was originally promoted to be. Practical use has shown that CDs do require some care. When removing a CD from its **jewel box**, put your thumb on the center spindle of the box and your index or middle finger on the rim of the disc. Press down with your thumb and gently pull up on the disc until it is free of the case. When handling a disc, continue to hold only the outer edges or the edges of the hole, and avoid bending the CD in any manner. Never touch the surface of the disc.

Some CD players will not track properly with dust or fingerprints on the surface, and scratches can also render a CD unplayable. CDs can be cleaned by gently wiping the side without the label with a lint-free cotton cloth from the

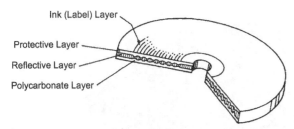

FIGURE 6.3 A CD is made up of several layers, as this cutaway view shows.

center hole of the CD directly toward the CD rim. Do *not* use a circular cleaning motion because this could cause a scratch that follows the spiral data on the disc, causing the disc to track improperly. Use a special CD-cleaning solution, isopropyl alcohol, or methanol to remove dirt or other material from the disc.

Bending a disc or touching it with oily fingers can damage the CD's protective coating and ultimately allow oxidation of the aluminum layer, which can ruin the CD. If your CD player has a tray that opens and closes, be sure to stop the CD *before* you eject the disc. Hitting the eject button while the CD is still playing can cause damage to the CD as it continues to rotate while the tray is opening.

If you must mark the disc, do not use pen, pencil, fine-tip markers, or adhesive labels; rather, use a non-solvent-based, felt-tip permanent marker on the label side of the disc. To get the maximum life span out of a CD, it should be stored in a plastic jewel box, or a binder made specifically to hold discs in an upright position, like a book.

6.4 THE CD RECORDER

CD recorders (see Figure 6.4) are found in most audio production studios. Although the first CD players were merely "turntables" for pre-recorded CDs, shortly after their development, manufacturers produced a recordable CD. Companies such as Tascam, Marantz, and Denon all offer CD recorders that work in the CD-R and/or CD-RW format.

A **CD-R (CD-recordable)** disc is based on what is known as the WORM design (write once, read many). The disc consists of a photosensitive dye layer and a reflective layer, encased in the normal protective clear polycarbonate. When heated by high-power laser pulses, the dye melts or "burns." This creates bumps and pits on the surface that have a different reflective nature, similar to those of a regular CD. The data is read off the disc by a lower powered laser. The recording laser follows a preformed spiral track so the data can be played back on any standard CD player. The recorder allows you to record several different sessions if you wish, or until the disc is full (usually 99 separate tracks, up to 80 minutes in total). Keep in mind, however, that you may *not* play the disc on a regular CD player until you *finalize* it and permanently write the table

of contents on it. A partially recorded CD-R *will* play on the CD-R deck at any time, but it will not play back on any other deck. You *must* finalize it first to do so.

The biggest drawback to the WORM design is the inability to record over old material. If you make a mistake in the recording process, you have ruined that disc or at least a section of it, and once you finalize the disc, you can't add any new material to it.

The **CD-RW (CD-rewritable)** disc is used in a similar fashion, however, the difference is the last recorded track can be erased before the disc has been finalized, and the space on the disc can be reused for recording other material. Some CD-RW discs that have been finalized can also be totally erased and rerecorded. The CD-RW format uses a phase-change technology, where the material on the recordable CD surface goes from crystalline to amorphous when heated during the recording process, which changes its reflective properties. During playback, the laser light is reflected differently from opposite-phase areas, allowing the player to distinguish binary zeros and ones and thus reproduce a digital signal. Be aware that an audio CD recorded in the CD-RW format may *not* play back correctly on every CD player. It will, however, play back on its own recorder, which will also play standard CDs.

6.5 DATA COMPRESSION

Data compression is possible with digital technology, and an absolute necessity in digital audio production, so it is important to understand its basics. Data compression is made possible by encoding only audible sounds. In other words, sound below the threshold of hearing or too soft to hear and sounds masked by louder sounds heard at the same time are excluded, according to a "perceptual coding" model. In addition, those sounds that are encoded use only the number of data bits necessary for near-CD-quality sound. This allows some digital recording technologies like the MiniDisc (see Appendix) and computerized portable recording formats such as MP3, WAV, and AIFF to cut down the amount of data that must be recorded by about one-fifth, with no apparent sound-quality change or loss.

There is no standard method of data compression, and different digital equipment will use various types of audio data reduction schemes. There has been some concern about the audible effects on sound that has been compressed more than once, for example, if an MP3 file is played through a signal processor that also uses data compression. Some tests have shown a degradation of the audio signal with some combinations of data compression; however, this doesn't appear to be a major problem with listeners, unless the degradation is extremely obvious.

Data compression can also be used with any digital media used in production. A CD-R or CD-RW disc, for instance, can hold a maximum of 80 minutes of uncompressed audio, but if those audio tracks were compressed

FIGURE 6.4   Most CD recorders found in the audio production studio can record in both the CD-R and CD-RW format. *(Image courtesy of Tascam®.)*

into MP3 format, 10 to 12 times that amount of audio could be stored on the same disc (approximately 13 hours). A standard DVD, with 4.6 gigabytes (GB) of space, could store over 66 hours of compressed, CD-quality audio. Storage space, then, is obviously not a problem. Ease of organization and utilization of files, however, does become an issue. While many compressed formats, such as MP3, can contain text and other data to help categorize and identify content, not every CD deck or computer can read and display that information. Additionally, some proprietary compression methods, such as Apple's.M4A format, require specific players or software for playback.

FIGURE 6.5   Many solid-state digital audio recorders fit easily into the palm of your hand and have become popular for news gathering and other field recording uses. *(Image courtesy of M-Audio.)*

6.6 COMPACTFLASH AND OTHER DIGITAL RECORDERS

Lately, the style of digital recorders that are enjoying the most prominent place in audio production work are **solid-state recorders** that record and store material on removable media, such as a **CompactFlash (CF)** or **Secure Digital (SD)** card. Portable, rack-mount, and mixer/recorder workstation units have all found their way into audio production work as well. The portable recorder shown in Figure 6.5 is one of many popular handheld recorders. The front panel of the unit shows an LCD screen and typical recorder controls. Most recorders allow both WAV and MP3 recording with built-in or external microphones. One-button recording is often featured, and additional controls are used to set various digital parameters, such as sample rate, data compression, or other effects.

Once recording is complete, basic non-destructive editing can be done on the deck itself by manipulating a waveform display on the LCD screen, or the audio files can be fed to a computer for editing or archiving. Perhaps the biggest plus for the portable digital recorder is that the recording system does not rely on moving parts and, once recorded, files can be directly transferred into an editing system via USB rather than having to rely on real-time transfers.

Other production recorders are designed to be rack mounted in the studio. In addition to providing multiformat

recording capability and normal transport controls, these recorders also feature an LCD screen for file manipulation or basic editing, and analog, digital, and USB ports. A 4-GB CF or SD card can hold over 4 hours of uncompressed stereo audio or up to 144 hours of MP3 mono audio.

Another type of portable recorder is HHB's FlashMic portable internal flash recording microphone. Shown in Figure 6.6, this digital recording microphone has a flash recorder with 1GB of recording memory built into the base of the microphone. A single button initiates recording with a high-quality Sennheiser omnidirectional or cardioid mic capsule or external line input. As the unit is entirely solid state, there is no internal noise that often plagues built-in microphones on portable units. The recorder also has a USB interface to export audio files for editing or adding effects to the recording. The advantage of this type of recorder is its efficient design as one single unit and its rugged design making it a durable, portable device for anyone recording in the field.

FIGURE 6.6 The HHB FlashMic is a professional digital recording microphone with a flash recorder built right into the base of the microphone. *(Image courtesy of HHB and Sennheiser Electronic Corporation.)*

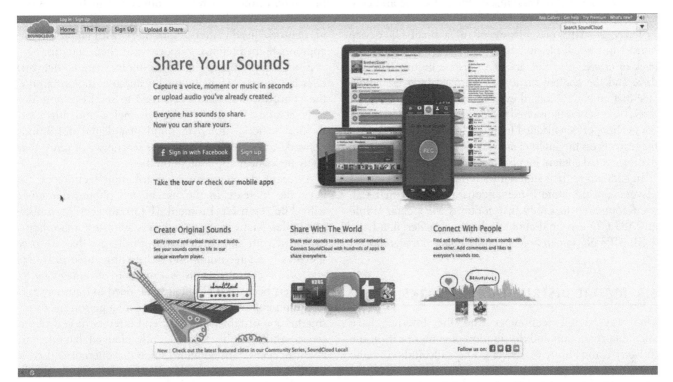

FIGURE 6.7 Recording and uploading audio to SoundCloud lets people easily store and share their material to blogs, sites, and social networks. SoundCloud can also be accessed anywhere using their official mobile and tablet apps.

6.7 STORAGE

A tremendous advantage of digital recording and processing is its ability for storage of content. It is not uncommon these days to find editing and playback computers with standard 500-GB or 1-terabyte (TB) hard drives. Additionally, external hard drives connected via USB can more than double a computer's storage space. USB flash drives that can hold up to 64 GB of data are also preferable storage and transfer options, especially compared to CDs or magnetic tape.

For production facilities with limited physical space, the option of online or cloud storage is becoming more popular. Online services such as SoundCloud.com (Figure 6.7) provide free or inexpensive storage solutions that can take the place of tape, discs, or hard drives. These storage solutions also offer flexibility for multiple producers by creating online interconnectivity and sharing between computers. An audio file produced in one facility can be uploaded to online storage and downloaded onto a computer, tablet, or mobile device practically anywhere in the world.

This option provides a low-cost alternative to sharing files among producers.

Perhaps the biggest disadvantages to online storage are file transfer rates, and the amount of storage space provided by an online host. While audio file sizes are much smaller compared to video, upload and download speeds can sometimes take more time than expected depending on the available bandwidth of a person's Internet connection. In regard to the amount of online storage space available, no free or inexpensive provider is going to offer 1TB of space. As of this writing, online storage services are good for quick transfer of smaller-size files among different computer devices. They are not the best option for long-term, large file storage.

6.8 MP3/PORTABLE AUDIO PLAYERS

One advantage of the CD (other than its digital capabilities) is its portability. At just under 5 inches in diameter, CDs store easily in cars and trucks, and were even played on portable devices in the 1990s. Although the portability of CDs may seem laughable to some these days, keep in mind they are still better as portable options compared to cassette tapes and vinyl records.

Portable MP3 players have been in existence since the late 1990s. It wasn't until 2001, however, that Apple's iPod made portable music listening easier and trendier. Continued developments in compression and playback capabilities have also made this method of music and content listening extremely popular.

Today, an MP3 player is essentially a small CF device housed in a small plastic box that can range in size from a pack of cigarettes to the size of a person's thumb. The CF drive and the content on it are controlled by a user interface that employs standard external buttons or even a touch screen. Content is transferred to a device via a USB computer connection, or downloaded from the Internet. Some devices that can access the Internet are also capable of streaming live or prerecorded content for playback.

In addition to their superior portability advantages, MP3 players can also store a tremendous amount of material, even compared to a CD that contains MP3 data. While an MP3 CD can hold about 66 hours of material, a basic 4-GB MP3 player can hold about 82 hours of material.

6.9 DIGITAL DISTRIBUTION NETWORKS

One way digital technology and the Internet have affected production and distribution of content *within* the industry is the ability to deliver CD-quality audio to radio stations using **digital distribution networks**. DG Systems

is one company that provides such a service (DG Fast Channel) by linking radio stations with ad agencies, production houses, and record companies through an online network.

In the past, radio spots were recorded onto audiotape masters, then dubbed to tape copies, and finally delivered by mail or courier to individual stations. A digital distribution network features PC-based servers in the ad agency, production house, or record company, which load, store, and share audio files. Audio is sent via satellite or Internet to network headquarters, and then, with approval and appropriate instructions, to receivers at client stations.

Since they employ digital technology, these networks offer fast delivery of digital-quality audio to stations around the country. They have been used to send out commercials, station IDs, newly released singles, and other production material. Keep in mind that digital distribution networks are "in-house," and that listeners will more than likely never hear any content on them.

A recent development in distribution technology has been the increase in the use of cell phones combined with **File Transfer Protocol (FTP)** services to transfer audio files. Many remote producers are using smartphones equipped with apps such as the iPush app, that are used specifically for recording and distributing audio to studios and on-air consoles. This is extremely advantageous for producers because it eliminates the need to transport a lot of equipment to a remote site, since all a person needs is a smartphone or tablet, the app, and internet or cell phone service (but just like always, fully charged batteries are also a necessity). In addition, since the internet is used as the distribution network, the files can be sent practically

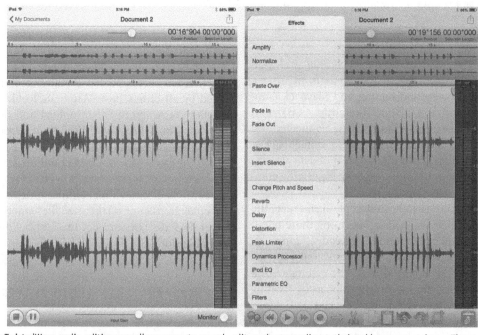

FIGURE 6.8 The TwistedWave audio editing app allows users to record, edit, and save audio on their tablet or smartphone. The app allows the user to record in stereo and includes a full suite of editing tools and features. *(Images courtesy of TwistedWave Software LTD, www.twistedwave.com; copyright 2015.)*

anywhere around the world. Finally, these easy-to-use tools allow greater flexibility in regard to scheduling and producing remote events.

PRODUCTION TIP 6A

Although editing is discussed in more detail in Chapter 3, this is a good point to mention the advent of high-quality audio editing apps for smartphones and tablets. Keep in mind that while successfully recording and distributing audio from remote settings is important, the ability to edit can also come in handy. Cutting dead air out of a recording, or using basic special effects can help make a remote production sound better. There are many audio editing apps available for free or for a very low cost, such as the TwistedWave app shown in Figure 6.8. This app can record in stereo up to a 96kHz sample rate, and includes a basic effects editor with reverb and echo effects and EQ controls. While there

is still no substitute for high-end studio editing, it is becoming easier for anyone to edit and produce high-quality audio from practically anywhere.

6.10 CONCLUSION

With the advent of digital audio equipment and the advantages in quality and convenience they offer, the use of older analog equipment continues to diminish and in many instances has been discontinued. Being comfortable working with various kinds of digital equipment, especially the recorders and players noted in this chapter, will be a necessary skill for most audio production people. It's also probable that new types of digital-based equipment, as well as new online storage options, will continue to be developed, and equipment not yet thought of will find a home in future audio production facilities.

Self-Study

1. What is one reason for using a professional-quality CD player rather than a consumer model in the audio production studio?

   a) Professional-quality players use a higher powered laser than a consumer model.
   b) Professional-quality players offer a better S/N ratio and less distortion than consumer models.
   c) Professional-quality players offer a greater frequency response and better dynamic range than consumer models.
   d) Professional-quality players are built for heavy-duty, continuous operation, and consumer models are not.

2. The lacquer coating on a CD makes it virtually indestructible in normal use.

   a) true
   b) false

3. Which format allows a CD recorder to record a blank CD only once?

   a) CD-RW
   b) FTP
   c) WORM
   d) DVD

4. CDs are best cleaned with a cloth wiped in a circular motion over the playing surface of the disc.

   a) true
   b) false

5. The standard for digital audio data compression is 192 kHz at 16 bits.

   a) true
   b) false

6. While many compressed data formats can contain text information to help make filing easier, not every CD deck or computer can read that information.

   a) true
   b) false

7. Once a CD-RW disc has been "finalized," it can't be erased and rerecorded.

   a) true
   b) false

8. Which of the following is *not* an advantage of the CD player?

   a) random access to material on the CD
   b) lack of physical contact between player and CD
   c) large signal-to-noise ratio
   d) ability to hold more information than a DVD

9. You know that a CD should be marked only on the label side, but which type of writing instrument should you use to mark a CD?

   a) ball-point pen
   b) No. 2 lead pencil

c) felt-tip permanent marker

d) none of the above; use an adhesive label

10. Which of the following is *not* one of the main internal components of the CD player?

a) drive motor

b) laser and lens system

c) CompactFlash card

d) tracking mechanism

11. Portable digital recorders that employ a CF or SD card have the ability to dub recordings using USB direct downloading rather than having to wait for real-time dubs.

a) true

b) false

12. Online cloud storage services allow users to store as much content as they want, for free.

a) true

b) false

13. Which of the following is *not* an effective consumer-based method of distributing or sharing digital audio files to more than one person online?

a) SoundCloud.com

b) Dropbox.com

c) email

d) digital distribution networks

14. When it comes to remote production, recording and distributing the audio files are all that matters.

a) true

b) false

15. Which of the following is not an advantage of using a smartphone or tablet for remote audio production?

a) content can be sent worldwide since the internet is the distribution network

b) producers do not have to pack and move a lot of equipment for a remote production

c) remotes are easier to schedule and produce

d) none of the above

ANSWERS

If you answered A to any of the questions:

1a. No. Both professional and consumer-model CD players use a similar laser system. (Reread 6.2.)

2a. No. Although the lacquer coating helps protect the CD, it is far from indestructible—fingerprints, dust, and scratches can damage CDs. (Reread 6.3.)

3a. No. This is a recordable/erasable format. (Reread 6.2.)

4a. Wrong. CDs should be wiped only in a straight line from the center hole toward the outer rim. (Reread 6.3.)

5a. No. There is no standard for data compression. (Reread 6.5)

6a. Correct. There are still some compatability issues in this regard, depending on the age of the digital file, operating system used, or software used.

7a. Wrong. CD-RW discs can be completely erased and rerecorded even after they have been finalized. (Reread 6.3.)

117

8a. No. Being able to instantly move from one track to another on the CD is a big advantage. (Reread 6.1.)

9a. No. You should not use a ball-point pen to label a CD. (Reread 6.3.)

10a. Wrong. The internal structure of a CD player centers on a drive motor. (Reread 6.2.)

11a. Correct. This feature is one of the main advantages of portable digital recorders.

12a. Wrong. You can store as much as you'd like, but not for free (Reread 6.7)

13a. Incorrect. SoundCloud.com allows consumers to upload audio files and share them online. (Reread 6.7)

14a. Incorrect. Editing content at a remote site can be just as important as recording and distributing (Reread 6.9)

15a. Wrong. This is an advantage of using smartphones or tablets for remote productions (Reread 6.9)

If you answered B to any of the questions:

1b. Wrong. All CD players have similar S/N and distortion characteristics. (Reread 6.2.)

2b. Yes. CDs require careful handling even though the lacquer coating helps prevent problems.

3b. No. FTP is a method used to distribute files online (Reread 6.9)

4b. Correct. Although a circular cleaning motion works for vinyl, it can damage a CD.

5b. Correct. There is no standard method of data compression.

6b. Wrong. Not all CD decks or computers can read text information on all digital files. (Reread 6.5.)

7b. Correct. CD-RW discs can be erased even after they have been finalized.

8b. No. Because a laser reads the CD, there is no wear on the CD and this is a big advantage. (Reread 6.2.)

9b. No. You should not use a pencil to label a CD. (Reread 6.3.)

10b. Wrong. The internal structure of a CD player centers on a laser and lens system. (Reread 6.2.)

11b. Wrong. This is a true statement and this feature is one of the main advantages of portable digital recorders. (Reread 6.6.)

12b. Correct. While you may be able to store up to a terabyte of content, it won't be for free.

13b. Wrong. Dropbox.com is free for consumers and has adequate storage for audio files. (Reread 6.7)

14b. Correct. Don't forget about editing!

15b. Wrong. This is an advantage of using smartphones or tablets for remote productions (Reread 6.9)

If you answered C to any of the questions:

1c. No. All CD players have similar frequency response and dynamic range characteristics. (Reread 6.2.)

3c. Yes. This is the "write-once, read-many" format used by CD-R machines.

8c. No. The larger the S/N ratio the better, so this is an advantage. (Reread 6.2.)

9c. Yes. If you must label a CD, use a soft, felt-tip permanent marker.

10c. Correct. The CompactFlash card is a removable recording medium for other digital recorders.

13c. Incorrect. Even though you can send an email to multiple parties, the attached file size can make it problematic to use. (Read 6.7)

15c. Wrong. This is an advantage of using smartphones or tablets for remote productions (Reread 6.9)

If you answered D to any of the questions:

1d. Yes. Most consumer-model CD players can't stand up long under the constant use of a broadcast facility.

3d. No. DVD is a different type of format. (Reread 6.4 and 6.6.)

8d. Correct. A DVD can hold more information than a CD.

9d. No. You should not use adhesive labels to mark a CD. (Reread 6.3.)

10d. Wrong. The internal structure of a CD player centers around a tracking mechanism. (Reread 6.2.)

13d. Correct. This is the best answer.

15d. Correct. These are all advantages of using smartphones or tablets for remote productions.

Projects

PROJECT 1

Prepare a report on a digital player/recorder or portable audio production software that is not discussed in this chapter.

Purpose

To keep you up to date on what is happening in this field.

Notes

1. Technological developments in this field are changing rapidly. Undoubtedly there will be new products available by the time you read this text that were not available when it was written. It behooves anyone in the audio production field to keep up to date. However, audio equipment also comes and goes, so don't be surprised if the player/recorder or software/app that you write about today is no longer available a year from now.
2. Try to select equipment that does more than play and/or record. For example, some current recorders and apps also allow for editing or mixing.
3. The Internet is an excellent place to find this information. The major manufacturers place a great deal of material about their products on their websites. Companies or software/app that are particularly active in this field include, Sony, Marantz, Roland, and Tascam. Be sure to look online for software and apps as well.

How to Do the Project

1. Find a piece of equipment or software that qualifies as a digital player/recorder or editor. As suggested, you can do this on the Internet or you can write to companies for their catalogs.
2. Research the characteristics of the equipment or software. If you know of a facility or station that uses this equipment or software, talk with the engineer or an operator who uses it. You could also call a distributor who sells the equipment for additional information.
3. Organize your report, taking into consideration some of the following points:
   a. What is the primary purpose of the equipment or software?
   b. How does it operate—with regular transport controls (play, record, stop, and so on)? Attached to a computer? On a tablet or smartphone?
   c. Is it designed to replace an older piece of equipment, or is it designed for a new application?
   d. What compression format does it use, if any?
   e. What type of recording medium does it use—tape, computer disk, flash card, flash drive?
   f. When was it introduced to the market?
   g. What does it cost?
   h. Are there optional features of this equipment?
   i. What is your assessment of whether this will be a successful product?
4. Write your report.
5. Turn in the report to your instructor to receive credit for this project.

PROJECT 2

Play and record several CD selections.

Purpose

To familiarize yourself with the operation of CD players.

Notes

1. For this exercise you will need to work with a microphone, audio board, and audio recorder, as well as a CD player.
2. Different brands of CD players have slightly different features, so you will need to learn the particular characteristics of your player.
3. You can play several selections from one CD or use several CDs. If you have only one CD player, you would be advised to use three cuts from one CD, as you will probably not have time to change discs. If you have two CD players, you would be better off playing selections from at least two CDs.

How to Do the Project

1. Make sure that your CD player is connected so that it can be faded out and will record onto an audio recorder.
2. Make sure that a microphone is available so that you can announce the title of the selections you choose.
3. Examine the CD player and practice with it so that you can cue, play, and pause it with ease.
4. When you feel familiar with the player and have decided on three selections to play, start the recorder.
5. Complete the project by doing the following:
   a. Announce the name of the first musical selection, bring it in, and then fade it out after about 30 seconds.
   b. Announce the name of the second selection, bring it in, and fade it out after about 30 seconds. (If you have two CD players, you can cue the second one while the first is playing. If not, you'll need to cue a second selection while you are introducing it. This is not particularly difficult, because most CD players enable you to cue easily.)
   c. In the same manner, announce the name of the third selection, bring it in, and then fade it out.
6. Label the assignment with your name and "CD Recording."
7. Turn in the assignment to your instructor to receive credit for this project.

PROJECT 3

Record a 5-minute interview with a classmate and edit the interview to 3 minutes, using a portable digital recording device

Purpose

To familiarize yourself with portable digital recording and editing.

How to Do the Project

1. Team up with a classmate and determine a time and place to conduct the interview (if your instructor requires you to complete the assignment out of class).
2. Before the interview, write a brief introduction (approximately 30 seconds) that lets the listener know who you are, who you are interviewing, and why you are interviewing that person (don't say its because you need to complete an assignment).
3. After your introduction is written, create a list of at least 10 questions to ask your guest. Be sure to write them down in the order you want to ask them. Do *not* let your guest know your questions beforehand.
4. Once you have your list of questions, write a brief outro/conclusion that thanks your guest for the interview, thanks your listener(s), and reminds everyone of who you are.
5. Check out the appropriate equipment from your department. Preferably you and your partner will use a portable recorder similar to those mentioned in this chapter. If one is not available,

it is fine to use another portable digital device with an internal microphone such as a tablet computer or MP3 device. Do *not* use a laptop or desktop computer for this assignment. Regardless of what device you use to record your interview, let your instructor know what you are using.

6. Conduct your interview in a non-studio setting but be aware of background noise! It might be a good idea to use headphones when you conduct your interview.

7. Once you have recorded 5 minutes worth of interview material (including your open and close), edit your interview down to 3 minutes, preferably using the editing options available in your recorder.

8. Once your interview (including your open and close) is edited to 3 minutes, submit it to your instructor as per his or her instructions to receive credit for this project.

7

MONITOR SPEAKERS AND STUDIO ACCESSORIES

7.1 INTRODUCTION

This chapter looks at some studio items that are often taken for granted, but are actually quite significant. **Monitor speakers** (see Figure 7.1) are used to listen to sound in the studio. They convert the audio signal, stored on recording media like computers, digital recorders, CDs, or tape, back into sound that can be heard. The sound that comes from them is the final product, so they are very important in determining a production's overall quality. What you hear on the monitor speakers should be the most accurate gauge of what you recorded and what the listener will hear.

Speakers are one of the various pieces of equipment used in the audio production studio. This chapter looks at various types of speakers and their **cables** and **connectors**.

7.2 TYPES OF SPEAKERS

You may want to review the sections on sound in Chapter 2, "The Studio Environment," as they will help

FIGURE 7.1 Studio monitor speakers help you judge the sound quality of your audio production. *(Image courtesy of Behringer North America, Inc.)*

you understand how speakers work. **Speakers** are transducers, but they work in a manner opposite to that of microphones. Instead of converting sound waves into electrical energy, speakers produce sound from an electrical signal by converting it into mechanical energy that causes vibrations to produce sound waves or audible sound.

The most common type of monitor speaker is a **dynamic speaker**. Also known as a **moving coil** or **electromagnetic speaker**, its transducing element (called a **driver**), produces sound by moving a flexible cone or diaphragm in and out very rapidly. The diaphragm, which may be made of paper, plastic, or metal, is suspended in a metal frame. Attached at the narrow end of the cone is a **voice coil** (a cylinder wound with a coil of wire), which is located between powerful circular magnets (see Figure 7.2). When an electrical current is generated in the voice coil, it creates another magnetic force that moves the coil and the cone back and forth, according to the electrical signal that is entering the coil. The cone vibration causes the surrounding air to move in a similar manner, which our ears pick up as sound.

Instead of a cone, some speakers use a dome, which is just another type of diaphragm that bulges out rather than tapers in. Other types of loudspeakers include **electrostatic loudspeakers** and **planar-magnetic loudspeakers**. However, these are considered to be fairly exotic and you aren't likely to run across them in a typical production studio.

As the modern audio production facility is typically an all-digital environment, several manufacturers offer digital monitor speakers (see Figure 7.3). The speakers have either digital inputs or digital and analog inputs. Most digital inputs can identify up to a 24-bit digital signal with a sampling rate as high as 192 kHz. When a digital signal is detected in a speaker with both analog and digital inputs, the speaker's internal digital-to-analog (D/A) converter becomes activated. There really is no monumental sound difference between an analog and a digital input; however,

123

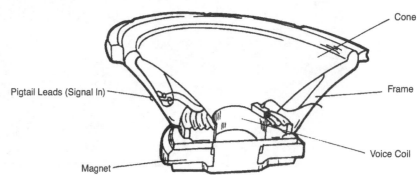

FIGURE 7.2 Cross-section diagram of a dynamic cone driver.

FIGURE 7.3 Digital monitor speakers feature an input that allows for direct digital hookup from an audio console or workstation. *(Images courtesy of Genelec, Inc., copyright 2012.)*

the convenience of attaching directly to the digital output of an audio console, digital audio workstation, or editing computer is a plus for the digital speaker. In addition, digital monitor speakers sometimes help eliminate line loss and possible hum that is associated with analog speakers.

7.3 BASIC SPEAKER SYSTEM COMPONENTS

The basic components of the typical monitor speaker are the **woofer**, **tweeter**, **crossover**, and **speaker enclosure**. "Woofer" and "tweeter" are names given to drivers or individual speakers used within a speaker system. Because no single speaker design can reproduce the entire frequency range of sound adequately, different speakers have been developed to handle different portions of it. A woofer is designed to move the large volume of air necessary to reproduce lower frequencies. The cone is usually larger in size (anywhere from 8 to 12 inches) and is able to make large movements. However, this bulk prevents the speaker from adequately reproducing the higher frequencies that

require rapid cone movement. The tweeter uses a lighter and smaller design; often a convex dome (1 inch or less in diameter) replaces the cone. There are also midrange speakers with cones from 3 to 5 inches in diameter that are designed to reproduce midrange, higher-bass, and lower-treble frequencies.

An individual monitor speaker is really a speaker system, in that many use at least a woofer and a tweeter driver. The crossover is used to divide the incoming electrical signal and send the proper frequencies to their respective driver. A crossover is a network of filters (mainly capacitors and inductors) between the input to the speaker and the individual speaker drivers. In a two-way system for example, an inductor would pass all audio *below* a certain frequency to the woofer. A capacitor in the same speaker would pass all audio *above* a certain frequency to the tweeter. Although there is no universal design for the crossover, most dividing points between the bass and treble frequencies are between 500 and 1,500 Hz. A speaker that has just a woofer, a tweeter, and a crossover is a **two-way speaker system**, like the one shown in Figure 7.4. A speaker that employs another driver (such as a midrange) is a **three-way speaker system**.

7.4 SPEAKER SYSTEM ENCLOSURE DESIGNS

Speaker drivers and crossover(s) are encased in a box (often called a **speaker enclosure**) that plays a role in how the speaker sounds. Every speaker produces sound both behind and in front of it: the back sound wave is exactly opposite of the one that goes into the forward listening space. If the two sound waves are allowed to combine naturally, they would be acoustically out of phase and cancel each other out. This would produce no sound at all or greatly diminished sound, especially in the bass part of the range.

The two most widely used speaker enclosure designs are the acoustic suspension and bass reflex systems shown in Figure 7.5. The **acoustic suspension** (also known as the **sealed-box**) design puts the speaker drivers and crossover(s) in a tightly sealed enclosure that produces an accurate, natural sound with a strong, tight bass. By containing and

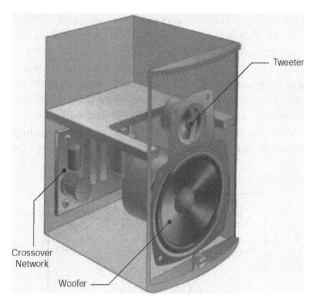

FIGURE 7.4 The basic two-way speaker system consists of a tweeter, a woofer, and a crossover housed within an enclosure.

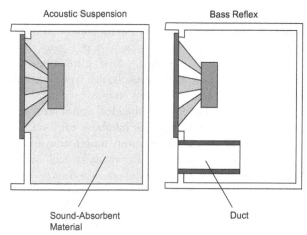

FIGURE 7.5 The most widely used speaker enclosure designs are the acoustic suspension and bass reflex systems.

absorbing the back wave in the enclosure, the acoustic suspension design prevents the rear sound from radiating and disrupting the main sound coming from the front of the speaker. Because half the sound energy is trapped and absorbed in the box, acoustic suspension speakers require a more powerful amplifier to drive them. The acoustic suspension design also demands a rather large, sturdily built physical enclosure to ensure accurate reproduction of the lowest bass notes. These are the main reasons the majority of monitors used in studio production are acoustic suspension.

On the other hand, the **bass reflex** (also known as the **vented-box**) design is quite efficient and produces a strong bass sound with less power required. The bass reflex speaker enclosure is designed with an opening (a vent, duct, or port) that is tuned to allow some of the rear sound (mainly the lower frequencies) to combine in phase and reinforce the main sound coming from the front of the speaker. A common place to see bass reflex speakers is on portable and tabletop stereo systems.

Some bass reflex design speakers have been criticized for not having tonal accuracy that is quite as good as the acoustic suspension design, and for adding a "boomy" quality to the sound. These problems, however, are often the fault of a particular speaker's tuning and construction and not of the bass reflex design itself. There are many different bass reflex designs, and most produce a clean, wide-ranging bass.

7.5 SPEAKER SOUND QUALITIES

There are a wide variety of monitor speaker systems to choose from, and as with some other audio equipment, the differences between various models may be minimal. One of the

important qualities that a good monitor speaker must have is excellent **frequency response**. Remember, humans are able to hear frequencies in the range of 20 Hz to 20 kHz, although most of us don't hear sounds quite that low or high. Top-line broadcast monitors often provide a frequency response range from 35 to 45 Hz at the low end, to 18 to 20 kHz at the high end. Increased use of digital equipment however, has made it necessary to have speakers that produce as much of this range as possible.

Another important quality for the broadcast monitor speaker is its ability to produce a flat frequency response, which is the speaker's ability to reproduce low, midrange, and high frequencies equally well while producing a natural sound. The speaker itself should not add *anything* to the audio signal that is heard. What is most important is merely how a speaker sounds. Among the combinations of driver types, speaker enclosure designs, and crossover frequencies, there is no one speaker configuration that produces the "best sound." Keep in mind, however, that a good speaker sound does not depend only on the speaker itself. How well a speaker sounds is also dependent on the content being played through it, the dimensions and acoustic properties of the room in which the speakers are located, the location of the speakers in relation to the listener, and the listener.

7.6 SPEAKER PLACEMENT

In the audio studio, there may not be many options when it comes to placing monitor speakers. Usually they are positioned on each side of the audio console, but exactly how they are positioned is open to various schools of thought. At the very least, there should be an acoustically symmetrical layout. For example, don't put one speaker in the corner next to a glass window and the other in the middle of a wall that's covered with acoustic tiles. One thought is to mount the speakers *in* the wall. As long as the speakers are isolated from the wall structure—using rubber shock mounts, for example—this flush mount creates an infinite baffle that prevents rear wave problems. Large speakers can be built into the wall, providing a loud, clear sound with plenty of "heavy" bass. Unfortunately, this type of installation is often not practical for many studios.

Some production people feel the ideal sound is obtained when speakers are even with or a bit above ear level. This is called **near-field** or **close-proximity monitoring** and can be accomplished by putting smaller speakers, such as those shown in Figure 7.3, on the audio console "bridge" or on short stands to the left and right of the console, about 3 feet apart. Keeping the speakers a couple of feet from the back wall will prevent any excessive bass boost from the wall. Since the speakers are so close to the listener, mostly direct sound is heard and there is no need to worry much about other effects. "Toeing in" the speakers (angling them toward the listener) or pointing them straight away from the wall will help control the treble, since speakers pointing toward the listener provide the greatest high-frequency response. Very clear detail and excellent stereo imaging are characteristics usually associated with near-field monitoring setups.

A greater concern with monitor speaker placement, especially in studios that employ near-field monitoring, has to do with the increased use of computer equipment in the studio. Unless the monitor speakers are magnetically shielded, they must be positioned far enough away from computer screens so that they don't distort their pictures.

Perhaps the most practical monitor installation for many production studios is having the speakers hung from the ceiling or attached to the wall behind the audio console, as shown in Figure 7.7. For the best sound dispersion, the speakers should be hung from the upper corners of the production room. While putting speakers

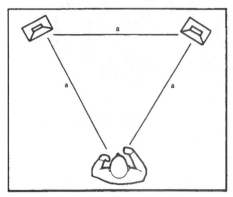

FIGURE 7.6 In this bird's-eye view, the most basic rule of speaker placement is displayed, showing that the distance between the speakers equals the distance from speaker to listener.

FIGURE 7.7 In this broadcast studio, note how the monitor speakers are located behind the console, and in the upper corners of the room. This setup gives the board operator an excellent environment to evaluate the quality of the sound as it is being produced. *(Image courtesy of Michael Parks, iHeartMedia-Harrisburg, PA.)*

totally into the upper corners may result in too much bass boost, this is not an uncommon location. Usually there is enough space between the back of the speaker and the wall for heavy bass frequencies to be dispersed without distorting the sound. To maximize the amount of space between the back of a speaker and the wall, speakers mounted in upper corners should be toed-in toward the console or main listening area. Speakers that are hung from walls also keep counter space available for other production equipment, which is an important consideration in many settings.

Regardless of where the speakers are placed, the location of the operator in relation to the speakers also plays a role in how well they sound, especially with stereo programming. Ideally, the operator is located directly between the two speakers and far enough back from them so that an equilateral triangle is formed if a line were drawn from speaker to speaker and from operator to speaker (see Figure 7.6 and Figure 7.7). If the layout of the production room positions the operator closer to one speaker than the other, the source of all the sound appears to shift toward that one speaker. As a production person, you may not have any control over speaker placement, so it's important to realize the effects of speaker placement on the sound you hear.

7.7 PHASE AND CHANNEL ORIENTATION

The concept of **phase** was previously mentioned in Chapter 4, "Microphones." Wiring monitor speakers incorrectly can cause phase problems. The sound signal is fed to each speaker from the audio console monitor amplifier (or sometimes an external amp) by a positive and negative wire. If the wires are reversed on one of the speakers (i.e., the positive wire is connected to the negative terminal), the two speakers will be out of phase. As the driver moves the cone of one speaker in and out, the driver on the other speaker is moving out and in, so that the two speaker sounds tend to cancel each other out and diminish the overall sound quality. This would be especially noticeable if the speakers were reproducing a monaural signal.

Another concern about speaker wiring is channel orientation. Most studio monitors are wired so that moving a balance control from left to right will shift the sound image from left to right *as you face the speakers*. In other words, while looking at the speakers, the left speaker is in line with your left hand. This setup is important for true stereo sound reproduction. For example, if you were listening to a classical music piece and the channel orientation was reversed, the violins would sound as if they were to the right of the sound stage, which would be contrary to normal symphony orchestra arrangement. However, since most speakers are wired by an engineer, phase and channel orientation should not be a problem in most audio studios.

7.8 MONITOR AMPLIFIERS

Recall that most audio consoles have an internal monitor amplifier that provides the signal to drive the monitor speakers. Although this is adequate for many studio applications, some production rooms and control rooms are set up with external monitor amplifiers. These are merely more powerful **amplifiers** that provide higher volume levels and clearer reproduction of the sound signal. Remember, the volume of the monitor speakers is for the operator's use only and has no relationship to the volume of the signal being broadcast or recorded.

7.9 SPEAKER SENSITIVITY

A speaker's **sensitivity** is the amount of sound (or output level) that can be produced from a given input level, much like the sensitivity of a microphone studied in Chapter 4. It is measured in decibels of sound-pressure level (db-SPL), and a good-quality broadcast monitor will usually have a sensitivity of more than 90 db-SPL measured at 1 meter with 1 watt of input level. There are speakers that range from low sensitivity to high sensitivity, but in reality, this characteristic of a speaker has little bearing on its quality. What it does mean is that a low-sensitivity speaker will need more power to drive it to any given volume level than a high-sensitivity speaker.

A speaker's size, sensitivity, and bass response are all related. To increase a speaker's low-frequency response, its enclosure can be made larger, or its sensitivity can be decreased. For many audio studio situations, a smaller speaker size offers a lot more flexibility in placement of the speaker and thus is preferred, even if it means using a low-sensitivity speaker that might require a more powerful monitor amplifier.

7.10 HEADPHONES

Headphones are tiny loudspeakers encased in a headset, which qualifies them as another type of monitor. Headphones are necessary in broadcast situations where studio monitor speakers are muted when a microphone is turned on, and the operator must be able to hear audio sources. For example, if a radio announcer is talking over the introduction of a song or a voice-over announcer is reading a commercial over the background of a music bed, headphones would allow him or her to hear both the other sound and the microphone sound so that he or she could balance the two or hit appropriate cues. Headphones are also portable, so sounds can be monitored when a standard monitor speaker might not be available.

Like regular monitor speakers, headphones come in a variety of designs and styles. Two of the most common types of headphones found in the production studio are closed-cushion headphones and open-air or hear-through cushion headphones. **Closed-cushion headphones**, also known as **circumaural headphones**, have a ring-shaped muff that rests on the head around the ear and not actually on the ear. The enclosure around the muff is solid or closed, as shown in Figure 7.8. These headphones are probably the most common, as they usually provide a full bass sound and minimize outside noise better than other styles. Closed-cushion headphones are also less likely to leak sound into the studio, however they're often heavier and more cumbersome than other styles.

Hear-through cushion headphones, also known as **supra-aural** or **open-air**, have a porous muff, instead of an ear cushion, that rests directly on the ear (see Figure 7.9). The enclosure around the muff typically has holes or other types of openings to give it the open-air design. Often made of very lightweight material, this design can be very comfortable for the wearer and provide superior sound reproduction. However, they are also more susceptible to feedback because the audio signal can leak out if driven at high volume levels.

Other headphone types include the tiny **earbud**, which is designed to fit in the ear; **electrostatic headphones**, which are extremely high-quality and require external amplification and special couplers to hook up; and **wireless headphones**, which operate similarly to wireless microphones by transmitting a radio frequency (RF) or infrared (IR) audio signal from the source to the headphones.

Unlike consumer headphones, many professional-quality headphones are purchased "barefoot," meaning they have no end connector on them. Although most audio equipment requires a standard $\frac{1}{4}$-inch phone connector for headphones, there are situations where barefoot headphones allow an engineer to wire them specifically if necessary.

The most appropriate headphone style is often determined by the personal taste of the user. Professional-quality headphones should always feature large drivers and full (but comfortable) ear cushions and a headband. One note of caution for all headphone users: *Listening at extremely high volume, especially for extended periods of time, can damage your hearing permanently.* Be sure to practice "safe sound" and listen only at a moderate volume for short periods of time.

7.11 HARDWIRING AND PATCHING

Audio equipment in the production studio is connected together by two methods: **hardwiring** and **patching**. Hardwired connections are usually permanent (such as a

FIGURE 7.8 Closed-cushion or circumaural headphones have a ring-shaped muff that rests on the head around the ear. *(Image courtesy of Audio-Technica U.S., Inc.)*

FIGURE 7.9 Hear-through or supra-aural headphones have a porous muff that rests directly on the ear. *(Image courtesy of AKG Acoustics.)*

FIGURE 7.10 A patch panel simplifies rewiring individual audio components into an audio console or studio. *(Image courtesy of Gentner Broadcast Systems.)*

CD player connected directly to the audio console) and can be soldered or wired by the engineer. Equipment that may be moved from one production area to another (such as a portable recorder) is often connected through male and female connectors known as plugs and jacks. More will be said about the typical audio connectors in the next few sections of this chapter.

Many pieces of audio equipment, and even separate production studios, can be connected together through the use of a **patch panel** or **patch bay**. Most patch panels are configured as two rows of 24 phone jacks in a one- or two-unit rack space as shown in Figure 7.10. The patch panel is located near the equipment and the audio console so that the input and output of each piece of equipment can be quickly and easily connected.

There are several modes that patch bays can be set to; however, in a typical setup, the *top row* of sockets on the panel is where the audio signal is *coming from*, and the *bottom row* of sockets is where the signal is *going to*. Putting a patch cord into the correct jacks in the panel allows an operator to interconnect and reconfigure various pieces of equipment or even separate studios. Patch cords are available as either a single-plug cord or a double-plug cord. A double-plug cord works well for stereo patch bays since it has two plugs at each end. One side of the plug casing will be marked (usually with a colored or ribbed edge), so that you can keep left and right channels correctly aligned on both the input and output sockets of the patch panel. Of course, single cords work fine, too—just remember to always plug in one cord from the top-row right channel of the patch panel to the bottom-row right channel before plugging in the left channel, so you won't cross channels. Crossing channels is easy to do if you plugged both left and right channels of the top row before plugging them in the bottom row.

Figure 7.11A shows a portion of a patch bay in which two CD players are linked to two channels on an audio console. With no patch cord put into the panel, CD 1 is linked to Channel 3 and CD 2 is linked to Channel 4. In patch panels, the top row is internally wired to the bottom row, and the audio signal flows in that manner. When it is unpatched like this, it's known as a "normalled" condition. Figure 7.11B shows another portion of a patch panel that links the output of the audio console to audio recorders. When normalled, the PGM (or Program) output goes to CD-R 1, and the AUD (or Audition) output goes to the Computer. The PGM signal could be sent to the Computer (as shown) by putting a patch cord from the PGM position on the top row to the Computer position on the bottom row. Now the signal flow has been changed, and that's the purpose of a patch panel—to allow flexibility in configuring the audio studio.

An alternative to the patch panel is the **audio routing switcher**. The router (see Figure 7.12) operates just like a patch panel by allowing several input sources to be switched to a single output or sometimes multiple outputs. The switching is done electronically by selecting the appropriate switches or buttons, rather than by using patch cords. Most audio routers are centered around a matrix of "x" number of inputs times "x" number of outputs. Any one input can be sent to all the outputs, or all the outputs can be sent to one single input, or any combination can be configured.

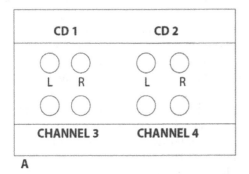

A

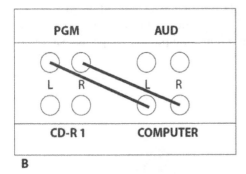

B

FIGURE 7.11 A portion of a patch panel that is "normalled" (A) and a portion of a patch panel that has been "patched" (B).

FIGURE 7.12 An audio routing switcher operates in a manner similar to a patch panel. *(Image courtesy of Lectrosonics.)*

7.12 COMMON AUDIO CONNECTORS

There are primarily four types of audio connectors: RCA, XLR, phone, and miniphone. With each type, the female, or receiving, connectors are called **jacks** and the male connectors are called **plugs**, but often the terms "plugs," "jacks," and "connectors" are used interchangeably.

The **RCA connector** is sometimes referred to by its old time name, **phono connector**, and "pin connector." Notice that it is "phono," not "phone." Most home stereo equipment and many professional-quality CD players use this type of connector. This connector is *always* a mono connector, so two of them are necessary for a stereo signal. Because of this, RCA connectors are often bundled in pairs and color-coded—red for right channel, and black or white for left channel. The male plug consists of a thin outer sleeve and a short center shaft that plugs into the female jack (see Figure 7.13). The female end is most often enclosed

in a piece of equipment so that the male end will just plug directly into the equipment. RCA connectors are used for unbalanced connections and are prone to picking up extraneous electrical noises, such as switch noises or humming.

The **XLR connector** is also known as the **Cannon connector** or **three-pin connector**. It's the most common microphone connector in audio production use and is often used as the input–output connection on audio recorders and loudspeakers. By convention, male XLR connectors are outputs and female XLR connectors are inputs. The three pins of the male plug fit into the three conductor inputs of the female jack. The guide pin on the female end fits into the slot for the guide pin on the male end so that the connector can't be put together improperly (see Figure 7.14). Like the RCA connector, the XLR connector is mono, so a stereo connection requires one XLR connector for the right channel and one for the left channel. The balanced three-conductor wiring of the XLR connector makes this a high-quality connection that is less likely to pick up noise through the cable.

The female XLR connection has a small spring lock on the outer casing that locks into the collar surrounding the male XLR connector. When connected properly, the XLR connection locks together and makes a snapping or clicking noise, letting you know your connection is strong. The connectors can only be unlocked by pressing the latch lock release.

The **phone connector** is also known as the **¼-inch phone**. Notice that it is called "phone," and not "phono." Most pro-quality headphones use a phone plug to connect to the audio console, and most patch bays have female phone jacks which rely on phone plugs. The **miniphone connector** (also called a **mini**) is most often used to connect portable audio recorders to other pieces of production equipment. The output on many portable recorders is a female minijack. The miniphone is a smaller version of the phone connector, and although there are various sizes of miniphone connectors, the most common is the ⅛-inch. Phone and miniphone plugs each have a tip and

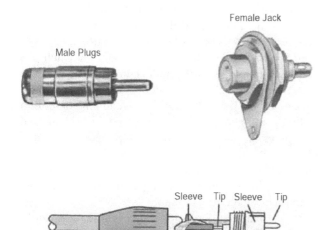

FIGURE 7.13 The RCA or phono plug is one of the common audio connectors. *(Images courtesy of Switchcraft, Inc., a Raytheon Company; and Mackie Designs, Inc.)*

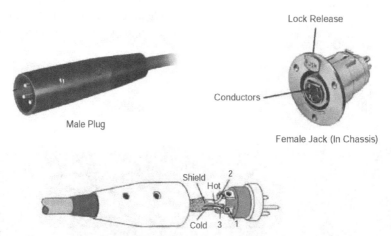

FIGURE 7.14 XLR or Cannon connectors feature a locking latch that prevents them from being accidentally disconnected. *(Images courtesy of Switchcraft, Inc., a Raytheon Company; and Mackie Designs, Inc.)*

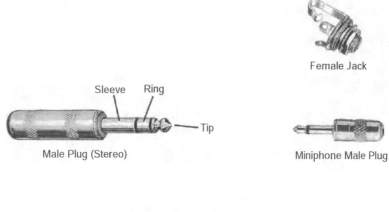

Female Jack

Sleeve   Ring

Tip

Male Plug (Stereo)

Miniphone Male Plug

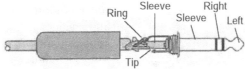

Ring   Sleeve   Right
Sleeve   Left

Tip

FIGURE 7.15 Phone and miniphone connectors can be either stereo or mono. *(Images courtesy of Switchcraft, Inc., a Raytheon Company; and Mackie Designs, Inc.)*

a sleeve, which go into a female jack. Male mono plugs have one insulating ring that separates the tip from the sleeve, and stereo plugs have two insulating rings, which actually define the ring portion of the connector (see Figure 7.15). If the signal is stereo, both the female and male connectors should be stereo. The female end is often enclosed in a piece of equipment, but inline jacks are available if needed.

7.13 OTHER CONNECTORS AND CONNECTOR ADAPTERS

Most of the connectors mentioned so far are intended for analog inputs or outputs; however. more equipment in the audio studio today has digital connections. Some digital connections utilize standard broadcast connectors, such as the XLR and RCA plugs and jacks, and other digital inputs and outputs use different types of connectors. When XLR connectors are used for digital audio, you may see an input or output labeled **AES/EBU**. This means the equipment meets the digital standards set by the Audio Engineering Society and the European Broadcasting Union. Digital RCA connectors follow the **S/PDIF** (Sony-Philips Digital Interface Format) standards. If these audio connectors are used for a digital audio connection, the connecting cable should also be designed for digital audio.

Two other connectors that are still being used for digital hook-ups in some studios include the **BNC** and **Toslink** connectors. The Toslink connector was developed by Toshiba and is the most common optical connector for digital audio. The connectors are usually molded plastic with a fiber-optic glass cable. BNC connectors (see Figure 7.16)

feature a locking design along the lines of the XLR and are designed for use with 75-ohm cable. Used extensively for years in video applications, BNC connectors have also been used specifically for digital audio equipment.

Other connectors commonly found in audio studios are based on the increased use of computer equipment in audio production. Many production studios are designed around **Ethernet** and **FireWire** cables and connectors (see Figure 7.17). The Ethernet or RJ-45 connector looks and operates like a larger version of the standard modular telephone jack and is commonly wired with CAT-5 cable. Many audio console control surfaces are now connected to their input/output frames with this connector, greatly decreasing the amount of wiring necessary for a typical audio studio. Apple gave the IEEE 1394b digital interface the name "FireWire," which is a tiny four- or six-pin jack or plug capable of very high speed digital data transfer. Although Apple has discontinued providing FireWire connections on its computers, some older editing machines using the technology are still in use.

A more common connector is the **USB** (**Universal Serial Bus**) which can link audio input/output equipment to computer recording and editing systems and has been used to directly connect microphones and loudspeakers to a computer. The increased practice of using USB flash drives for storage and playback of audio content has also helped solidify its use in the modern studio environment.

Something else that is very handy to have in any audio production studio is a supply of **connector adapters**, which enable you to change a connection hookup from one form to another. Let's say, for example, that you need

FIGURE 7.16 Some digital connections require the use of connectors not normally associated with the wiring of audio equipment. *(Image courtesy of Neutrik USA.)*

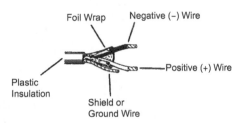

FIGURE 7.18 Typical audio cable is designed for a balanced wiring scheme. *(Image courtesy of Cooper Industries, Belden Division.)*

FIGURE 7.17 Connectors common to the computer industry are being used more and more in audio facilities and equipment: (A) Ethernet, (B) FireWire, (C) USB.

to connect an RCA output to a phone connector input, but the only cable you can find has an RCA connector at both ends. You can convert one of the RCA connectors to a phone connector with an adapter. This is a single piece of metal, which in this case houses a female RCA input at one end and a male phone output at the other. When the male RCA connector is inserted into the female end of the adapter, the signal is transferred from the RCA connector to the phone connector, and from there it can go to the phone input. Adapters usually come in handy in emergency situations when some connecting cable fails; therefore having a variety of them available is good production practice.

7.14 BALANCED AND UNBALANCED LINES

The standard monitor cable most often used in analog audio production consists of two stranded-wire conductors that are encased in plastic insulation along with a third

uninsulated shield wire, all encased in a foil wrapping and another plastic sheathing (Figure 7.18). For most wiring practices, the inner wires are designated 1 (red) and 2 (black), and the uninsulated wire is the shield or ground wire. The audio signal is carried on the positive and negative conductors. This type of cable is referred to as three-wire, or **balanced cable**, and often requires the XLR connector, since that is the one best designed to connect three wires in addition to being the one most utilized by many types of audio equipment.

Another type of cable is two-wire, or **unbalanced**. In this configuration, the negative wire also acts as the ground. Since there is no plastic or foil shielding, an unbalanced cable is more susceptible to unwanted audio interference, such as that created by nearby electric sources. Unbalanced lines are also more likely to suffer from signal degradation and attenuation as they get longer. If that were not enough, equipment that utilizes unbalanced cable also generally outputs a lower level signal than equipment that employs balanced wiring. Because of these shortcomings, unbalanced cable is more often used in home stereo systems than production studios. Ideally, balanced and unbalanced cable should not be mixed in the same audio setup. However, sometimes this can't be avoided, because different pieces of equipment require different cabling.

7.15 MICROPHONE, LINE, AND SPEAKER LEVELS

Equipment inputs and outputs can be considered in terms of one of three audio *levels*: microphone, line, or speaker. Think of these levels as very *low* for microphone (which usually must be preamplified), *normal* for line level (most equipment will use line levels), and *very high* for speaker levels (designed to drive a speaker only). Problems arise when various levels are mismatched. For example, if you tried to feed an audio recorder from a speaker-level source on a console or patch panel, you would probably distort the recording, because the speaker-level source would be too loud, and there is no control to turn it down. Another problem would occur if you fed a microphone-level signal into a line-level input. In this case, the signal would be too low because microphone levels must be preamplified to a usable level. Most audio equipment inputs and outputs are clearly designated as microphone, line, or speaker level, and good production practice dictates connecting them properly.

7.16 STUDIO TIMERS

Although many audio consoles have built-in studio timers, it is not uncommon to find a separate timer in the audio production room (see Figure 7.19). Because the timing of production work is so important, an accurate timing device is crucial. Most studio timers are digital, showing minutes and seconds, and include at least start, stop, and reset controls. Many timers can be interfaced with other equipment (such as computers, recorders, and CD players) so that they automatically reset to zero when that piece of equipment is started. While shorter timers (10 minutes) are usually adequate for audio production work, 24-hour timers are also often found in the studio.

7.17 TELEPHONE INTERFACE

A simple **telephone interface**, or **telephone coupler** (see Figure 7.20), is a piece of equipment designed to connect telephone lines and cellular networks to broadcast or recording equipment. In a basic configuration, the telephone goes through the interface and comes into the audio console on its own channel. The caller volume is controlled with that channel's fader, in addition to a caller volume control on the interface. Once a call is taken by the announcer and the interface is switched on, the announcer talks to the caller through the studio microphone and hears the caller through the headphones. The

interface electronically maintains an isolation between the studio "send" signal and the caller "return" signal, providing a high-quality, clear telephone signal.

A popular portable telephone interface is the Comrex system, which is a compact, handheld unit that allows for connection to a number of different wired and wireless data circuits. Although units like those shown in Figures 7.21 and 7.22 can connect to standard telephone lines or POTS (Plain Old Telephone Service), they can also send audio over DSL, cable, Wi-Fi, 3G and 4G cellular, Internet, satellite, and other communication networks. It is also possible to use the Comrex system with an app such as iPush (see chapter 6). This portable IP (Internet Protocol) audio codec makes field remotes possible in situations where setting up a rack of telephone interface equipment can't happen.

Keep in mind that each of these interface devices can be used with cell phones as well as land lines. The biggest concern for a studio operator when it comes to recording or broadcasting cell phone calls is that the quality of the recording or broadcast relies on the quality of the cell phone connection. A poor connection will lead to a poor production.

7.18 CONCLUSION

Often, monitor speakers and connectors are given little or no thought. Some production people are concerned only with making sure sound comes out of the speakers and that some accessory items are available, but the purpose of the monitor speaker in audio production is not as minor as one might initially believe. Although much of the equipment mentioned in this chapter may never be installed or adjusted by the studio operator, it is critical to not take speakers, their connectors, and other interface equipment for granted. Always remember that loudspeakers in any form are the only tool you have to adequately determine the final quality of an audio production.

FIGURE 7.19 An audio studio timer provides accurate timing for production work. *(Image courtesy of Radio Systems.)*

FIGURE 7.20   A telephone interface allows mixing an audio signal and a telephone signal. *(Image courtesy of Comrex Corporation.)*

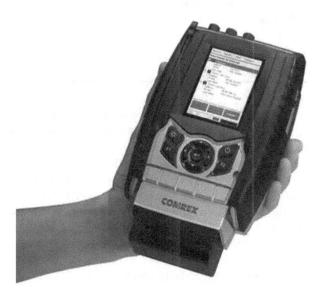

FIGURE 7.21   This portable IP audio codec can interface with DSL, cable, Wi-Fi, 3G cellular, satellite, POTS, and other communication services. *(Image courtesy of Comrex Corporation.)*

FIGURE 7.22   CBS Sports NFL Analyst Rich Gannon uses the portable Comrex system to produce his live call-in program on Sirius/XM Radio from a remote location. *(Used with permission.)*

Self-Study

1. What components make up a two-way speaker system?

   a) tweeter, woofer, and midrange speaker
   b) tweeter, woofer, and crossover
   c) tweeter and woofer
   d) tweeter, woofer, and two crossovers

2. What is the transducing element of a speaker called?

   a) tweeter
   b) crossover
   c) woofer
   d) driver

3. Which speaker enclosure design utilizes a tuned port to provide a highly efficient system with a full bass sound?

   a) acoustic suspension
   b) bass reflex
   c) bass boom
   d) sealed box

4. Which individual driver or speaker is designed to reproduce higher frequencies?

   a) woofer
   b) crossover
   c) tweeter
   d) bass reflex

5. For proper stereo sound, the listening angle formed between the speakers and the listener should be 90 degrees.

   a) true
   b) false

6. Which broadcast connector has a guide pin?

   a) RCA
   b) phone
   c) XLR or Cannon
   d) phono

7. Which is the most practical place to locate monitor speakers in a production room?

   a) near the upper corners, close to the wall
   b) on the counter
   c) not in the room at all, but in an adjoining room
   d) in the middle of the wall, close together

8. Which is true if two speakers are out of phase?

   a) The bass sounds will be generated at the rear of the cones.
   b) Both negative wires will be connected to negative terminals.
   c) Both positive wires will be connected to positive terminals.
   d) The cone of one speaker will be moving out while the cone of the other speaker is moving in.

9. Which broadcast connector is always mono?

a) phone
b) miniphone
c) ¼-inch phone
d) RCA

10. Which type of monitor speaker will most likely be found in the audio production studio?

a) ribbon speaker
b) electrostatic loudspeaker
c) dynamic loudspeaker
d) condenser loudspeaker

11. Which component of a speaker system divides the incoming audio signal into different frequencies and sends the proper frequencies to the appropriate driver?

a) pigtail leads
b) tweeter
c) woofer
d) crossover

12. Which type of headphone is designed with a porous muff that rests directly on the ear?

a) closed-cushion headphone
b) circumaural headphone
c) earbud
d) supra-aural headphone

13. Unbalanced audio cables are more susceptible to interference than balanced cables.

a) true
b) false

14. Having a high-power, external monitor amplifier in your production studio will allow you to record or broadcast a louder signal than using the internal monitor amp in the audio console.

a) true
b) false

15. Small speakers set on short stands, placed left and right of the audio console so the listener hears mostly direct sound at ear level, are known as which type of monitors?

a) dynamic
b) near-field
c) acoustic suspension
d) out-of-phase

16. When a patch panel is normalled, a patch cord is used to link broadcast equipment assigned to the top row of the panel to the equipment assigned to the bottom row.

a) true
b) false

17. Which broadcast connector has a sleeve, ring, and tip?

a) RCA
b) CAT-5
c) XLR or Cannon
d) phone

18. Which connector is most likely to be used for a patch bay?

   a) phone
   b) miniphone
   c) RCA
   d) multipin connector

19. What is a connector adapter used for?

   a) to transfer a signal in a patch bay
   b) to change a connector from one form to another
   c) to make a balanced line unbalanced
   d) to change a telephone signal to an audio signal

20. Which configuration describes a balanced cable?

   a) two wires
   b) three wires
   c) three ground wires
   d) two ground wires

21. The normal outputs of a CD player produce which level of audio signal?

   a) microphone
   b) line
   c) speaker
   d) none of the above

22. Practically all consumer models of headphones can be purchased "barefoot."

   a) true
   b) false

23. By convention, male XLR connectors are outputs and female XLR connectors are inputs.

   a) true
   b) false

24. Which production room accessory is used to connect telephone lines directly to broadcast equipment?

   a) audio routing switcher
   b) telephone coupler
   c) patch panel
   d) XLR connector

25. What would be the best type of monitor when you need to use a microphone in the production studio to record a voice over music?

   a) headphones
   b) tweeter
   c) acoustic suspension
   d) bass reflex

ANSWERS

If you answered A to any of the questions:

1a.   No. This speaker complement would be in a three-way system. (Reread 7.3.)
2a.   No. A tweeter is a speaker designed to produce high frequencies. (Reread 7.2 and 7.3.)

3a. No. The acoustic suspension design is relatively inefficient. (Reread 7.4.)

4a. Wrong. The woofer is designed to reproduce the lower frequencies. (Reread 7.3.)

5a. No. This would put the listener directly in front of one of the speakers, and all the sound would appear to be coming out of that speaker. (Reread 7.6; check Figure 7.6.)

6a. No. (Review Figures 7.11 to 7.14; reread 7.12 and 7.13.)

7a. Correct. This gives the operator good sound and also leaves the counter clear for other equipment.

8a. No. You're confusing this with speaker enclosures. (Reread 7.4 and 7.7.)

9a. No. The phone connector can be either stereo or mono. (Review Figures 7.12 to 7.14; reread 7.12 and 7.13.)

10a. No. Ribbon speakers are generally too exotic in design and too expensive for broadcast use. (Reread 7.2.)

11a. Wrong. You may be confused because this is a part of an individual speaker that receives an input signal, but only after it has been divided into the proper frequencies. (Reread 7.3.)

12a. No. This is a headphone with a ring-shaped ear cushion designed to encircle the ear and rest on the head. (Reread 7.10.)

13a. Right. This is a true statement because one wire conducts the signal and also acts as a ground, so unbalanced cables are more likely to pick up unwanted noise.

14a. Wrong. A monitor amp has no relation to the broadcast or recorded signal; it only controls the volume of the monitor speakers. (Reread 7.8.)

15a. Although they may be dynamic speakers, this term usually refers to the speaker driver. There is a better answer. (Reread 7.2 and 7.6.)

16a. Wrong. A normalled patch panel is actually unpatched, so no patch cords are used. (Reread 7.11.)

17a. No. The phono connector only has an outer sleeve and tip. (Review Figures 7.12 to 7.14; reread 7.12 and 7.13.)

18a. Right. A male phone plug is the most likely connector to use with a patch bay.

19a. No. You would be very unlikely to use an adapter with a patch bay. (Reread 7.11 and 7.13.)

20a. No. Two wires would be unbalanced. (Reread 7.14.)

21a. Wrong. Microphone level is a low output level, which must be preamplified to be usable, and is primarily produced by microphones, as the name implies. (Reread 7.15.)

22a. Wrong. Most consumer headphones come with a plug already attached. Most professional headphones can be purchased "barefoot." (Reread 7.10.)

23a. Correct. This is a true statement.

24a. Wrong. An audio router can be involved in selecting various inputs and outputs, but it can't connect a telephone line by itself. (Reread 7.11 and 7.17.)

25a. Yes. Headphones would prevent feedback and allow the operator to hear the music when the speakers are muted because the microphone is on.

If you answered B to any of the questions:

1b. Correct. These are the basic components of a two-way speaker system.

2b. Wrong. The crossover divides the electrical signals and sends them to the speaker drivers. (Reread 7.2 and 7.3.)

3b. Right. This answer is correct.

4b. No. The crossover is not a speaker but an electronic device for sending various frequencies to different speaker drivers. (Reread 7.3.)

5b. Correct. This is false because an angle of about 60 degrees should be formed between the listener and the speakers for the best stereo sound.

6b. No. (Review Figures 7.11 to 7.14; reread 7.12 and 7.13.)

7b. No. The sound can be good, but the speaker takes up counter space that could be used for something else. (Reread 7.6.)

8b. No. One negative wire connected to a positive terminal would put them out of phase. (Reread 7.7.)

9b. No. The miniphone connector can be either stereo or mono. (Review Figures 7.12 to 7.14; reread 7.12 and 7.13.)

10b. No. Electrostatic loudspeakers are generally too exotic in design and too expensive for broadcast use. (Reread 7.2.)

11b. Wrong. This is a part of a speaker system that reproduces high frequencies. (Reread 7.3.)

12b. No. This is another name for the closed-cushion headphone, which has a ring-shaped ear cushion designed to encircle the ear and rest on the head. (Reread 7.10.)

13b. Wrong. This is a true statement. (Reread 7.14.)

14b. Correct, because monitor amps have no relation to the broadcast or recorded signal.

15b. Yes. Some production people think that the best sound is heard through near-field monitoring.

16b. Yes. When a patch cord is put into a patch panel, it is no longer normalled.

17b. Wrong. (Review Figures 7.12 to 7.14; reread 7.12 and 7.13.)

18b. No. You're warm but not correct. (Reread 7.11 and 7.12.)

19b. Correct. It transfers the signal so that another form of connector can be used.

20b. Right. There are three wires: positive, negative, and ground.

21b. Right. Line-level outputs are standard for most broadcast production equipment, such as CD players and audio recorders.

22b. Correct. This is the best answer.

23b. No. This is a true statement. (Reread 7.12.)

24b. Correct. This is the best answer.

25b. No. This is only part of a monitor speaker. (Reread 7.3 and 7.10.)

If you answered C to any of the questions:

1c. No. You're close though, but you've left out one component. (Reread 7.3.)

2c. No. A woofer is a speaker designed to reproduce low frequencies. (Reread 7.2 and 7.3.)

3c. Wrong. There's no such enclosure design. (Reread 7.4.)

4c. Correct. The tweeter is the speaker designed to reproduce high frequencies.

6c. Right. The XLR or Cannon jack has a guide pin that prevents it from being connected incorrectly.

7c. No. You couldn't hear them if they were in another room. (Reread 7.6.)

8c. No. One positive wire connected to a negative terminal would put them out of phase. (Reread 7.7.)

9c. No. A $\frac{1}{4}$-inch phone or phone connector can be either stereo or mono. (Review Figures 7.12 to 7.14; reread 7.12 and 7.13.)

10c. Yes. The dynamic loudspeaker is found most often in the production studio.

11c. Wrong. This is a part of a speaker system that reproduces low frequencies. (Reread 7.3.)

12c. No. This is a type of headphone that is designed to fit into the ear. (Reread 7.10.)

15c. Wrong. Although they may be acoustic suspension speakers, this term is usually associated with the speaker enclosure. There is a better answer. (Reread 7.4 and 7.6.)

17c. No. (Review Figures 7.12 to 7.14; reread 7.12 and 7.13.)

18c. No. (Reread 7.11 and 7.12.)

19c. No. Adapters are not related to balance. (Reread 7.13 and 7.14.)

20c. No. Although there are three wires, they can't all be ground wires. (Reread 7.14.)

21c. No. Speaker level is quite high and is designed to drive a monitor speaker. Most production equipment, like a CD player, would have to have its output signal amplified to reach speaker level. (Reread 7.15.)

24c. Wrong. A patch panel could be involved in wiring a telephone line to an audio console, but it can't connect a telephone line by itself. (Reread 7.11 and 7.17.)

25c. No. You're confusing monitors and enclosures. (Reread 7.4 and 7.10.)

If you answered D to any of the questions:

1d. Wrong. Only a single crossover is required in a speaker system. (Reread 7.3.)

2d. Yes. The driver transforms electrical signals into mechanical energy and thus audible sound.

3d. No. This is another name for an acoustic suspension speaker, which is relatively inefficient. (Reread 7.4.)

4d. No. Bass reflex describes a speaker enclosure design, not an individual speaker. (Reread 7.3 and 7.4.)

6d. No. Phono is just another name for the RCA connector. (Review Figures 7.11 to 7.14; reread 7.12 and 7.13.)

7d. Wrong. This would really limit sound dispersion and any stereo imaging. (Reread 7.6.)

8d. Right. The sounds will be fighting each other when this happens.

9d. Correct. You must use two RCA connectors for stereo, one for each channel.

10d. No. You might be thinking of a microphone; there is no such loudspeaker. (Reread 7.2.)

11d. Correct. A crossover is a network of filters that divides the audio signal into different frequencies and sends it to the proper individual speaker.

12d. Yes. Also known as open-air or hear-through cushion headphones, supra-aural headphones are designed to rest directly on the ear.

15d. No. Out-of-phase speakers are miswired, and this has nothing to do with speaker placement. (Reread 7.6 and 7.7.)

17d. Correct. The miniphone connectors also have them.

18d. No. (Reread 7.11 to 7.13.)

19d. No. There are devices to do this, but they are interfaces, not connector adapters. (Reread 7.13 and 7.17.)

20d. Wrong. (Reread 7.14.)

21d. No. There is a correct answer and this isn't it. (Reread 7.15.)

24d. Wrong. You're quite confused if you chose this answer. (Reread 7.12 and 7.17.)

25d. No. You're confusing monitors with speaker enclosure designs. (Reread 7.4 and 7.10.)

Projects

PROJECT 1

Compare speaker/listener placement.

Purpose

To make you aware of how sound can change as the relationship between speaker and listener changes.

Notes

1. The most important thing for your drawings will be to show the relative dimensions of the studio and the position of the speakers.

How to Do the Project

1. Make three sketches of your production room, showing where the speakers are located.
2. On the first drawing, put an *X* where the production person usually sits. On the second drawing, put an *X* at another spot in the control room where you can stand and listen to the monitors. Do the same for the third drawing, but in a different location in the room.
3. Play some music through the monitor speakers, and position yourself in each of the three places where you have placed *X*s. Listen for any differences in the way the music sounds at the three locations.
4. Write a short report detailing how the music sounded at each position.
5. Turn in your drawings and your report together with your name and the title, "Speaker/Listener Relationship," to your instructor to receive credit for this project.

PROJECT 2

Identify common connectors found in the audio production studio.

Purpose

To familiarize yourself with audio connectors and how they are used in the production studio.

Notes

1. If you have trouble identifying them, review Figures 7.12 to 7.16.

How to Do the Project

1. Draw four columns on a sheet of paper and label them "XLR/Cannon," "RCA," "Phone," "USB," and "Other." Count the number of different connectors you see and log that number on your sheet. Be sure to include the equipment they are connected to and also be sure to include as many pieces of equipment and connectors as you can.
2. Indicate whether this connector is mono, stereo, or can be both.
3. Next, see how many instances you can find where each connector is used and note this on your list. Provide specifics on how the connector is being used. For example, "the output of the CD player uses RCA connectors for the left and right channels."
4. Try to find at least one instance of use for each connecter, but be aware that some connectors may be used numerous times in the studio.
5. Remember to note how they are being used in the studio.
6. Label your list with your name and the title, "Audio Connectors." Turn it in to your instructor to receive credit for this project.

8

SIGNAL PROCESSING AND AUDIO PROCESSORS

8.1 INTRODUCTION

Signal processing, or audio processing, is nothing more than the process of altering the sound of audio—such as an announcer's voice or a CD. We've already seen a few forms of audio processing in the equalization capabilities of some audio consoles and the "bass roll-off" feature of some microphones (this is actually a form of processing, specifically equalization), but most signal processing is accomplished either by the use of separate audio components or software. Modern audio equipment, especially audio editing software and digital audio workstations, includes features that allow for **digital signal processing (DSP)**. In many of today's audio production situations, any audio processing is accomplished in a postproduction environment.

Several signal processing devices will refer to the terms "wet" and "dry." The incoming, or unprocessed, audio signal is considered a dry signal, and the outgoing audio signal that has been processed is considered a wet signal.

This chapter focuses on the capability of audio editing software to process signals and also looks at other commonly used electronic audio processors. However, be aware that the amount of equipment available for signal processing in any production facility can vary widely, from essentially none to a veritable smorgasbord of electronic black boxes or software effects to choose from.

8.2 WHY USE SIGNAL PROCESSING EFFECTS?

Signal processing usually involves the manipulation of the frequency response (tonal balance), stereo imaging, or dynamic range of the sound signal. As noted earlier in the text, frequency response refers to the range of all frequencies (pitches) that an audio component can reproduce and we can hear, that is usually around 20 Hz (low bass) to 20 kHz (high treble). Imaging refers to the perceived space

between, behind, and in front of monitor speakers and how we hear individual sounds within that plane. Dynamic range refers to the audible distance between the softest sounds that can be heard over noise and the loudest sounds that can be produced before distortion is heard. Most audio processing used in the production studio is to accomplish one of three effects: to alter recordings to suit individual taste, to create special audio effects, or to reduce various forms of audio noise.

Audio processing can be used to alter the sound of a CD (or any music) or an announcer's voice to suit an individual's taste, and changing the tone with an equalizer is one of the most common kinds of audio processing. For instance, in some production work, you may need to accentuate the bass of an announcer's vocal track. Boosting frequency settings between 50 and 100 Hz certainly would help to do so. Perhaps a background music bed seems overly "bright," so you attenuate frequencies in the 10- to 18-kHz range to lessen their impact. As another example, reverb is sometimes added to an announcer's microphone to give it a more "live" sound or to simulate a different acoustic environment. Remember, however, that audio processing is a subjective process and what sounds great to one person may not sound so great to another. However, some specific adjustments like those just mentioned would usually be noticed as a positive improvement if used properly. The key to good signal processing is to use any effect moderately and to experiment to find just the right sound you're looking for.

Another reason for using signal processing is to achieve a special audio effect. For example, cutting down most of the lower frequencies below 1 kHz leaves a very tinny-sounding vocal that might be perfect as a robot voice. You could create an effect that approximates an old-fashioned, poor-quality telephone sound by boosting the frequency bands at 640 and 800 Hz and reducing the frequency bands on either side of this range. Many such effects are achieved by experimenting with various settings for the equalizer. Using a flanger effect (where one audio signal is mixed with an identical signal that is slightly delayed a continuously changing amount of time) can give a voice an otherworldly sound, and a chorusing effect can easily make one voice appear to be several.

Audio processing is also used to deal with unwanted noise such as "hiss," one of the common forms of noise in production work. **Hiss** describes a high-frequency noise problem inherent in analog equipment. Hiss sounds exactly as you would imagine and is heard by the human ear within the frequency range of 4 to 15 kHz. Using an equalizer to cut or turn down the frequency settings closest to this range can attenuate hiss to a less noticeable level. An audio filter is sometimes used to combat "hum," which is another form of noise in production work. **Hum** is a low-frequency problem associated with leakage of the 60-Hz AC electrical current into the audio signal. A poor or broken ground in any part of the electrical circuit can cause hum, and it, too, sounds exactly as you would imagine. Attenuating frequencies right around 60 Hz would lessen the problem.

However, even with careful adjustment, it's often impossible to eliminate noise entirely. Remember, you also affect the program signal as you equalize, so any audio processing is usually a compromise between less noise and a still-discernible signal. Digital recording technique has lessened hiss and hum problems, but in audio production you are often working with prerecorded audio that may already have such inherent problems.

More audio processing uses will become apparent later in this chapter as we look at the various signal processing devices.

8.3 SOFTWARE OR BLACK BOX SIGNAL PROCESSING

As mentioned previously, the majority of audio processing today is accomplished with software programs. Computer-based recording and mixing programs, like Adobe® Audition® CC or Avid Pro Tools, feature the ability to utilize a multitude of audio processing effects, including equalization, echo chamber, hiss reduction, chorusing, pitch shift, reverb, flanging, and noise reduction, to mention only a few.

The basic process with most audio editing programs is to select the audio you wish to process and then select the effect you wish to apply. Usually, you are able to preview the effect to make sure it is doing what you wish and, in most cases, you will be able to adjust numerous parameters about the effect to fine-tune and get the precise sound you're trying to achieve.

Originally, signal processing was accomplished with electronic "black boxes" or individual units designed to produce one particular effect. So if you wanted to add reverb, you needed a reverb unit (a physical piece of equipment), and if you wanted to change the tone of a voice, you needed to have a graphic or parametric equalizer. The audio signal was routed through the device and then to a recorder, public address system, or to be broadcast over the air. Each device was generally rack mounted, and most featured a number of controls that would allow the operator to adjust the effect and achieve the exact sound desired.

Studios might employ a whole rack of various units to have available the audio processing that might be required; however, eventually, multi-effects boxes were developed to offer, as the name implies, several effects in one unit.

Software and black box devices continue to coexist, and there is no real agreement over which approach is better. In fact, you'll find proponents of both who will argue long and hard which systems work best.

8.4 EQUALIZERS

One of the most commonly used signal processors is the equalizer. The term **EQ** refers to the general process of equalization when an equalizer is used to cut or boost the signal level of specific frequencies; essentially, it is just **tone control**. You're familiar with bass and treble tone controls, because most stereo systems have them. When you turn up the treble, you increase the volume of the higher frequencies. Unfortunately, a treble control turns up *all* the higher frequencies. An equalizer, on the other hand, offers greater flexibility and allows the operator to differentiate, for example, between lower high frequencies and upper high frequencies, and to make different adjustments to each.

Most equalizers found in the audio production studio are passive or **unity-gain** devices. The sound signal that has passed through the equalizer's circuits is no stronger or weaker overall than the input sound signal was. In other words, if the processor's controls are set at zero, they should have no processing effect on the incoming audio signal. Of course, if the signal is processed by increasing the level of several different frequencies, then the overall volume of the outgoing signal will be louder than the unprocessed signal. The two main kinds of equalizers found in audio production are the graphic equalizer and the parametric equalizer.

8.5 THE GRAPHIC EQUALIZER

The **graphic equalizer** is more common and derives its name from the rough graph of a sound's altered **frequency response** formed by the slider control settings on the equalizer's front faceplate (see Figures 8.1 and 8.2). A digital graphic equalizer may have a SELECT button for each band and RAISE and LOWER buttons to adjust the volume of each band. As shown in Figure 8.1, you'll still see the frequency response curve formed by your settings, but on multiple LED bar-graph meters instead of the positions of the slide controls.

Graphic equalizers come in different designs, but all of them divide the frequency response range into separate frequency bands, arranged left to right in order of increasing frequency. A complex design would be a stereo 13-octave equalizer with 31 controls; a simple mono 10-band octave equalizer would feature ten controls on 1-octave intervals. If the first band of a full-octave equalizer was at 31 Hz, the second would be at or near 62 Hz, the third at 125 Hz, and so on. Obviously, the more bands you have to work with,

FIGURE 8.1 The graphic equalizer, whether digital or analog, can boost or cut the volume at various frequencies and provide a visual representation of the EQ being provided. *(Image courtesy of Alesis.)*

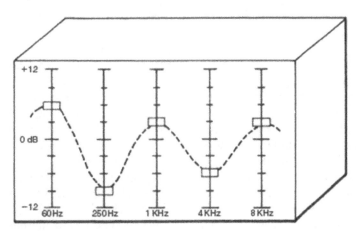

FIGURE 8.2 A simple 5-band graphic equalizer.

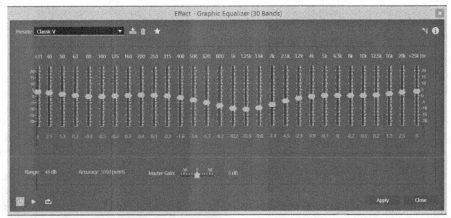

FIGURE 8.3 The graphic equalizer screen from an audio editing program. *(Adobe® Audition® CC screenshot reprinted with permission from Adobe Systems Incorporated.)*

the greater tone control you have, but it also becomes harder to correctly manipulate the equalizer. Most broadcast-quality graphic equalizers feature 15 or 31 band channels. Each band has a volume control that's "off" in a middle or flat position and can move up to increase ("boost") or down to attenuate ("cut") the volume at that particular frequency. This volume range varies, but 112 to 212, or 12 dB of boost or cut, is a common configuration.

To understand equalization techniques, look at the simple 5-band equalizer shown in Figure 8.2. Note how the slider setting would demark a frequency response graph, indicating how the original sound was being manipulated. Typical equalizing procedure starts with all bands set "flat" or at the mid-position. Then each band is adjusted by increasing or decreasing the volume of various frequencies. If we were playing rock music through this equalizer, the boost at 60 Hz would give the drums extra punch; cutting

back at 250 Hz would help minimize bass "boom." The increase at 4 or 5 kHz would add brilliance to the voices, and the slight decrease at 6 to 10 kHz would minimize the harshness of the sound. Finally, the boost at 10 to 13 kHZ would add presence to the highs. There is no one correct setting for an equalizer, and we could just as easily have had all different settings and yet produced an excellent sound. There are, however, settings that could make the sound very poor, and the production operator needs to beware of altering any sound too much. By adding a "little here" and taking away a "little there," you can usually improve any poor-sounding audio.

As previously noted, audio editing software often includes signal processing capability. Figure 8.3 shows the screen for Adobe® Audition® CC graphic equalizer, which allows you to select a 10-, 20-, or 30-band unit. Clicking on the virtual slider control lets you cut or boost the volume of the specific frequency. The position of the sliders and the top portion of the screen give a visual representation of what is happening to the audio signal. Equalization procedures are exactly the same as with any other graphic equalizer. Clicking on the PREVIEW button lets you hear the equalized or "wet" signal and actually change settings as it is playing back. When you have the sound you're looking for, click on the OK button and the original audio will be changed according to the EQ settings.

8.6 THE PARAMETRIC EQUALIZER

The **parametric equalizer** gives the operator more flexibility with three basic controls: one to choose the actual center frequency to work on, one to control the bandwidth of frequencies selected, and one to control the volume of the specified frequencies. In other words, the parametric equalizer is used for finer control of the frequencies within a sound than the broad control offered by graphic equalizers. For example, the 5-band graphic equalizer mentioned previously had fixed frequency bands at 60, 250, 4,000, 8,000, and 12,000 Hz, but a parametric equalizer allows the operator to select an exact frequency. For example, instead of a set band at 1,000 Hz, you might choose a band at 925 Hz or 1,200 Hz. If the graphic equalizer were increased at 1 kHz, not only would that frequency get a boost, but so would the adjacent frequencies according to a preset bandwidth determined by the manufacturer (perhaps from 500 to 1,500 Hz).

Parametric equalizers allow the operator to adjust that bandwidth. For example, still using the 1-kHz center frequency, the parametric equalizer operator could select a bandwidth of 800 to 1,200 Hz or a narrower bandwidth of 950 to 1,050 Hz to be equalized. Most equalizers also have a bypass switch that allows the sound signal to pass through the equalizer unaffected, or not equalized at all, so you can compare the original sound with the equalized sound. Figure 8.4 shows one type of parametric equalizer.

8.7 AUDIO FILTERS

An audio filter is a simple type of equalizer which affects a whole range of the audio signal. For example, a **low cut filter** cuts, or reduces, frequencies below a certain point,

say 500 Hz. Instead of a normal frequency response of 20 to 20,000 Hz, once the signal goes through this filter, its response is 500 to 20,000 Hz. A **low pass filter** allows frequencies below a certain point to pass or remain unaffected. In other words, a cut filter and a pass filter are opposites in their actions. A **band pass filter** eliminates frequencies except for a specified band. It has a low cut point (maybe 1,000 Hz) and a high cut point (maybe 3,000 Hz). When a signal is sent through this filter, only that portion of the signal between the two cut points is heard. A **band reject filter** is just the opposite of the band pass filter in that it allows frequencies to pass except for a specified frequency range. A **notch filter** is a special filter that completely eliminates an extremely narrow range of frequencies or one individual frequency, such as AC hum at 60 Hz.

Usually filters are used to correct a specific problem. Of course, filters eliminate both the problem *and* the actual program signal, so careful use is necessary to maintain a good audio signal after filtering has taken place. Figure 8.5 shows a digital audio editing program that has a filtering function. This screen illustrates the use of a low pass filter so that low frequencies are allowed to pass through the filter unaltered. The cutoff point has been set at 661.5 Hz, so the high frequencies above this point are rejected or removed from the original audio. The graph gives a visual representation of what is happening to the audio.

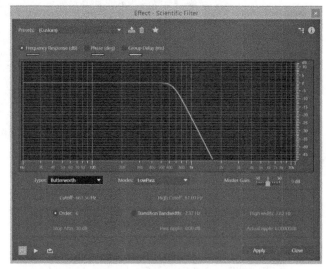

FIGURE 8.5 Digital editing software can replicate an analog filter. *(Adobe screenshot reprinted with permission from Adobe Systems Incorporated.)*

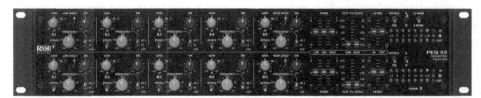

FIGURE 8.4 A parametric equalizer offers greater control of various frequencies within a sound than the graphic equalizer. *(Image courtesy of Rane Corporation.)*

8.8 NOISE REDUCTION

We've mentioned the problem of noise that is inherent in the production process, especially with tape-based or other analog recording. Signal processing devices known as **noise-reduction** systems, have been devised to help prevent noise. Keep in mind that these electronic devices can't get rid of noise that already exists in a recording; their job is to prevent noise from being *added* to a recording. Digital recording does not need to use external noise-reduction devices such as EQs because the noises they are meant to eliminate are those inherent in the analog recording process.

Dolby and dbx® are two systems commonly used to help with noise reduction in analog recording. Dolby noise-reduction is often found in cassette and reel-to-reel recorders and you'll find the dbx® Type III noise-reduction circuitry in some of their EQ products. Both systems are **companders**—their general operation is to compress, or reduce the dynamic range of an audio signal during recording and then expand the signal during playback—but they are not compatible systems, however.

There are several **Dolby** noise-reduction systems in use in recording studios, broadcast stations, home stereo equipment, and movie theaters, but the Dolby B, C, S, and SR systems are most likely to be found in production equipment. The first part of the Dolby process consists of encoding the signal before recording begins so noise cannot be introduced into the recording. In the second part of the process during playback decoding, levels that were increased are decreased to their original levels, and any noise that may have been introduced during the recording process is also reduced so that it seems much lower in relation to the program level. Depending on the system employed, up to 25 decibels of noise reduction can be attained using the Dolby process. As with all other signal processing equipment, careful use is required to achieve the results the operator wants. Dolby SR (for spectral recording) and Dolby S are more sophisticated noise-reduction systems; Dolby S is designed to be an improved version of the Dolby B and Dolby C systems.

Unlike Dolby, the **dbx®** system compresses the signal over the entire frequency range during recording. Noise buildup usually introduced in the recording process is dramatically reduced (noise reduction up to 40 decibels can be attained); however, as with all noise-reduction systems, any noise present in the original audio signal is not reduced.

Although digital audio recording will not generally introduce new noise, most software programs have features for noise reduction so that audio recorded outside the studio and then brought in can be improved. For example, a sound effect or piece of music from a vinyl record or tape may contain pops, clicks, or hiss from excessive use or inherent machine noise. Some software programs include a feature that eliminates this noise by first analyzing the audio to determine where the pops and hiss are and then repairing or reconstructing the audio at that point. There may be several parameters to set, and a bit of trial and error is usually required to get the correct settings, but salvaging just the right piece of music may be critical to a specific production project. Other noise-reduction features may help eliminate or reduce 60-cycle AC hum, microphone background noise, or other noise problems heard in the recording.

Generally speaking, the more consistent the noise you want to remove is, the easier it will be to dispose of. For example, the consistent hum of a refrigerator is easier to remove than the sound of birds in the background, as each bird's whistle or chirp will have its own frequency. Some noise-reduction software can help with this by allowing you to sample a short section of unwanted background noise, and then removing the noise from a longer section of audio. Using this process, the hum of a refrigerator behind a conversation can virtually be eliminated very simply, and without having to experiment with frequencies. Be careful however, because noise-reduction software does not discriminate between frequencies. In the previous example for instance, if the frequency of a person's voice is the same as the frequency of the running refrigerator in the background, both frequencies will be eliminated.

Noise reduction is a trade-off between how much noise level you can decrease and how much loss of audio quality you can afford. Depending upon the sophistication of the software program you may have available, there is a great deal that can be done to improve a noisy audio signal.

8.9 REVERB AND DIGITAL DELAY

A signal processor that affects the "imaging" of sound is the reverberation unit, probably the second most common type of processor. It manipulates the sound signal to artificially produce the sound of different acoustic environments. As we've mentioned, reverb is reflected sound that has bounced off two or more surfaces. Sound heard (or produced) in a small studio "sounds" different than sound produced in a large hall or auditorium, and reverb is the main characteristic that audibly creates the difference.

The modern production studio is likely to have a **digital reverb** unit that is either an electronic device (see Figure 8.6)

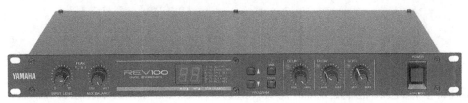

FIGURE 8.6 A digital reverb unit can simulate various acoustic environments by altering the "image" of the sound. *(Image courtesy of Yamaha Corporation.)*

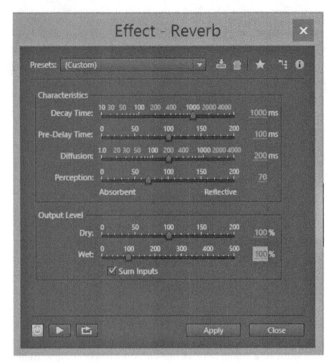

FIGURE 8.7 A digital reverb program, within audio editing software, can have various preset acoustic environments, ranging from a coat closet to a concert hall. *(Adobe® Audition® CC screenshot reprinted with permission from Adobe Systems Incorporated.)*

from fractions of a second to several seconds. Although there are analog delay units, most systems are digital, and if the incoming signal is analog, it is converted from its analog form to digital for processing and back to analog after processing. In a radio station on-air studio, a delay unit designed to prevent profanity (see Figure 8.8) is often used in conjunction with phone-in talk shows. The program signal is sent through the delay unit to provide a delay of 7 to 80 seconds before it is sent to the transmitter. If something is said by a caller that should not be broadcast, the operator has time to "kill" the offending utterance before it is actually broadcast. The delay system will "dump" the delayed signal and revert to the live signal or an earlier signal to do this. Listeners who call a program using a delay system are asked to turn their home radio down if they're talking to an announcer on-air, because the sound they hear on the radio is the delayed sound and not the words they are actually saying into the telephone. It's extremely difficult to carry on a conversation when you can hear both the live and delayed sounds.

In the audio production studio, delay units are used to create special effects similar to reverb. Set for an extremely short delay, the units can create an effect that sounds like a doubled voice or even a chorus of voices. These types of effects are most commonly included in a multi-effects processor, which will be mentioned shortly, but also may be available as one of the special effects of digital audio software programs.

or a feature of an audio editing program. The original signal is fed into the unit and electronically processed to achieve the reverb effect; then the altered signal is sent out of the unit. Audio editing software may also have reverb, echo, or delay functions. Figure 8.7 shows a reverb screen from one audio editing program. In addition to building your own reverb sound by setting various parameters, note that several preset acoustic environments can be chosen, such as a medium-sized auditorium or a tiled bathroom shower. Merely selecting the appropriate preset and clicking on the OK button will transform the original audio so that it sounds as if it had been recorded in that environment with the corresponding reverb effect.

A **digital delay** unit can be used in both the production studio and the on-air control room. As its name implies, this signal processor actually takes the audio signal, holds it, and then releases it to allow the signal to be used further. The time the signal is held or delayed can be varied

---

PRODUCTION TIP 8A
World Wide Web Effects

The Internet can be a source for some interesting DSP effects. This isn't pirated software, but legitimate programs that are available for free or nominal cost. Most are plug-ins that you can add to your digital audio software program, but some you can just install on your computer on their own.

One free program you might want to check out is Izotope's Vinyl. It's a DirectX plug-in and you'll need to go to their website at www.izotope.com and click on the VINYL icon. Then you'll have to supply some registration information that will allow them to email you back the necessary install information. Once you've completed the download, you'll need the install information the first time you use the plug-in with your audio software.

FIGURE 8.8 A delay unit is often used in conjunction with radio talk shows to prevent material that should not be broadcast from going out over the air. *(Image courtesy of Eventide, Inc.)*

Vinyl is a vinyl record simulation. It lets you take an audio file and process it so it sounds like it is being played back on a record player. There are several presets, and you can also make a number of parameter adjustments, including adding hum (electrical noise) or turntable rumble (mechanical noise). You can add dust and scratches to your recording, as well as warp to give you that great off-speed or off-center "wow" sound. As with most signal processing devices, you'll have to experiment to find the right sound you're looking for, but Izotope provides a handy "10 Steps to Lo-Fi" help screen to get you started. There are, no doubt, other programs on the Web, so some search time might be beneficial. However, remember that one small effect used as part of a production can be very effective, but too many special effects take away from their uniqueness.

8.10 DYNAMIC RANGE

When we mention **dynamic range** in an audio production context, we're referring to the range of volumes of sound that equipment can handle. This intensity of a sound is measured in decibels (dB). One **decibel** represents the minimum difference in volume that we can hear, but a change of 3 decibels is often necessary before we actually do perceive a difference in volume. The dynamic range goes from 0 decibels at the **threshold of hearing** to 120+ decibels at the **threshold of pain**. A whisper is around 20 decibels, normal conversation is near 60 decibels, and shouting is about 80 decibels. Average music-listening levels are between 30 and 80 decibels, but some rock concerts have been measured above 110 decibels. Dynamic range also relates the volume of one signal to another, such as signal to noise or input to output.

We should note that dynamic range is measured on a logarithmic scale. For example, to hear one audio signal twice as loudly as another, you would have to increase the volume of one by 10 decibels in relation to the other. A 60-decibel dynamic range was once considered quite adequate for high-quality production equipment; today's digital equipment, however, offers an increased dynamic range and a 120 to 140 decibel range is now common.

Why would you want to reduce dynamic range? Even modern equipment can have trouble handling extremes in dynamic range. If equipment is set to handle typical volume levels, when extremely loud levels are encountered the equipment overloads. If everything is set to handle extremely loud levels, when you hit a quiet section nobody can hear it.

8.11 COMPRESSORS, EXPANDERS, AND NOISE GATES

The compressor is one of the two most common signal processing devices used to affect the dynamic range of the audio signal; the limiter is the other. Although they are most often used to process an audio signal between the studio and the transmitter in broadcasting, they are also used in the production room to process the signal before it's sent to an audio recorder.

One type of **compressor** operates as an automatic volume control or automatic gain control (AGC) and reduces the dynamic range of an audio signal put through this unit. In the contemporary production studio, this type of compressor is used if the audio signal is too loud to automatically lower it. Several adjustments on the compressor determine its actual operation. The threshold of compression is the setting of the level of signal needed to activate the circuit. As long as the audio signal stays below this point, the compressor doesn't do anything, so the compressor really needs to work only some of the time. If the input is too low, the threshold of compression will never be reached, and if the input is too high, the compressor will severely restrict the dynamic range. The compression "ratio" determines how much the compressor works or how much it will turn down audio above the threshold. A ratio of 5 to 1 means that if the level of the incoming signal increases to 10 times its current level, the output of that signal from the compressor will only double.

Compressors also have settings for attack time (how quickly volume is reduced once it exceeds the threshold) and release time (how quickly a compressed signal is allowed to return to its original volume). The release time adjustment is very important because too fast a setting can create an audible "pumping" sound as the compressor releases, especially if there is a loud sound immediately followed by a soft sound or a period of silence.

An **expander** is essentially the opposite of a compressor and is used to expand the dynamic range of an audio signal. Today's expanders are set to operate only below a threshold level on low-volume audio so they make quiet passages quieter. As was noted earlier, the process of noise reduction uses compressor and expander technology. A **noise gate** is a type of expander used to reduce noise by turning the level of an audio signal that falls below a set threshold point way down. For example, it could be set so that pauses in a piece of music would be attenuated to "zero" volume so that background noise would not be recorded. Audio editing software often includes some type of dynamics processor that can emulate stand-alone compressors, limiters, expanders, and noise gates.

8.12 LIMITERS

A **limiter** is a form of compressor with a large compression ratio of 10 to 1 or more. Once a threshold level is reached, a limiter doesn't allow the signal to increase anymore. Regardless of how high the input signal becomes, the output remains at its preset level. If they're adjustable, attack and release times on limiters should be quite short. Both the limiter and the compressor can be rather complicated

FIGURE 8.9 The compressor/limiter is used to process the dynamic range of an audio signal. *(Image courtesy of Symetrix, Inc.)*

to adjust properly. Too much compression of the dynamic range makes an audio signal that can be tiresome to listen to, and pauses or quiet passages in the audio signal are subject to the pumping problem mentioned earlier.

Because they're often associated with the transmitter only, compressors and limiters are usually the domain of the engineer. If you do have access to them in the production facility, it may take some experimenting to get the kind of processing that you're looking for because there are no standard settings for signal processing devices. Figure 8.9 shows a system that combines both a compressor and limiter in a single unit.

8.13 OTHER SIGNAL PROCESSORS

A **flanger** is another processor for producing a specific special effect. This unit electronically combines an original signal with a slightly delayed signal in such a way as to cause an out-of-phase frequency response that creates a filtered swishing sound. Flanging, also called phasing, was originally accomplished by sending an audio signal to two reel-to-reel tape recorders and then physically slowing down one of the reels by holding a hand on the flange, or hub, of the reel. As the two signals went in and out of sync with each other, the audio would cancel and reinforce, creating a psychedelic phase-shift effect. Figure 8.10 shows the flanger screen from an audio editing software program. The Original–Delayed slider on the top left side of the screen controls the mix between the dry (original) audio and the wet (flanged) signal. Portions of both must be mixed to achieve the effect of slightly delaying and phasing the signals. As with many of the other signal processing effects using audio software, there are several other adjustments that can be made to create just the effect you're looking for, or you can just choose a preset effect by clicking on one of the descriptive names on the right side of the screen.

A **de-esser** is an electronic processor designed to control the sibilant sounds without affecting other parts of the sound signal. Many microphone processor units include a section that is essentially a de-esser. A **stereo synthesizer**, as the name implies, is a processor that takes a mono audio signal (input) and simulates a stereo signal (output). Some processors use a form of delay to provide separation between the left and right channel outputs; others use a form of filtering to send certain frequencies to one

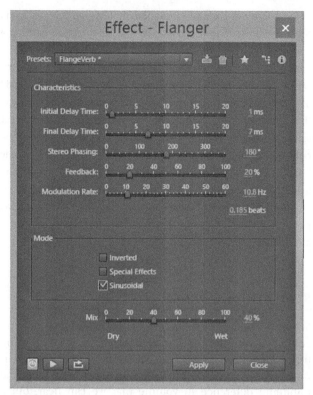

FIGURE 8.10 Flanging—a psychedelic, swooshing sound—is a phase-shifted, time-delay effect easily created with audio editing software. *(Adobe® Audition® CC screenshot reprinted with permission from Adobe Systems Incorporated.)*

channel and other frequencies to the other channel to provide a synthetic stereo effect.

8.14 MULTI-EFFECTS PROCESSORS

Rather than using several individual signal processor "boxes," many signal processing devices are designed so that they perform more than one function—that is, more digital effects can be created from a single black box. For example, Eventide, Yamaha, and Lexicon all provide popular signal processing tools for the production studio (see Figure 8.11). These devices offer a variety of audio effects in one unit, including the ability to alter the pitch of an incoming audio signal, time compression and expansion, delay, natural reverb effects, flanging, time reversal, and repeat capabilities. Any production person should find it enjoyable to experiment with the variety of creative effects that can be produced with a multi-effects processor.

In addition to offering improved editing capabilities, most workstations or software programs have provisions for adding special effects to the audio signal. Often you can add an effect to the audio by mouse-clicking on a preset effects button. Otherwise, you may have to set various parameters to create the signal processing effect that you desire, and that can range from being fairly easy to

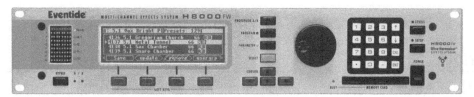

FIGURE 8.11   This multi-effects processor, an Eventide Harmonizer, puts several signal processing effects in a single "black box." *(Image courtesy of Eventide, Inc.)*

extremely complex. Regardless of this fact, editing software or workstation systems are still, in many cases, simpler to set up and operate than their black box counterparts.

8.15 CONCLUSION

This chapter is not intended to be a complete guide to signal processing equipment. There are other units in use in production facilities, and there are other effects available through audio editing software programs. Neither is this chapter designed to make you a professional operator of such equipment. What is intended is that you become aware of a number of the more common processors and that you have an understanding of their basic purpose. The actual operation of most of this equipment will take some trial-and-error work in your production facility. Use signal processing in moderation, and bear in mind that a lot of great production has been produced using no signal processing equipment at all.

Self-Study

1. The equalizer processes an audio signal by altering which of the following?

   a) volume
   b) imaging
   c) dynamic range
   d) frequency response

2. An audio signal that has been equalized would be called a "dry" signal.

   a) true
   b) false

3. Which type of equalizer can select an exact center frequency and bandwidth as well as alter the volume at that frequency and bandwidth?

   a) graphic
   b) parametric
   c) dielectric
   d) full-octave, 10-band

4. What type of filter would most likely be used to attenuate or eliminate a 60-Hz hum in a recording?

   a) low pass filter
   b) band pass filter
   c) notch filter
   d) low cut filter

5. The 60-Hz hum mentioned in Question 4 could also have been eliminated by the use of either Dolby or dbx® noise reduction.

   a) true
   b) false

6. Which signal processor affects the imaging of a sound?

   a) equalizer
   b) noise-reduction unit
   c) de-esser
   d) reverb unit

7. How could you create a tinny voice using signal processing equipment?

   a) Cut out most of the lower frequencies.
   b) Cut out most of the higher frequencies.
   c) Eliminate the EQ.
   d) Increase compression.

8. Which noise-reduction system is most likely to be found in the audio production studio?

   a) Dolby A
   b) Dolby S
   c) Type I
   d) Flanger

9. Which is a true statement about the Dolby system of noise reduction?

   a) Volumes of certain frequencies are increased during recording and decreased during playback.
   b) The dbx® is increased with a calibrated tone so that it attains the level of 30 decibels.
   c) All frequencies pass through, except ones that have been preset by the notch filter.
   d) Once a threshold level is reached, the signal isn't allowed to increase any more.

10. A noise gate is which type of signal processing equipment?

    a) limiter
    b) equalizer
    c) expander
    d) imager

11. Which is true about a compressor?

    a) usually has a compression ratio of 10 to 1
    b) lowers a signal that's too loud and raises one that's too soft
    c) doesn't operate unless it's connected to a digital delay unit
    d) produces an out-of-phase, filtered, swishing sound

12. Any signal processing equipment that's labeled a unity-gain device would amplify all frequencies of the signal going through that equipment an equal amount.

    a) true
    b) false

13. Which of the following is *not* a type of Dolby noise reduction?

    a) Dolby S
    b) Dolby B
    c) Dolby C
    d) Dolby D

14. Which signal processing device is most likely to offer a variety of effects, such as pitch change, time compression, reverb, and flanging?

    a) dbx® Graphic Equalizer
    b) Eventide Broadcast Delay
    c) dbx® Compressor/Limiter
    d) Eventide Harmonizer

15. Which signal processing device inputs a mono signal and outputs a simulated stereo signal?

    a) digital delay
    b) band cut filter
    c) flanger
    d) stereo synthesizer

16. Today's digital equipment offers an increased dynamic range, and a range of how many decibels is now common?

    a) 153
    b) 120 to 140
    c) 60
    d) 3

17. A digital audio workstation usually includes digital reverb and delay effects but rarely any other type of signal processing capability.

    a) true
    b) false

18. Which graphic equalizer setting would most likely be used to add "brilliance" to an announcer's voice?

    a) boost at 60 Hz
    b) cut at 250 Hz
    c) boost at 1 kHz
    d) cut at 8 kHz

19. Which signal processor is designed to help control sibilance in an announcer's voice?

    a) compressor
    b) limiter
    c) de-esser
    d) digital reverb

20. Which of the following is the opposite of a compressor?

    a) equalizer
    b) expander
    c) flanger
    d) limiter

ANSWERS

If you answered A to any of the questions:

1a. No. You're headed in the right direction. Equalizers do increase or attenuate volumes at specific frequencies, but they do this to alter another sound characteristic. (Reread 8.1, 8.2, and 8.4.)

2a. No. A dry audio signal is an unprocessed signal. (Reread 8.1.)

3a. No. Center frequencies and bandwidths are usually preset on graphic equalizers. (Reread 8.5 and 8.6.)

4a. No. This type of filter allows lower frequencies to pass and would not eliminate noise at 60 Hz. (Reread 8.7.)

5a. No. Noise-reduction units, regardless of brand name, can't eliminate noise that already exists in a recording. They only prevent noise during the recording process. (Reread 8.8.)

6a. No. Equalizers affect frequency response. (Reread 8.2, 8.4, and 8.9.)

7a. Correct. Cutting the bass will give a tinny sound.

8a. No. (Reread 8.8.)

9a. Correct. It's a two-step process.

10a. No. In a way it is the opposite. (Reread 8.11 and 8.12.)

11a. Wrong. That's a limiter. (Reread 8.11 and 8.12.)

12a. No. A unity-gain device doesn't amplify the overall level of the incoming signal at all. (Reread 8.4.)

13a. Wrong. This is one of the Dolby noise-reduction systems. (Reread 8.8.)

14a. No. This doesn't really exist. It's a mixture of signal processing. (Reread 8.5, 8.8, and 8.13.)

15a. No. You may be thinking about a stereo synthesizer that uses a form of delay. (Reread 8.9 and 8.14.)

16a. No. You may be getting this confused with the threshold of pain. (Reread 8.10.)

17a. No. Most digital audio workstations have multi-effect capability. (Reread 8.1 and 8.13.)

18a. No. This would punch up the bass. (Reread 8.5.)

19a. Wrong. This is a form of automatic volume control used to affect dynamic range. (Reread 8.11 and 8.14.)

20a. No. You might be confusing frequency response and dynamic range. (Reread 8.2 to 8.6, 8.10, and 8.11.)

If you answered B to any of the questions:

1b. No. Imaging can be affected by other signal processors. (Reread 8.1, 8.2, and 8.4.)

2b. Correct. This is the right response.

3b. Yes. The parametric equalizer gives the operator the greatest control over the EQ process.

4b. No. This type of filter is usually used to allow a range of frequencies to pass, not to eliminate a single frequency. (Reread 8.7.)

5b. Yes, because you can't eliminate existing noise with noise-reduction units.

6b. No. Noise-reduction units affect dynamic range. (Reread 8.8 and 8.9.)

7b. No. That is where the tinny sound would be. (Reread 8.2.)

8b. Correct. It's one of the newer common ones.

9b. No. (Reread 8.8.)

10b. No. A noise gate has nothing to do with frequencies. (Reread 8.4 and 8.11.)

11b. Yes. It lowers and raises signals in that manner.

12b. Yes. Unity-gain devices don't amplify the overall level of the incoming signal.

13b. Wrong. Dolby B is a common noise-reduction system found on cassette recorders. (Reread 8.8.)

14b. No. This isn't correct. Delay is a single type of audio processing, often used with radio talk shows. (Reread 8.9 and 8.13.)

15b. No. You may be thinking about a stereo synthesizer, which uses a form of filtering. (Reread 8.7 and 8.14.)

16b. Correct. This is the right response.

17b. Yes. Most digital audio workstations have multi-effect capability.

18b. No. This would minimize bass boominess. (Reread 8.5.)

19b. Wrong. This is a form of automatic volume control used to affect dynamic range. (Reread 8.12 and 8.14.)

20b. Correct. Because the expander increases dynamic range, it is the opposite of a compressor.

If you answered C to any of the questions:

1c. No. Dynamic range can be affected by other signal processors. (Reread 8.1, 8.2, and 8.4.)

3c. Wrong. There's no such thing. (Reread 8.5 and 8.6.)

4c. Correct. This type of filter allows all frequencies to pass except a specified one, which we could specify at 60 Hz to eliminate the hum.

6c. Wrong. A de-esser is an electronic processor designed to control sibilant sounds. (Reread 8.9 and 8.14.)

7c. Wrong. That wouldn't really be possible. (Reread 8.2.)

8c. No. (Reread 8.8.)

9c. No. You are confusing this with filters. (Reread 8.7 and 8.8.)

10c. Correct. A noise gate is a type of expander.

11c. No. They have nothing to do with each other. (Reread 8.9 and 8.11.)

13c. Wrong. Dolby C is a common noise-reduction system found on cassette recorders. (Reread 8.8.)

14c. No. This isn't correct. (Reread 8.8 and 8.11 to 8.13.)

15c. No. You're thinking about another signal processing device. (Reread 8.14.)

16c. Wrong. You may be getting this confused with the level of normal conversation. (Reread 8.10.)

18c. Yes. This would add brilliance to a voice.

19c. Right. This is the correct answer.

20c. No. Flanging concerns phase-shift and time delay, but not dynamic range. (Reread 8.10, 8.11, and 8.13.)

If you answered D to any of the questions:

1d. Correct. Equalizers allow you to adjust selected frequency volumes and thus alter the audio signal's frequency response.

3d. No. This would be a type of graphic equalizer. (Reread 8.5 and 8.6.)

4d. Although it could eliminate a 60-Hz hum, low cut filters normally are designed to eliminate all frequencies below a certain point. There's a better response. (Reread 8.7.)

6d. Yes. This is what reverb units do by electronically changing the apparent acoustic environment in which we hear the sound.

7d. No. A compressor acts as an automatic volume control affecting the dynamic range of an audio signal. (Reread 8.2 and 8.11.)

8d. No. This is not a noise-reduction unit. (Reread 8.8 and 8.13.)

9d. Wrong. This happens in a limiter, but not during noise reduction. (Reread 8.8 and 8.12.)

10d. No. A noise gate does not deal with imaging. (Reread 8.2 and 8.11.)

11d. No. You're thinking of a "flange" effect. (Reread 8.11 and 8.14.)

13d. Right. This is not a current Dolby noise-reduction system.

14d. Yes. This is a multi-effect processor.

15d. Correct. As the name implies, a stereo synthesizer makes a mono signal into a "fake" stereo signal.

16d. Wrong. You're thinking about the amount of change in volume that's usually necessary before we actually hear a difference in level. (Reread 8.10.)

18d. No. This would decrease the highs, if anything. (Reread 8.5.)

19d. Wrong. This is a type of processor used to affect imaging of an audio signal. (Reread 8.9 and 8.14.)

20d. Wrong. A limiter is a type of compressor, not the opposite of it. (Reread 8.10 to 8.12.)

Projects

PROJECT 1

Record a commercial spot that uses a signal processing effect.

Purpose

To develop skill in incorporating a signal processing effect into an audio production.

Notes

1. This project assumes that you have enough familiarity with your studio equipment to accomplish basic recording and production techniques.
2. The production incorporates a single announcer voice, a music bed, and some type of signal processing technique.
3. You will need to write a simple script that can be read in about 20 seconds. Write the copy about a store and use the phrase "One Day Sale!" several times in the script.
4. There are many ways to accomplish this project, so don't feel like you must follow the production directions exactly.

How to Do the Project

1. Record the script (voice only) onto an audio recorder. Each time the "sale" phrase is read, add some type of signal processing effect (reverb, flanging, echo, EQ, and so on). Depending on your studio, you may be using a black box processor or a signal processing effect available in your audio editing software.
2. Select a music bed that is appropriate for the style of the spot. You might find one on the website that came with this text.
3. You can play back the music bed directly from a CD, or you may have to record it on another recorder from the website.
4. Set correct playback levels for both the vocal track and the music bed. Both will start at full volume. Then cue both to the beginning sound.
5. Mix the vocal track and music bed to complete the spot.
6. Start the music bed at full volume. Start the vocal track and simultaneously fade the music bed slightly so the vocal track is dominant.
7. As the vocal ends, bring the music bed back to full volume and then quickly fade it out as you approach 30 seconds.
8. It may take you several attempts to get the spot to come out correctly. If you need to "do it over," just cue everything and try again.
9. On the completed project, write your name and "Signal Processing Spot," and turn it in to your instructor to receive credit for this project.

PROJECT 2

Use multitrack recording to create a chorusing effect.

Purpose

To practice the multitrack techniques of chorusing and bouncing while keeping a track sheet.

Notes

1. This project assumes you have a multitrack recorder or an editing software system with at least four tracks.
2. Your system should be set up so that you can record from a microphone as well as a source (such as a CD player) that can play music.

3. You can use any dialogue you want, but the first line of "Mary Had a Little Lamb" works well.
4. You may find music appropriate for this project on the website that came with this text.

How to Do the Project

1. Record your selected dialogue on Track 1.
2. While playing back Track 1, overdub the dialogue again on Track 2.
3. As you're doing your recording, keep a track sheet similar to the one shown in Chapter 3.
4. Bounce Tracks 1 and 2 onto Track 3.
5. While listening to Track 3, overdub the dialogue again on Track 1.
6. Bounce Tracks 3 and 1 onto Track 4.
7. Listen to Track 4. It should sound like a chorus of four voices.
8. Record a music bed on Track 1.
9. Mix down Track 1 and Track 4 so that you have a spot with a vocal track over a music bed.
10. Dub the finished project, and turn in the recording to your instructor to receive credit for this project. Make sure you include the track sheet with your name and label it "Chorusing Production."

PROJECT 3

Restore an audio clip using noise-reduction software.

Purpose

To make you aware of some of the characteristics of noise-reduction signal processing.

Notes

1. You'll need audio editing software that includes noise-reduction capability to complete this project.
2. Adobe® Audition® CC is being used as an example; however, you should be able to complete the project with other similar programs.
3. The audio clip which you can "restore" is on the website that came with this text or your instructor may provide a different clip for your use.

How to Do the Project

1. Using Adobe® Audition® CC in "edit view," select "file" (from the top menu bar) and then "extract audio from CD." You could also record the audio clip from a CD player into Audition.
2. With the audio clip waveform on the screen, select and highlight a small section of "noise." Usually there is something just before the music or voice starts that can be grabbed.
3. Select "edit" (from the top menu bar), then "restoration," and then "capture noise-reduction profile." (If you get a pop-up window, just click "OK.")
4. Select "edit" (from the top menu bar), then "select entire wave."
5. Next, from the top menu bar, select "effects," then "restoration," and then "noise-reduction (process)."
6. In the pop up window, select "preview."
7. Listen to the processed audio. If you don't like how it sounds, you can try changing some of the various adjustments, but usually the presets will give you very good results.
8. Click "OK" to have the noise reduction applied to the audio clip.
9. Record both the original audio clip and processed audio clip onto a CD-R or other medium.
10. Listen to both clips several times.
11. Write a short paper that describes the differences you noted. Label the paper "Noise-Reduction Differences."
12. Turn in this paper and the recording you made to your instructor to receive credit for this project.

9

PRODUCTION SITUATIONS

9.1 INTRODUCTION

The equipment that has been discussed in previous chapters can be used for many types of studio audio productions, including commercials, image enhancers, on-air shows, music recording, news, sports, traffic, weather, talk shows, drama, and variety. Although many of these program forms relate to radio, some can be used for other audio-based material, such as music CDs, club disc jockeying, and demo recordings for job applications. Entire books have been written about each of these types of material; this chapter will not give you the in-depth knowledge you need to perfect any of the program forms. It will, however, get you started in the right direction. Experience and advanced training can then provide you with more specialized skills.

9.2 PRODUCING COMMERCIALS

People who work for advertising agencies and radio stations are likely to be involved with producing commercials. At some stations, dealing with commercials is a full-time position, and at many stations the disc jockeys spend a portion of their workday doing commercial production work. At their simplest level, commercials consist of straight copy that the announcer reads over the air. At their most complex level, they are highly produced vignettes that include several voices, sound effects, and music. In between are commercials such as those that feature celebrity testimonials, dialogue between two announcers, or an announcer reading over a **music bed**.

Commercials, because they are inserted within other programming, are usually an exact length such as 15 seconds, 30 seconds, or 60 seconds. Otherwise part of the commercial might be cut off if it runs long, or yield **dead air** if it runs short. At many stations, commercials are brought in and taken off by computers, which are very unsympathetic to anything that's too long or too short.

Anyone reading commercial copy should use a natural, sincere style. Reading commercials in a condescending manner is definitely uncalled for. Commercials pay station salaries and should be treated with respect. If you're to read commercial copy live, you should read it over ahead of time to avoid stumbling on words. If the commercial involves both reading live and playing a prerecorded segment, make sure you rehearse the transition between the two. Sometimes a prerecorded spot sounds like it's ending, but it actually contains additional information—perhaps new store hours or a bargain price. If you're to read material after the recording ends, you want to make sure you don't read on top of the recorded information.

Probably the most basic form of commercial you'll produce involves an announcer reading over a music bed or sound effects. The usual format for a commercial with music is that the music bed fades in or begins at full volume for a few seconds and then fades under and holds as the announcer begins reading the spot. The **voice-over** is read on top of the music bed, and then the music bed is brought up to full volume at the end of the voice-over for a few seconds until it fades out or ends cold (see Figure 9.1). Although this seems simple enough, it may not come easily until you've practiced. Not only must you be concerned with timing, you also have to determine how much music to use to establish the spot, balance the levels between voice and music bed, and correctly manipulate the broadcast equipment. Often you do all this at the same time, although it is possible to record the voice-over and mix it with music at a later time.

Reading over sound effects is similar to reading over music, in that the sound should fade up and down. Sound effects are usually short; they often need to be **looped** (recorded over several times) in order to cover the length of the commercial. This task is easy to accomplish with computer editing. The sound can simply be copied and laid on the **time line** as many times as needed.

Highly produced spots usually take a relatively long time to prepare. **Minidramas**, which often start with music and

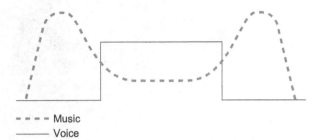

- - - - Music
———— Voice

FIGURE 9.1 A drawing that illustrates a commercial that consists of an announcer reading over a music bed that fades in and out.

involve several people bantering to the accompaniment of sound effects, involve a great deal of preproduction, rehearsing, mixing, and editing. They are among the most challenging, creative products that an audio production person handles. Often different parts of them are recorded at different times. The talent record their parts and then leave, especially if they're high-paid talent. Then the production person mixes the talent skit with music and sound effects.

Known as "spot production," the attributes of commercial radio production can be applied to the creation of commercials (that promote a good or service), station promotional announcements (promo), and public service announcements (PSA). After the concept for the announcement has been determined, a script is prepared and production of the spot takes place. At times, the client for an advertising spot will have to be convinced as to what is needed to get the message across. Like all production elements (see the different imaging materials described below), these announcements should fit the format of the station so that they do not detract from the "sound" of the station, or even worse, serve as tune-out factors. Having a good working knowledge of both production techniques and advertising attributes will aid in producing effective messages. (See Sauls, S. J. (2007). *Basic Audio Production: Sound Applications for Radio, Television, and Film*, 2nd ed., Thomson Custom Solutions, Publisher, pp. 9-7 - 9-8)

PRODUCTION TIP 9A
Music Punctuators

For radio spots that consist of a mixture of the announcer's voice and a background music bed, the voice should be dominant, because it conveys the important information of the spot. The music is in the background and helps convey the mood of the spot. You can make this basic radio production more interesting for the listener by raising the volume level of the background music slightly during natural pauses in the vocal track. This step will move the music from the background to the foreground momentarily. The listener will pay attention to this change, because our ears "follow" movement and will focus on the subtle shift. Make sure that you turn down the background music when

the voice starts again. You can punctuate key phrases or concepts in your spot with this technique, and your production work will sound more lively than just keeping the music constantly in the background. Remember that music, like all production elements, should support the intended message.

9.3 ENHANCING IMAGE

Imaging (sometimes called **branding**) defines the station as a product so that the listeners know what to expect when they tune in. It involves many elements, some of which are printed material such as posters or billboard ads. But imaging is also heavily incorporated between program elements such as music, commercials, or news stories. People who can produce short, catchy material to help establish the brand are valuable to the station.

There are many types of imaging materials. They have been given names, some of which are:

Bumper: A prerecorded audio element that consists of voice-over music that is used as a transition between different forms of content. It might, for example, consist of music and the disc jockey saying his or her name in an unusual manner.

Jingle: A produced programming element that includes the singing of call letters, a station slogan, disc jockey names, or other imaging elements.

Liner: A sentence or sentences that a disc jockey says over the intro to a song or during a break between songs. An example might be "It's always golden here at KAAA."

Promo: Short for promotional spot, this is an announcement, usually the length of a commercial, that promotes an upcoming station event such as a concert or a contest. A promo for a contest, for example, might include excited reactions of past contest winners, the contest rules, the prize, and how this contest relates to the station programming.

Slogan: A short pithy group of words that help listeners remember the main image a station wants to convey (see Figure 9.2). An example is "All news all the time."

Station ID: An identification of the station. The FCC requires that all stations identify themselves by call letters and city of license at least once an hour. Many stations also include how the station is known within their ID—for example, "WQSU, The Pulse, 88.9 FM" or "KNTU, 88.1 FM, The One." Also many stations do station IDs more often than once an hour, but they do not need to include all the required elements when they do so.

Stinger: A sound effect or musical effect that punctuates so that it will capture attention. A station might, for example, play four chimes accompanied by its call letters at the top of every hour.

Sweeper: A recorded element of voice, voice-over music, or a sound effect that bridges two songs together

RADIO STATION SLOGANS
B1 with the Music
Goodtime Oldies
More Stimulating Talk Radio
Radio Jalapeno
Timeless Classics
Today's New Country
The Voice of Pittsburgh
Where Ministry Blesses Many
The World's Greatest Radio Station
Your Peace and Justice Community Radio Station

FIGURE 9.2 Here are ten slogans. Which do you think are most creative? Most effective? Most descriptive?

FIGURE 9.3 The radio disc jockey manipulates all the broadcast equipment and adds the element of live announcing within the style of the radio station's format. *(Image courtesy of Michael Parks, iHeartMedia-Harrisburg, PA.)*

or creates a transition from commercials back to music or vice versa.

Teaser: A short segment, usually broadcast before a commercial break, the intent of which is to keep the listener turned to the station. And example would be "Coming up—how you can win two free tickets to the Cowboys game."

As you can tell, many of these concepts are similar or overlapping. In addition, the terms are not uniform. What one station calls a bumper, another might call a sweeper. The exact definitions are not nearly as important as the need for all these elements to convey something to the listener that will distinguish the station from all the other stations on the air and attract the **target audience** that the station wants to attract.

All these imaging elements are short. Although a station cares a great deal about how it is perceived, it can't devote too much of airtime to establishing itself or it will lose its audience. If you are producing something to enhance a station's image, you must make it compact, memorable, and unique.

Remember that very successful spot productions are sometimes very simple. Complex does not necessarily mean creative. Easy production and microphone techniques can make for great audio production.

9.4 ANNOUNCING MUSIC

Music constitutes the largest percentage of radio station programming and is usually introduced and coordinated by a **disc jockey (DJ)**. If you become a disc jockey, you'll probably be spending most of your time in the on-air studio doing your production work live (see Figure 9.3). On-air broadcasting is fast-paced, pressure-packed, and, for most people, a lot of fun.

Although the main element of the programming is music, the main duty of the disc jockey is talking. Much of this talk involves introducing music. For this, your announcing style must fit the format of the radio station. For example, fast-paced, high-energy, rapid-fire speech is

not appropriate for a classical music or big band format but may be required at a contemporary hit or rock radio station. Develop a variety of ways of getting into and out of the music. Many beginning announcers latch onto one introduction and use it over and over ("Here's a classic from the Beatles…"; "Here's a classic from Bob Dylan…"; "Here's a classic from…"). If you have trouble thinking of clever material, read the **liner notes** on the CD or research the artist on the Internet. This will often give you ideas for something to say that's unusual or informative.

Your station may have certain policies regarding what you say and how you say it. For example, some stations require you to talk over the beginning and ending of every record. Other stations may require you to say the station call letters every time you open the microphone.

As a disc jockey, you'll also be talking about things other than music. For example, you may need to give the time, temperature, commercials, weather, news, or traffic reports. Or you may introduce other people, such as the newscaster, who will give some of this information. Station policy will probably dictate whether you must be formal about these introductions or have the latitude to banter with the other person.

Regardless of what's going on within the live production situation, always assume that the microphone is open. Don't say anything that you wouldn't want to go out over the air. This includes personal conversations and, of course, indecent language. Many studios have an on-air light inside the studio as well as the one outside the door. This inside light is to alert the announcer to the fact that the mic is live, but the best rule is to assume that the microphone is always on.

Obviously, as a disc jockey, you need to be proficient at operating the equipment so that you can cue up and play the music. You need to think ahead, because you are normally operating equipment and talking. For example, if you have to bring in a feed from a network, you must know exactly when and how to do it. It's also good practice to have an alternative if something goes wrong. If the CD player you want to play doesn't start, have a backup plan prepared in advance. A good announcer can overcome most miscues so that the listening audience doesn't even know that anything went wrong.

Make sure you have previewed your music, especially to know how songs begin and end. This will ensure that you avoid **walking over** (beginning to "outro" a song before it's really over) a false ending. Previewing also helps you know how much instrumentation there is at the beginning of the record before the vocal starts so that you can talk over the instrumental but not the vocal (referred to as "stepping on"). There is no excuse for a disc jockey playing music on the air that he or she isn't completely familiar with.

Plan how you'll get from one piece of music to the next if you play them consecutively. You might want to review the sound transitions mentioned in Chapter 5, "The Audio Console." Remember to use a variety of ways rather than the same method time after time. Listen to the on-air monitor frequently, if not continuously. Most audio consoles allow the DJ to hear the program line, the audition line, or the on-air signal; however, only the on-air signal allows you to hear exactly what the listener hears.

Sometimes disc jockeys are not in an on-air studio announcing music live. Instead, they are **voice tracking**. Voice tracking is simply when an announcer prerecords the vocal portion of his or her air shift—such as song introductions, generic time checks, and so on—that will be mixed with the music, commercials, and other programming elements later. Often computer software that allows a station to automate its programming also facilitates voice tracking. In this way, rather than doing a 4-hour air shift in real time, the DJ can record in a much shorter period of time—often about one-quarter of the time. In addition, voice tracking allows announcers to record several shifts both for the local station and for stations in other markets. It is not uncommon for an air personality to voice track a number of different formats for stations in several markets in one session.

Voice tracking has been around for quite a while; the **countdown** shows heard on radio have long been voice tracked. But technology and radio consolidation of the last several decades has made it more prevalent and more controversial because the jocks who create this "virtual radio" have replaced live and local DJs. Using "cheat sheets" listing local names, places, and events, a voice-tracking announcer can sound like he or she is in Portland, Maine, when the announcer is really in Dallas, Texas. Many listeners may not even realize their favorite "local" announcer has never been to their market. One result of voice tracking is that DJ jobs are being lost because one announcer can cover several air shifts. On the other hand, top-quality DJs from larger markets can be heard on smaller-market stations, and stations can save money through voice tracking. There are pros and cons to this method of creating radio content, but it is a practice that anyone wishing to become a radio announcer should be familiar with and become proficient at.

No matter what the format is of the station, the goal of all on-air talent is to maintain the sound of the station. Of course, the number one rule is to follow the station format. If everyone on the air "did their own thing," you would have a different sound every time a new talent came on-the-air. A word of caution: the quickest way to lose your job as a DJ is to not follow format. You are talent, not a programmer.

Learn to work in different formats. Chances are that over time, you will work in a variety of formats. Learn the techniques of different formats when it comes to delivery and announcing style, as well as specifics contained in different formats. The more versatile you can be, the more employable you will become. For example, don't limit yourself to just one style of music. With today's industry consolidation in the radio business, you never know where you will be needed. Flexibility will go a long way in placing you in positions.

Today, music on most radio stations is on computer, and the computer programs now accomplish some of the tasks that DJs used to do on their own. This would include showing on the computer screen the talk-up time before the lyrics start in a song, the countdown of the song to its end, the fading from one song to another, the ability to easily preview a song, etc.

All of the techniques for on-air personalities described can be applied across the many distribution methods being employed today. So, it doesn't matter if you are on terrestrial radio, cable only radio, Internet radio, satellite delivered radio, or what is digital audio broadcasting (DAB/HD radio), learn techniques that will give you the professional edge. (See Sauls, S. J. (2007). *Basic Audio Production: Sound Applications for Radio, Television, and Film*, 2nd ed., Thomson Custom Solutions, Publisher, pp. 9-6 - 9-9)

9.5 RECORDING MUSIC

Another type of job open to people with audio skills is that of recording musical groups. Most of these jobs involve recording and mixing songs to be distributed on CDs or over the Internet. Sometimes groups are recorded (or aired live) in radio station studios where the disc jockey interviews the group and then has them perform.

Recording music is a fairly complicated form of audio production that often involves numerous microphones and long hours of recording. You can record the whole musical group at once or you can record individual instruments and mix them at a later time. (Obviously, if the performance is live, you must send out the whole group at once.)

When you record a whole group, you must place microphones very carefully so that you achieve the proper balance. If the group is fairly small, you can place a mic by the sound hole of each instrument and, of course, mic the vocalists (see Figure 9.4). But if you place a mic the same distance from a trumpet sound hole as from a violin sound hole, the trumpet will drown out the violin—and probably everything else. There are no set rules on how to position the mics; it takes experimentation and experience. In general, for recording music, you want **cardioid condenser** mics with a large

FIGURE 9.4 This is an effective setup for a person who plays guitar and sings. One mic is by the person's mouth and the other is by the guitar sound hole.

frequency response and **dynamic range**, and you want to position them on stands (see Chapter 4, "Microphones"). It helps if the output of each mic is sent to a separate input into an audio board. Then several audio technicians working at the console can make minor changes in the volume of various instruments as the sound is on its way to the recorder.

Better yet, in terms of quality and control, is to record each instrument separately. You might, for example, start by recording the guitar player performing the entire song. Then, the singer, while listening to a playback of the guitar, would perform. The singer would be recorded on a different track than the guitar and you would not need to adjust for the differing volume levels because you will be mixing them together later. But you do need to make sure you have good sound quality in terms of being **distortion**-free, **on-mic**, rich in frequencies, and so forth. By the same manner, you could lay down a drum track, a clarinet track, and so on. Later you could mix the sounds together starting and stopping and adjusting as needed to obtain the best recording possible. You can also add effects, such as **reverberation** and **equalization** (see Chapter 8, "Signal Processing") at a later time.

Recording each instrument at a different time, however, may not allow for the energy and flow that musicians in a band, for example, might exchange during a performance. This is especially true in the case of jazz music, where riffing off of and interacting between the various musicians is a hallmark of the form. However, the different musicians can be sonically isolated from each other, allowing for each instrument to be recorded separately, while the musicians can, by way of headsets, be fed the live mix of all the performers so as to be able to interact and react to the composition as it ebbs and flows among the musicians. Many studios are equipped so as to allow sonically isolated musicians to see each other through glass walls.

If you are recording a large group, such as a symphony orchestra, you can't mic each instrument. Instead, you

mic various sections of the orchestra (violins, drums, trombones), usually with mics that are hung from overhead. As with a smaller group, technicians at an audio console set and adjust volume levels, but it is also wise to record each mic on a separate track so that changes can be made in the recording after the fact.

There are many other possible steps related to recording music. Sometimes music recorded from real instruments is mixed with computer-created music. Sometimes several versions made from the same recording session are released. In general, the recording process is a very important aspect of the music business.

PRODUCTION TIP 9B
Miking a Guitar

If you are into guitar music, you might like to try this guitar-miking method used by professional sound engineer Bil VornDick. It is for someone who is just playing the guitar and not singing, but it uses two microphones. Bil uses different microphones, depending who the artist is, but one of his favorites is the Neumann KM84, a condenser cardioid mic that operates on **phantom power**. It has a switch that allows it to be used close to loud instruments, handling 130 dB without distortion.

Bil places one mic toward the right side where the neck joins the body of the guitar pointed between the hole, arch, and neck—the area where the higher transients are. The other mic looks down from where the musician's right shoulder is, because most guitar players play to the right ear. This mic is pointed toward the sound hole and covers the area between the wrist and shoulder. It emulates what the guitarist is hearing and has a deeper tonal **timbre** than the first mic. He has to be careful about **multiple-microphone interference**, or keeping the mics far enough apart that they do not cancel each other's sound.

9.6 PREPARING AND ANNOUNCING NEWS

If you're involved with radio production, you're likely to do news work at some point. In some instances, the disc jockey may just **rip and read** from the news service on the hour or half hour. Of course, most radio news is now read directly off a computer screen so "rip 'n' read" just refers to the practice of reading a newscast cold or with no rewriting or preparation. As we'll note in a moment, this is not good broadcast practice. At the other extreme are all-news stations, which have many people involved with the news, often including two on-air anchors at any one time (see Figure 9.5). Whatever situation you find yourself in, some knowledge about newscasting will prove useful.

Even if you're expected to simply read a short newscast each hour, you should give some time and thought to

FIGURE 9.5 This setup at an all-news stations is made for two anchors. KRLD Afternoon News Anchors Tasha Stevens and Chris Sommer both have their own audio controller and monitors. Note the adjacent studio. *(Photo courtesy of Dan Halyburton, CBS Radio Dallas, Texas State Networks.)*

your presentation. First, you must decide which news to present. When assembling the radio newscast, you must choose the stories and story order in terms of importance, immediacy, and geography. The first story becomes the "lead" story. You probably have a station news format to follow, but you should also try to select those items that are most likely to be of interest to your listeners. Thus, local items are quite often paramount.

You may find that you need to do some rewriting of the wire service news and write some transitions to take the listener from one story to the next, or to localize the story. Timing is important on a newsbreak. Beginning newscasters sometimes run out of news to read before the newscast time is up. To prevent this, you should pad your newscast with some extra stories that you can cut if you have to, but that also provide a cushion if you need extra material. You'll usually need to get commercials in at an appropriate time during a newscast, so make sure that you know when this happens and be prepared for it. Don't read a newscast cold. Read it over first so that you're familiar with the material. Rewrite anything that isn't natural for you, such as tongue-twisting phrases that you might trip over or long sentences that make you run out of breath. Avoid too many numbers or facts jammed into a single sentence. Whenever something is unclear, rewrite to make it simple and easy to understand. Remember, broadcast news should be conversational and written for the ear, not the eye.

When a news story includes the actual voice of the person in the news, such as the mayor commenting on the new city budget, that segment is called an **actuality**. Most radio news operations strive to include many actualities

within a newscast, because these bring life to the news. It's more interesting to hear the mayor's comments than the voice of an announcer telling what the mayor said. Many of these actualities are gathered by field reporters (see Chapter 10, "Location Sound Recording"); however, you can also make use of the phone to gather them. Many small radio stations have one-person news departments, and in these cases the phone actuality is especially crucial. At some stations, a **telephone interface** is used to semipermanently hook up phones to an audio recorder or audio console. But you can also use a microphone that's specifically made to attach to the phone. You must obtain permission to record someone for broadcast use, and your station newsroom probably has specific guidelines to follow for doing this. A recorded segment with only the reporter speaking is called a **voicer**. A "wrap-around" type of segment that includes both the reporter and the voice of the person in the news is known as a **voicer-actuality (V/O)**.

Actualities generally need to be edited. For this, you should use all the editing techniques presented in Chapter 3, "Digital Audio Production," and include one very important rule: Make sure that when you edit, you don't change the meaning of what someone has said. Ethical news procedures dictate that a great deal of care be taken in this area because the elimination of a single word can significantly alter a news report. When you're editing and need to eliminate part of what a person said, either because it's too long or it's irrelevant, try to match voice expressions where the statement leaves off and where a new one begins. For example, don't make an edit that jumps from a fast-paced explanation a speaker was making to a measured response he or she was formulating. Edits are usually best if made at the end of thoughts, because a person's voice drops into a concluding mode at that point. Be careful not to edit out all the breaths the person takes, because doing this will destroy the natural rhythm. Try to maintain a constant background level throughout the actuality. To do this, you may have to mix in background noise from one part of the recording to another.

9.7 REPORTING SPORTS, TRAFFIC, AND WEATHER

At some radio stations, the same person who reads the news also reports on other related elements such as sports, traffic, and weather. At other stations, especially all-news stations, there are separate reporters to cover these topics.

Sports announcers need to know sports in general, as well as the particulars of the sports for the area in which they are reporting. Unlike other reporters, they are allowed to show a bias for the home team and are expected to have an upbeat, hyper vocal presentation. At some stations the same people who report sports on a regular basis from the studio also cover **play-by-play** at various sports venues. The equipment setup at a sports box at a stadium is similar to that in a studio, in that there are microphones and a small audio board, but there are also differences, which will be covered in Chapter 10.

In cities with heavy congestion, traffic reports have increased in importance, with some major market stations reporting traffic 24 hours a day. But traffic reporting can be very expensive if it involves a helicopter and reporters driving the streets to report on conditions (see Chapter 10). In some cities, specialized companies handle traffic reports for a number of stations. Reporters change their tone and banter as they switch from reporting for a rock station to an all-news station to an easy listening station. In addition to obtaining details from reporters in the field, employees of traffic companies or radio stations can obtain traffic information by listening to police radios, by encouraging listeners to call in with traffic jam information, and by looking at Internet site maps that are tied to sensors in the highways that indicate how fast cars are progressing on the road. Someone involved with traffic reporting needs a knowledge of the area in order to know which traffic jams are most likely to impact listeners.

Weather is an important subject in most communities; particularly so in the morning, as people prepare to leave for the day and in the evening as they are heading home. Some stations require their forecasters to have meteorology degrees so that they can explain weather conditions and the weather forecast to the listeners. Other stations have someone who calls the weather service or checks one of the government or private websites that constantly update local and national weather (see Figure 9.6). At the simplest level, the person giving the weather report can simply look out the window.

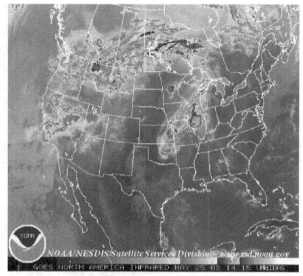

FIGURE 9.6 This is one of many maps that the National Oceanic and Atmospheric Administration (NOAA) provides on the Internet.

Some stations also have a person who deals with business news, regularly reporting on the stock market and related items. There are specialists who report on other topics such as art shows, fashion, education, health, or movies, and there are people who give commentaries on current events. Often these people work on a contract basis for syndication companies who supply audio material to a host of stations nationwide.

9.8 HOSTING TALK SHOWS

There are many types of talk shows—**shock jock** morning shows, public affairs programs that deal with local or national issues, sports talk programs, psychology and self-help programs, religious talk, right-wing and left-wing political talkers, and on and on. Some talk shows consist of one person talking, others involve a host and a knowledgeable guest, and many include call-ins from listeners (see Figure 9.7).

Not everyone is cut out to handle hosting a call-in talk show. You must be fast on your feet and able to ad lib in an entertaining and effective manner. For many programs, you're expected to have more than broadcast production knowledge. For example, a sports program or radio psychology show requires a host with some expertise in those areas. Sometimes the call-in talk show host has an engineer handling the equipment and a producer screening the calls so that the host can concentrate on dealing with the callers, but this is not always the case, especially in smaller-market radio. A call-in show host has to be able to handle people tactfully (or, in some cases, abrasively, if that's the style of the program).

The host must remain in charge of the program. You should keep the program moving, trying not to give too much time to any one caller. If you have a guest to whom

FIGURE 9.7 This public affairs talk show involves a discussion between the program host and a studio guest. *(Image courtesy of Michael Parks, iHeartMedia-Harrisburg, PA.)*

people are posing questions, you should give information about the guest and redirect questions if they're not understandable.

Research is crucial to talk shows. Even if the show is one host spouting his or her opinions, the audience will sense when the host doesn't know what he or she is talking about. A good host will research the subject and the guest before doing a program that involves a guest. Not only will this provide background, but it should also enable the host to generate a list of questions to ask.

Asking the right questions really means asking good questions. For example, ask questions that require more than a simple yes or no answer. Rather than asking, "Do you agree with the mayor's new policy regarding the police?" ask, "What do you think of the mayor's new policy regarding the police?" Ask short, simple, and direct questions. The question, "Given the salaries of employees and the possible raises they will receive, what do you think the effects will be on the social security system and the GNP?" will most likely get a response of "Huh?" Break complex questions down into a number of questions such as, "How do you think increasing salaries 5 percent will affect the GNP?" Ask questions that don't require very long answers. Don't ask, "What would you do to improve the city?" Instead ask, "What is the first thing you would do to improve the city?" Don't bias your questions, such as "You do believe that the mayor's new policy is correct, don't you?"

Asking good questions also means knowing how to handle the answers, whether they come from a guest or a listener. For example, if the answer is too wordy, ask the person to summarize the response. If the answer is muddy or unclear, ask the question over again in smaller parts. If the answer is evasive, come back to it later, or ask it again from a different angle. If the response gets off track, redirect. And if the response goes on and on, interrupt politely and redirect.

Listen carefully to what the guest or call-in person says so that you can ask appropriate follow-up questions. Sometimes interviewers become so engrossed in thinking about the next question that they miss an important point that could lead to something significant. Although you should organize your program ahead of time and jot down some questions or points you want to cover, you aren't required to stick to your questions. In all probability, when you ask question one, the guest will also answer questions three and seven, for example, so you must constantly flow with the conversation. The more you can lead off what the guest or a listener says, the more natural the whole show will appear. But make sure you do get the information you want. Most talk shows are aired live, but if the material is recorded, it may be necessary to edit. As with news, be certain that you do not change the meaning of what someone said.

When you have a number of different guests, identify them frequently, because the listener may have difficulty

FIGURE 9.8 *Prairie Home Companion* starring Garrison Keillor (right) involves many production elements, including sound effects performed live by Tom Keith (left). (Copyright Prairie Home Productions. Used with permission. All rights reserved.)

keeping track of the various voices. Even a single guest should be reintroduced several times during a 30-minute program, particularly after returning from commercial or sponsorship breaks. Not only does this remind listeners who your guest is, but it introduces him or her to those listeners who joined the program while it was in progress.

9.9 PERFORMING DRAMA AND VARIETY

Drama isn't produced very often on radio anymore. Sometimes there are comedic skits within variety shows that also include music, jokes, quizzes, and other elements. But variety shows, too, are rather rare, occurring primarily on public radio (see Figure 9.8). When drama or variety shows are performed in a studio, they require some unique production skills. Usually actors perform several parts, so an actor needs to be able to create a number of distinct voices. Otherwise audience members will not be able to distinguish one character from another.

Sometimes sound effects are performed live or at the same time that the drama or variety show is being recorded. The people performing them are skilled in using their bodies, voices, and various implements to create a wide variety of sounds. Along with music, these sound

effects establish locale, tell the time, create mood, indicate entries and exits, establish transitions, and add humor. Sound effects and music can also be prerecorded or taken from a CD and edited into the production at a later time. Dramas and variety shows are fun to produce, but they do not have the broad audience appeal they had during the 1930s and 1940s before television was available.

9.10 CONCLUSION

This chapter has been devoted mainly to audio productions that occur within the studio environment. But there are many times when sound is gathered at a field location, broadcast from a helicopter, played at a club, used in conjunction with video, or connected to a variety of media forms such as the Internet and podcasting. The following chapters will explore these varieties of sound recording. When you finish the book you should have a well-rounded view of a large number of audio recording possibilities and an appreciation for the fact that audio production can be an exciting career. It is certainly possible to have a great deal of variety in your day-to-day occupation and you will have numerous opportunities to be creative and innovative.

Self-Study

1. How would a radio DJ "walk over" a false ending of a song?

   a) by leaving the microphone open when it should be off
   b) by beginning the outro of a song before it is really over
   c) by recueing a CD
   d) by researching an artist on the Internet

2. What is the reason sound effects sometimes need to be looped?

   a) They are too short.
   b) They need to be recorded after the talent is recorded.
   c) Then need to punctuate a minidrama.
   d) They need to be created live for a variety show.

3. Which of the following might be used before a commercial break?

   a) a sweeper
   b) a teaser
   c) a bumper
   d) any of the above

4. Which of the following describes good radio talk show technique?

   a) Give each caller all the time he or she wants.
   b) Ask all your questions in the order you have them written down.
   c) Ask complex questions so that you cover the topic thoroughly.
   d) Ask follow-up questions based on the interviewee's response.

5. A fast-paced announcing style would be most appropriate for a disc jockey working at which type of radio station?

   a) classical music format
   b) big band format
   c) all-news format
   d) contemporary hit radio format

6. Which of the following best describes what should occur for a commercial that consists of an announcer reading over a music bed?

   a) The announcer will read with a condescending tone.
   b) The commercial will always be cut off because it is not the prescribed length.
   c) The music bed will begin at full volume then fade under as the announcer begins talking.
   d) The balance between the voice and music bed will favor the music bed.

7. Approximately how long should it take an announcer to record a 4-hour air shift by voice tracking?

   a) one-quarter hour
   b) 1 hour
   c) 2 hours
   d) 4 hours

8. Which of the following are you least likely to do when you record each instrument of a band separately rather than recording them all together?

   a) adjust the volume of each instrument so it is at a different level
   b) record the guitar and then record the drums
   c) make sure your recording is not distorted
   d) mix the sounds together when you are finished recording and then add reverb

9. What does the term "actuality" refer to?

   a) the voice of the news announcer
   b) a pad for a newscast
   c) wire service copy
   d) the voice of a person in the news

10. Which of the following is true about the content of radio news copy?

    a) It should contain as many facts and numbers as possible.
    b) It should be written in simple, easy-to-understand sentences.
    c) It should be sonically isolated.
    d) It should be written using long, explanatory sentences.

11. Which of the following are you more likely to do if you are recording an orchestra than if you are recording a string quartet?

    a) place mics by the sound hole of each instrument
    b) place mics by sections rather than individual instruments
    c) place mics carefully to achieve proper balance
    d) record each mic on a separate track

12. Which of the following would not be a good reason for editing a news actuality?

    a) to cut out a cough
    b) to cut out material that is irrelevant
    c) to change the meaning of the story
    d) to shorten the length of the story

13. Which type of announcer has the most latitude to be hyper and biased?

    a) a sportscaster
    b) a traffic reporter
    c) a stock market reporter
    d) a news anchor

14. Which of the following is the best definition of a "jingle"?

    a) a sentence that a disc jockey says over the intro to a song
    b) a short imaging element that includes singing
    c) a promotional spot that explains a contest
    d) a sound effect that punctuates

15. Which of the following would be the best radio interview question?

    a) Do you favor capital punishment?
    b) What do you think will happen regarding an amendment against capital punishment being added to the Constitution after it has been discussed by the state legislature in light of the case pending in Florida at the present time and the one recently decided in Illinois?
    c) What do you think will be the outcome of the present attempt to outlaw capital punishment?
    d) You favor capital punishment, don't you?

ANSWERS

If you answered A to any of the questions:

1a. No. This is something that shouldn't be done. (Reread 9.4.)
2a. Yes, this is the correct answer. They need to be recorded over and over to cover the length of something, such as a commercial.
3a. This answer is correct, but it's not the best answer. (Reread 9.3.)
4a. No. This could get very boring. (Reread 9.8.)
5a. Definitely not. (Reread 9.4.)
6a. No. This should never happen. (Reread 9.2.)
7a. No. This is too short. (Reread 9.4.)
8a. Correct. You would adjust the volumes later.
9a. No. You have the wrong voice. (Reread 9.6.)
10a. Wrong. Audience members won't be able to comprehend all of them as they go by quickly. (Reread 9.6.)
11a. No. That would take way too many microphones. (Reread 9.5.)
12a. Wrong. This would be a legitimate reason to edit. (Reread 9.6.)
13a. Correct.
14a. Wrong. This is more likely to be the definition for a liner. (Reread 9.3.)
15a. Wrong. This question could be answered "yes" or "no." (Reread 9.8.)

If you answered B to any of the questions:

1b. Correct. That's the definition of "walk over."
2b. No. It wouldn't make any difference. (Reread 9.2.)
3b. This answer is correct, but it's not the best answer. (Reread 9.3.)
4b. No. You don't want to stick slavishly to your questions. (Reread 9.8.)
5b. No, that wouldn't be appropriate. (Reread 9.4.)
6b. Wrong. You can certainly make it the prescribed length. (Reread 9.2.)
7b. Yes. Usually an announcer can produce an air shift in about one-quarter of the actual time by voice tracking.
8b. Wrong. You would be likely to record one instrument and then another when recording separately. (Reread 9.5.)
9b. Wrong. (Reread 9.6.)
10b. Definitely correct.
11b. Yes. When there are that many instruments you place the mics by sections.
12b. No. That is one of the main reasons for editing. (Reread 9.6.)
13b. Wrong. Traffic reporters aren't biased. (Reread 9.7.)
14b. This is correct.
15b. Wrong. This question is much too convoluted. (Reread 9.8.)

If you answered C to any of the questions:

1c. No. This has nothing to do with cueing. (Reread 9.4.)
2c. No. This would have nothing to do with looping. (Reread 9.2.)
3c. This answer is correct, but there is a better answer. (Reread 9.3.)
4c. Wrong. Complex questions are not called for. (Reread 9.8.)
5c. Wrong. There aren't even disc jockeys for this format. (Reread 9.4.)
6c. This is the correct answer.
7c. No. This is too long. (Reread 9.4.)
8c. Wrong. You want to check distortion regardless how you are recording. (Reread 9.5.)
9c. No. It has nothing to do with wire service copy. (Reread 9.6.)
10c. No. Sonically isolated refers to recording music. (Reread 9.5 and 9.6.)

11c. No. You always want to place mics carefully. (Reread 9.5.)

12c. This is correct. You definitely don't want to change the meaning of the story.

13c. No. A stock market reporter should be serious. (Reread 9.7.)

14c. No. This is more likely to refer to a promo. (Reread 9.3.)

15c. Yes. This is the best-stated question.

If you answered D to any of the questions:

1d. Wrong. This has nothing to do with "walking over." (Reread 9.4.)

2d. Wrong. If they are live, they definitely don't need to be looped. (Reread 9.2.)

3d. Yes. Many of these imaging terms have interchangeable uses.

4d. Correct.

5d. Right. This format is fast-paced.

6d. No. If anything, it's the other way around. (Reread 9.2.)

7d. Wrong. This is as long as the shift. (Reread 9.4.)

8d. No. This would be something you do when you record separately. (Reread 9.5.)

9d. This is the correct answer.

10d. No. The audience needs shorter sentences. (Reread 9.6.)

11d. Wrong. You'd be more likely to do this with a string quartet. (Reread 9.5.)

12d. No. Shortening is a legitimate reason to edit. (Reread 9.6.)

13d. Wrong. News anchors should not be hyper or biased. (Reread 9.7.)

14d. No. This is the definition of a stinger. (Reread 9.3.)

15d. Wrong. This is a biased question. (Reread 9.8.)

Projects

PROJECT 1

Record an air-check tape.

Purpose

To instruct those interested in doing on-air broadcasting in how to make an audition or demo recording, something required when applying for a job.

Notes

1. To apply for on-air jobs in broadcasting, you may send your résumé and an air-check recording to many stations. An air-check is a recording of less than 5 minutes that shows how you handle on-air broadcast situations.
2. Ideally, an air-check is an edited-down sample of your actual on-air work, but if you aren't on the air on a regular basis, a simulated air-check can be put together in the production studio.
3. Try to make the recording as general as possible so that it could be sent to several different types of stations.
4. Put those things you do best at the beginning. Many potential employers don't have time to listen past the first 30 seconds and will rule you out if they don't like the beginning. Don't structure the recording so that it builds to a climax, because probably no one will listen that far.
5. Feel free to use things that you've done for other projects for this assignment.
6. Keep the pace moving. Don't do any one thing for too long.
7. The website contains a sample air-check. You may want to listen to it before starting your own.

How to Do the Project

1. Plan what you intend to include in your recording. An air-check format might include ad lib introductions to a few songs (either fade out the music after a few seconds, or edit to the end of the songs so that the listener doesn't have to hear the whole song), some production work (commercials, station promos, and so on), and a short newscast. If you can do play-by-play sports, you might want to include that. There is no standard format, so do whatever showcases your talent best.
2. Plan the order of your recording. Make it sound like a continuous radio show as much as possible.
3. Record the project and listen to it. Redo it if it doesn't present good broadcasting skills.
4. On the recording, write your name and "Air-Check Project." Turn in the completed recording to your instructor to receive credit for this project.

PROJECT 2

Record a 5-minute radio interview show in which you are the interviewer.

Purpose

To prepare you for this common type of broadcasting situation.

Notes

1. Your interview must be exactly 5 minutes. Meeting the exact time without having an awkward ending will probably be the hardest part of the project, but it's a lesson worth learning because broadcasting is built around time sequences.

2. Don't underprepare. Don't fall into the trap of feeling that you can wing this. In 5 minutes you must come up with the essence of something interesting, and you can't do this unless you are organized. You'll also only be able to record the interview once. You can't redo this project, so you need to get it right the first time.
3. Don't overprepare. Don't write out the interview word for word. It will sound stilted and canned if you do.
4. Five minutes is actually a long time; you'll be amazed at how much you can cover in this time.
5. As the interviewer, don't talk too much. Remember, the purpose is to convey the ideas of your guest to the audience, not your own ideas.

How to Do the Project

1. Select someone to interview. If you're taking a course, it will probably be easier to do this project with someone in class.
2. Decide what the interview will be about. You may select any subject you wish. You could talk about some facet of a person's life or his or her views on a current subject, or you could pretend the interviewee is a famous person.
3. Work up a list of questions. Generate more than you think you'll actually need, just in case you run short.
4. Think of a structured beginning and ending for the show, because those will probably be the most awkward parts.
5. Discuss the interview organization with your guest so that you are in accord as to what is to be discussed, but don't go over the actual questions you're going to ask.
6. Record the interview, making sure you stop at 5 minutes. Listen to it, and check that it has recorded before your guest leaves. You are finished with the project once the interview is recorded. Even if it didn't come out as you had hoped, do not redo this project.
7. On the recording write "Interview Project" and your name. Give the interview to your instructor to receive credit for this project.

10

LOCATION SOUND RECORDING

10.1 INTRODUCTION

Sometimes you need to undertake audio production beyond the confines and comforts of a studio. You lose some control when you enter the "real world," because you no longer have a room designed to maximize proper acoustics and you don't have sophisticated equipment to aid your production. But what you lack in control, you often make up in adventure once you are out in the field.

10.2 TYPES OF FIELD PRODUCTION

One very common type of field production involves gathering news. Radio networks, news agencies, all-news radio stations, and many other radio stations have reporters in the field interviewing newsmakers and members of the public, and reporting on what is happening locally, nationally, and internationally. In a similar manner, producers of public affairs programs often venture into the field to obtain **actualities** and other audio for their programs (see Figure 10.1).

Traffic reporters travel surface streets, sit in fixed-wing aircraft, or ride in helicopters to report traffic conditions. They have become increasingly important as cities become more congested and are often the reason people turn to a particular station. They must respond quickly to changing conditions.

Play-by-play sportscasters, by definition, need to be at the location where the sports event is taking place. They must keep up the chatter to avoid **dead air**, and they must describe and convey visual information that the audience can't see. They must be aware of crowd noises and sounds from the game itself that will enhance (or detract from) the broadcast. Sports announcers often work as a team, with one announcer providing the play-by-play while another offers color commentary, along with game statistics.

Concerts and other live events are often recorded in the field at the same time they are being performed in front of an audience. Miking such events so that everyone in the audience can hear the material in a properly balanced fashion can be a challenge (see Figure 10.2). On a smaller scale, audio technicians are often asked to set up sound equipment for company activities, such as sales or promotional events, which involve people speaking and also using various pieces of audio/visual equipment.

There are many other types of location situations. Sometimes you may need to record background noises, such as monkeys chattering for a feature about a zoo or incoherent sounds of people talking to use as background for a commercial. Disc jockeys who play at clubs (see Figure 10.3) need to project their voices into the crowd. They also make use of "play" audio equipment, such as turntables, in novel ways more akin to a musician playing an instrument than a sound technician capturing, editing, or outputting audio.

FIGURE 10.1 An example of recording an interview in the field. *(Image courtesy of iStockphoto, Bonnie Jacobs, Image #5803709.)*

175

FIGURE 10.2 This live location event featuring a mariachi band needed a fair amount of audio equipment so that it could be heard by the people in attendance and also recorded.

FIGURE 10.3 This club disc jockey uses a turntable as part of her performance. *(Image courtesy of iStockphoto, Lise Gagne, Image #5512525.)*

10.3 COMMON LOCATION SOUND PROBLEMS

The main sound problems you are likely to encounter when you are on location are those that come from everything else that is happening at the location. You are likely to hear people talking in the background, air conditioners operating, lawn mowers cutting grass, airplanes flying overhead, sports fans cheering a home run, or jackhammers digging up streets. You can use these sounds to your advantage if they are elements that fit with your production and if they are not so loud that they drown out your voice, but it is more likely that they are going to be a nuisance.

Microphones do not pick up sound in the same manner as people's ears and brains. When you are talking to a friend you are often not aware of airplanes flying overhead or birds chirping in the trees; you are able to concentrate on your friend's words—a phenomenon known as **selective attention principle**. Microphones and recording equipment do not abide by this principle. If your recording includes an airplane sound that you didn't notice, the sound will come through loud and clear on playback.

Other sound problems on location are related to the recording process. Talent and crew members who are out in the field are often not as disciplined about keeping quiet as when they are in the silent environment of a studio. If you are covering a news story, there will probably be other reporters there who aren't going to stand around quietly while you record your material. Recording equipment that is close to the microphone can itself be a source of noise. Sensitive microphones pick up wind noises, sometimes even ones that are too gentle to be heard with the ear. Your equipment can also be in conflict with other electronic equipment resulting in a **hum**.

PRODUCTION TIP 10A
How to Get Rid of a Hum

Hum is a low rumble, usually at a frequency of 60 Hz that can be picked up by recording equipment even though you do not hear it. It is usually caused by electric equipment such as fluorescent lights, vacuum cleaners, or power saws. The easiest way to get rid of it is to find the offending equipment and turn it off. But that may not be possible—a factory owner is not going to shut down the factory because you are getting hum. Wearing headphones is paramount when in the field to check for any hum, buzz, or other errant noise being picked up by the microphone(s).

Another option is to make sure your equipment cables are not running parallel to the power cables of the offending equipment. Cables that run parallel are more likely to pick up hum than those that run perpendicular. If that doesn't work, try moving to a different location. Also, some smartphones used as recording devices can inject a hum into your recording.

If you absolutely have to live with the hum then try not to record anything else at the same frequency—anything that sounds like it at the same pitch as the hum. When you go back to the studio, you may be able to use signal processing to filter out the 60-Hz frequency where the hum is and leave the rest of the recording undisturbed.

Weather is another problem that causes location sound problems. Thunder and lightning can cause static or just plain noise. Some equipment won't work in very cold or very hot conditions. Some locks up if there is too much humidity. If there is rain, you must have some means for keeping the equipment covered.

Other problems relate to the way that people treat you. If someone makes the effort to come to a studio, they are, in all probability, going to talk to you. But if you are on location, especially if you are reporting news, people may refuse to talk to you because they don't want to be part of the story—they may not want to be associated with something they feel could reflect on them negatively. On the other hand, people may talk too much, trying to break in to your coverage or production with opinions of their own. Disc jockeys who work at clubs often have problems with people drowning them out.

And then, of course, you may be at least partially responsible for loud noise that annoys the neighbors who call the police. This is particularly likely if you are setting up sound for a rock band.

10.4 SITE PLANNING FOR LOCATION RECORDING

Some of the problems common to location sound recording can be solved by surveying the location ahead of time. (Figure 10.4 lists elements you might want to consider when you scout a location.) If you know that the air conditioning is loud and intermittent, you may be able to secure permission to turn it off for the period of time you will be recording. If the site is near an airport, try to determine how often the planes take off, so you can plan to stop recording occasionally. If there are fluorescent lights (which can cause hum), ask whether you can turn them off. If you are miking a band or performing in a club,

---

LOCATION SURVEY CHECKLIST

Type of material being recorded:
Potential location of shooting:
Principal contact person:
Address:
Phone number:
Email address:

Sound:
    Are there background noises, such as air conditioning, phones ringing, airplanes, lawn mowers, that may interfere with audio? If so, how can they be corrected?
    Where can the microphones and cable be placed?
    Are any particular microphone holders or stands needed?
    What types of microphones should be used?
    Will microphone cable be needed? If so, how much?
    Will it be necessary to plug equipment into wall outlets? If so, where are they located?
    How many, if any, extension cords will be needed?

General:
    Where is parking available?
    If passes are needed to enter the premises, how can they be obtained?
    Where are the nearest places to eat?
    Where are the nearest washrooms?

FIGURE 10.4   A location survey form.

check out the acoustics ahead of time and think through microphone needs. Determine how much crowd noise wafts up to the sports announcers' booth; you may want to enhance it by placing an omnidirectional mic over a section of the crowd or you may want to mitigate it by using a more directional announcer's mic or a shotgun mic.

Of course, you can't always survey a location ahead of time, especially if you are covering news. But even then, you can survey places where news occurs frequently—city hall, a police station, a community center—and determine the best places to position microphones. Occasionally producers hire guards to keep people away from sites where material is being recorded. That keeps down the people noises, but the guards can't keep birds from singing in the trees.

Sometimes you need a permit from a city government in order to record, especially if music is involved. Even if you don't need a permit, it is a good idea to have owners of

facilities (such as restaurants) sign a permission form (see Figure 10.5) just in case someone tries to stop your audio recording.

Parking is another consideration; make sure there is enough parking for all the people involved. Reporters have special permits that enable them to park close to the scene of stories, but these have exceptions. For example, they are not allowed to park in spaces reserved for the handicapped. Carry along a good map or GPS in case you get lost; have a radio so you can listen to weather reports (and your radio station, if you work for one).

Even if you haven't been able to check a site ahead of time, make sure you check your equipment carefully before you leave the studio area. Students are well advised to actually record and play back some material before leaving for a location so that they know the equipment is working properly.

LOCATION PERMISSION FORM
Note: This is not intended as an authentic legal document. It is just a representation of a location release. Check with your organization's lawyer before using.

The undersigned Owner or authorized representative of the Premises located at:

_____

_____

_____

hereby grants to the Production Company:

_____

its licensees, successors, assigns, and designees, an irrevocable license to enter upon and record at the Premises in connection with the production and distribution of the audio production tentatively entitled:

_____

This permission includes, but is not limited to, the right to bring cast, crew, and equipment onto the Premises. All things brought onto the Premises will be removed at the end of the production period. The Production Company shall leave the Premises in as good order and condition as when entered, reasonable wear and tear excepted. The Production Company shall indemnify the Owner for any loss and liability incurred as a direct result of any property damage or personal injury occurring on the Premises caused in connection with the Production Company's use of the Premises.

The Production Company shall own all rights in and to recordings made on or about the Premises forever and for all purposes. The Production Company shall have no obligations to pay any fee or other consideration for the exercise of the rights granted.

This Agreement may not be modified except by another writing signed by the parties.

Signature_____

Printed Name_____

Phone Number_____

Email Address _____

Date_____

FIGURE 10.5   A sample permission form.

10.5 USING MICROPHONES

Out in the field you will probably want to use a **dynamic mic**, because it is more rugged than a **condenser mic** and will not require a battery. **Cardioid** is better than **omnidirectional** for interviewing, and if you are recording sounds in the distance—an animal in a tree, a fire in the forest—you may want to use a **super-cardioid**. If you are going to be recording crowd noises, you will want an omnidirectional mic. A hand mic is best for news and interviews (see Figure 10.6); you position it near your mouth when you are talking and near the interviewee when he or she is talking. Except in unusual circumstances, don't let people you are interviewing hold the mic; they probably won't hold it correctly and you will lose control of the interview. If you need to record moving talent, you should use a **wireless lavaliere mic** (see Chapter 4). If at all possible, carry at least one more mic than you will need, just in case.

When you are outside, it is a good idea to have a **windscreen** on your mic (see Figure 10.7A). Mics pick up even gentle breezes as a hissing sound and windscreens cut down on this noise. Hand-held microphone recorders (as shown in Figure 10.7B) allow for high-quality, close-proximity recording.

Miking musical groups is particularly interesting and can be complex. If the performance is just for people in the vicinity, sometimes you need to mic only the vocalist because the instruments will carry well enough (see Figure 10.8). Other

FIGURE 10.6 Reporter Pete Demetriou of Los Angeles station KFWB using a Sennheiser MD40 cardioid hand mic to interview a woman leading a group of picketers.

FIGURE 10.7 A microphone with a built-in windscreen (A). *(Image courtesy of iStockphoto, Kryzysztof Kwiatkowski, Image #3428747.)* A Zoom microphone recorder (B). *(Image courtesy of Michael Parks, iHeartMedia-Harrisburg, PA.)*

FIGURE 10.8 In this situation, where the band is performing live, only the vocalist has a mic. *(Image courtesy of iStockphoto, A-Digit, Image #4693200.)*

times, especially if you are recording a band, you should mic the vocalist and the sound holes of the instruments. Refer to Figure 10.2 and notice how the mics are pointed at the string instrument sound holes, how the singer is miked, and how the trumpets—which are louder than other instruments—are not miked. If you have a large band or orchestra, you can mic individual sections (the violins, the trombones, etc.) and send each mic to a separate audio **channel**. Technicians at an audio board then mix the sounds so that they are at the desired volumes.

10.6 USING RECORDERS

The main recorders used for location work are the small portable ones that record on **hard drive** or **memory cards** (see Chapter 6, "Digital Audio Players/Recorders"). The most valuable recorders are those that can be used to undertake at least minimal editing. News reporters, in particular, need to edit interviews and other sounds so that they can send their stories back to their stations from the field. It is also helpful if the recording medium is compatible with the studio editing system—in other words, it works best when the memory card or hard drive can be inserted into the studio equipment so that the material on it can be edited into a production without a great deal of downloading and complicated transfer of data. You might, for example, record cars buzzing by on a freeway, then go back to the studio and give the memory card to a person editing a feature on the history of transportation. The easier it is for that person to incorporate the car sounds into the feature, the better.

Batteries are the main potential problem related to portable recorders. You should make sure that a recorder's battery is fully charged before you take it out, and you should bring along extra batteries. Most vehicles that are used regularly for audio recording (such as news vans) have built in charging capabilities and reporters should charge batteries that have run down. You should replace the battery when it registers 25 percent full; if you don't, it is bound to go dead just as you are recording your most important interview. At the end of your recording session, make sure you charge the battery for the next user. About once a week, take the battery all the way to empty and then recharge it in order to maximize its performance. At some colleges, technicians handle recharging batteries, but in other situations students check out battery chargers and are expected to return the batteries fully charged.

10.7 USING MIXERS

Most audio **mixers** used in the field are simple ones that allow you to adjust the volume for several inputs (see Figure 10.9). News vans often have small audio boards

FIGURE 10.9 A portable mixer being used in a location situation.

built into them so the reporter can mix together recorded sounds and his/her voice to build a story.

More complicated mixers are used for club performances and mixing band or orchestra music. These have more inputs and possibilities for **signal processing** (see Chapter 8). If the band is performing live, but someone also wants to record it, that person might connect a recorder to **house sound**—the sound generated for the people at the venue where the band is performing (see Figure 10.10). If you're taking a house console feed to record, make sure the audio engineer sends all the necessary audio to cover the entire ensemble.

If you record house sound, you must pay careful attention to the quality and volume of the sound entering your recorder. Concerts are often loud and the mix going to the crowd gathered at the venue will result in distortion when recorded. Don't trust your **VU meters**; they may not be set to take into account the loudness of the music. See whether someone at the venue can send you tone from a **tone generator** so that you can calibrate your sound. Listen to the output from your recorder on **headphones** and lower your volume input controls if it is distorted.

10.8 USING HEADPHONES

When you are on location it is important to have high-quality headphones, preferably **circumaural** (see Chapter 7, "Monitor Speakers and Studio Accessories"). You want something that shuts out external noises so that you can hear clearly what you are recording and be aware of unwanted sound. Headphones can be used when you are recording, while you are listening to material to edit it, and after you have it in the finished form (see

FIGURE 10.10 This reporter is plugging a portable Marantz recorder into a mixer. This mixer is used to amplify and combine the sounds from microphones picking up the announcer, singers, and musical instruments.

FIGURE 10.11 Monitoring the live audio mix with headphones during a remote broadcast. (*Image courtesy of Mark Lambert, KNTU program/operations/news director.*)

FIGURE 10.12 These sports announcers (A) who work as a team, with one announcer providing the play-by-play, while another offers color commentary and game statistics, wear headphones with microphones attached so they can work hands-free. (*Image courtesy of Alan R. Stephenson.*) In (B), headphone–microphone combination headsets play a critical role in radio coverage of NASCAR racing events. (*Image courtesy of Rob Albright, PRN, Performance Racing Network.*)

Figure 10.11). You can also use them to **cue** up material when you have an audience and don't want them to hear the cueing sound. When it is inappropriate to wear headphones or earbuds (such as when you are interviewing someone), make sure ahead of time that you have your recorder levels set properly so that the interview will not be too loud or too soft.

Sportscasters and helicopter traffic reporters wear headphones in part so they can hear instructions from the station ("time for a commercial break," "head over to Route 58"). Usually there is a microphone attached to the headphone assembly so that they can operate hands-free — a necessity for someone flying a helicopter and a convenience for sportscasters who are often juggling statistics while they report (see Figure 10.12). Headphone–microphone combinations are usually referred to as a **headset**. Headsets also come in handy for radio station live broadcasts, remotes, and at concerts or promotional events.

10.9 GETTING THE SIGNAL BACK TO THE STUDIO

Sometimes you record material in the field and carry it by hand back to the studio where you or others edit it and prepare it for broadcast. But in many instances, it is necessary or desirable to send the material to the station so that it can be edited or broadcast immediately. The most obvious example is on-the-spot news. A radio news station or network would not want a reporter to cover a fire by driving back to the studio with each update. People want the news as it is happening. In the same manner, play-by-play needs to get from the venue to the airwaves as the game is in progress.

Audio facilities use various methods to transmit a signal from a location to the station and from there to the station transmitter. One of the oldest methods, still in use to some degree, is a plain old telephone line. A microphone or a recorder is attached to a phone jack and the sound is sent through the phone switching system in the same manner that your voice is sent when you call a friend on a landline phone. There are phone systems, such as **ISDN** (Integrated Services Digital Network) and **ATM** (Asynchronous Transfer Mode), that yield higher sound quality than your home phone, but they operate through wires in much the same manner as a home phone. **Telex** is another system that has been around for awhile. It is basically a telegraph system that was adapted for shortwave radio. It uses airwaves rather than wires.

Another traditional system that has been used for many years is called a **remote pickup unit (RPU)** which uses a band of frequencies that the FCC set aside for electronic news gathering and other remote broadcasting. This system uses special equipment often referred to as "Marti," because that is the name of the main company that provides the gear (see Figure 10.13). For this method, a small transmitter at the field location must line up with an antenna that a station has placed somewhere in the local vicinity, usually on a high hill. In order for the transmission to work it must be in "**line of sight.**" That means the transmitter must be able to "see" the antenna—not literally, but in such a way that the signal can get from the location to the antenna without being cut off by buildings or hills.

Today, more and more radio stations are using smart phones and Skype for getting the remote signal back to the station. One such method being employed is iPush, as described in Chapter 6 and shown in Figure 10.14.

FIGURE 10.13   A Marti RPU. *(Image courtesy of Broadcast Electronics.)*

FIGURE 10.14   iPush allows producers to record and program content from remote locations. *(Image courtesy of Michael Parks, iHeartMedia-Harrisburg, PA.)*

FIGURE 10.15   Equipment needed for wireless broadband transmission.

Wireless **broadband** (see Figure 10.15) is another way that audio facilities receive information from the field. Companies such as Sprint sell to stations and networks secure transmission services that travel to and over the Internet. In addition, with special equipment or applications, reports from the field can be sent through cell phones or smartphones (see Figure 10.16), and there are also ways to send an audio report as an attachment to an email.

If you are sending audio back to a station, it is important to know the characteristics of the particular system you are using. If it is essential that the audio feed actually gets to its destination (as is the case in news reporting), then the station should provide redundancy so that if one system doesn't work, another can. For example, if your primary system involves a Marti, you should know where in your town there are **dead spots**—places where the signal will be obstructed. In those places you may need to use a cell phone or an ISDN line instead.

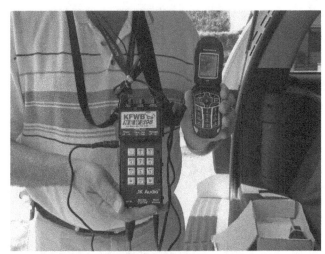

FIGURE 10.16 Equipment that can be used to send audio with a cell phone.

FIGURE 10.17 A reporter writing a story in his "studio." Notice the mixer and editing system that are next to him. The transmitting equipment is farther to the right.

People at locations also need a way to talk to people at the home base, so that they can find out when they need to send a feed, where they should go next for breaking news, and other logistical information. Through the years, two-way radios and walkie-talkies have been used for this purpose, but now cell phones are the main way that people communicate.

Whenever possible, conduct a trial feed to make sure that the people at the station or network can receive your field report before you send it. A feed that cuts off in the middle is annoying to listeners. Also, all the transmission equipment runs on batteries, so you must make sure they are properly charged in the same manner as your recording equipment batteries. Reporters sometimes go out to cover stories by themselves, and so they are responsible for operating all the equipment. The best story in the world will be useless if it can't get from the location to the listener.

10.10 HANDLING VEHICLES

If you are going to a location, you obviously need to have some way to get there. As a student, this probably means that you will drive your own car, although for some location recording it is possible to use public transportation and carry your equipment with you. **Stringers** (people who search for news stories that they can cover and then sell the reports to stations, networks, or news agencies) also use their own transportation.

Most radio stations and other news gathering organizations provide vehicles for reporters (or pay mileage reimbursement), commonly a news van that is outfitted with equipment for editing and transmitting. Sometimes the vans have **scanners** so that reporters can listen to police and fire department radios. Reporters are often issued laptops so they can write their stories (see Figure 10.17) and also search the Internet if they are in an area where they can access it. Having paper around to make notes and to

FIGURE 10.18 The same reporter taking care of his studio's vehicle by keeping the tank three quarters full.

use as a backup when the computer inevitably crashes is still a must. Laptops, too, have batteries that need to be checked and charged.

A vehicle serves as a mobile office/studio—the place where the audio person writes, records, mixes, and transmits. It is the responsibility of the person using it to keep it in good condition. If you are a reporter, keep the gas tank three quarters full (see Figure 10.18), because you never know when you will need to suddenly travel a long distance to cover a story. Fill up at the end of your shift so the next person using the van will have a full tank. Check the vital signs—oil level, tire pressure, and, of course, batteries. Most news gathering organizations have a mechanic who keeps the various vehicles operating, but the reporter needs to undertake routine maintenance and report any malfunctions to the mechanic.

There also are news and traffic helicopters (see Figure 10.19) that people reporting from the air operate. Obviously, these are very expensive vehicles that require special training. Although it is somewhat dangerous, one

FIGURE 10.19 A news helicopter. *(Image courtesy of iStockphoto, Dan Barnes, Image #5288449.)*

person can fly the helicopter and also report the traffic or news. Stations with helicopters often hire a person who is first and foremost a helicopter pilot and then train that person to be a reporter. Sometimes several news agencies will share one helicopter reporter. Traffic services tend to rent helicopters and pilots don't do on-air work. However, TV stations often have pilot/reporters.

10.11 PROVIDING FOR YOUR OWN NEEDS

When you are recording on location, you may not have the comforts of a studio, so you need to make sure that you will be safe and able to carry out your duties. If you are only going out for an hour or two to record someone in an office building, you probably do not need to take any extra precautions. But if you are going to be out in the sun for several hours, you may at least want to take along some water and suntan lotion and find out where the nearest food and washroom facilities are located.

If you are a news reporter, you should have more supplies with you, because you never know where you might wind up during the course of a day (see Figure 10.20). You might be covering a fire or bank robbery where you can accidentally find yourself in harm's way. You should think through all the possibilities in your part of the country and plan your supplies accordingly.

PRODUCTION TIP 10B
How to Pack a Survival Bag

What are some of the items in Pete's bag and why might he need to use them?

- Jacket with many pockets: This item allows him to place equipment and supplies in the jacket and operate hands-free.
- Bulletproof vest: Issued by his radio station, he wears this whenever he is covering a story, such as a bank robbery or

FIGURE 10.20 Reporter Pete Demetriou holds the bag (A) of supplies that he keeps with him when reporting. In (B), he shows some of the contents of the bag. (Note: The big cases are not part of the bag's contents; they are there just to help display the contents.)

a hostage-taking, where there might be firearm action that results in stray bullets.
- First-aid kit: Similar to any first-aid kit, this is something that all car owners would be well-advised to carry in their vehicles.
- Solution for cleaning out eyes: If covering a fire or chemical spill, this is beneficial in order to keep eyes from being damaged. The main ingredient is Johnson's baby shampoo.
- Hard hat: This, too, is issued by the radio station and is needed for any construction area or any situation where there could be danger to the head.
- Fireproof clothing: Because Pete reports in an area that has numerous fires, his station issues all reporters a fireproof shirt and pants and a fireproof covering for the head.

- Gas mask: Another station-issued item, a gas mask is needed if there are toxic chemicals or gases.
- Boots: Keeping a pair of old boots in the bag comes in handy for covering stories in rugged terrain.
- Flares: Like flares that people should carry in their automobiles, these can be lit to warn of danger.
- Binoculars: These come in handy for covering stories where reporters are not allowed near the scene. He can look through the binoculars to get a better idea of the situation and then report more thoroughly.
- Food and water: Energy bars come in handy when it is not possible to go to a restaurant, and water can be used for drinking and for washing off anything that may need it.
- Flashlight: For covering stories at night, a flashlight is indispensable.

Protect yourself in other ways, too. Keep in shape, because gathering news can, on occasion, be physically rigorous and require long hours. Make sure to eat at regular intervals whenever you can. Keep your own "gas tank" three quarters full, in case you need to cover a story where you will not be able to eat a regular meal. If someone relieves you when you are on an extended story, make sure that you convey to them not only the latest events related to the story but also the possibilities of danger and the status of the equipment. Expect someone to do the same for you and ask specific questions if they don't.

10.12 POSTPRODUCTION CONCERNS FOR LOCATION RECORDING

As already mentioned, it is possible that you will do a little editing at the location, especially for short items. But if you are going to be recording material that will become part of a lengthy feature or documentary that is highly edited, there are some special considerations you may wish to think about.

You should try to record everything **flat**—that is, everything should be recorded at the same volume level and with no effects or EQ (equalization) added. During editing, you or someone else will no doubt raise and lower the volume of different elements, but if they were all recorded at the same level, this will be an easier task than trying to compensate for a variety of levels.

You may need extra audio referred to as **ambient sounds** or **wild track**. Sometimes ambient sounds are broken into atmosphere sounds, room tone, and walla walla.

Atmosphere sounds add authenticity and interest to your production. For example, you might record sounds of a brook for a feature about the environment.

Room tone involves recording about 30 seconds of "silence" in the location (a city park, a concert hall) where you have been recording your program material (an interview, a musical group). Cast and crew remain silent while you record but you pick up the general sound of the location, using the same mic that you used for your main audio. Each location has a particular sound because of its acoustics and elements within it, such as air conditioning or the whir of a computer fan. Room tone is used during editing to smooth audio transitions. For example, if you cut from a recording where a person was talking in a machine room full of operating equipment to a recording of that same person talking in the studio, you may find that the cut is too abrupt, because the machine noise disappears too quickly. To solve the problem, you mix some room tone from the machine room with the studio audio.

For **walla walla**, you record people's voices so that you can't understand actual words that they are saying. Walla walla is useful when you have important foreground noise but you want to fill it in with background noise. For example, if you are producing a commercial for a restaurant where you record the restaurant owner in a studio, you might want to mix in walla walla to give the impression of a crowded restaurant. Walla walla got its name because sometimes the best way to record it is to have a group of people say "walla walla" over and over at different speeds and with different emphasis.

Anytime you are engaged in postproduction, you should keep ethics uppermost in your mind. During editing, it is very easy to change the meaning and intent of what someone said. Don't succumb to this temptation just because it makes your editing easier. Always aim to make something sound as close as possible to what actually occurred.

10.13 CONCLUSION

As you can see, location recording can be quite different from studio recording. When you are on location, you have more technical responsibility. You must also be more flexible. As a reporter, you may be pulled off one story and sent to another. Even though you have undertaken careful preplanning for an audio documentary, you may find when you arrive that crucial elements have changed in the last few hours. You need to do the best job you can under the circumstances and, whenever possible, enjoy the experience.

Self-Study

QUESTIONS

1. Which of the following is most likely to use a helicopter?

   a) a play-by-play sportscaster
   b) a traffic reporter
   c) a disc jockey working in a club
   d) a rock band

2. Why should a reporter keep a station news van gas tank three quarters full?

   a) to conserve gasoline
   b) to make sure that he/she will have enough gas to leave quickly to cover a story that is at some distance
   c) to keep the battery chargers in the van from discharging
   d) to keep the scanner operating

3. Which of the following should you do when you are scouting a location that you plan to use for recording?

   a) Park in a handicapped parking space to see if you get a ticket.
   b) Give your location survey form to a musician so he or she can arrange to record a song about it.
   c) Bring along some malfunctioning equipment and try to get someone at the location to fix it.
   d) Have the owner of the location facility sign a form that gives you permission to record there.

4. Which of the following should you be sure to do if you are using house sound?

   a) Listen with your headphones to the output of your recorder so you can tell whether your sound is distorted.
   b) Cue up the sound carefully so the audience can't hear it.
   c) Always use a DAT recorder.
   d) Bring your own audio mixer.

5. Which of the following is most likely to have a problem caused by low battery power?

   a) a microphone
   b) a memory card
   c) a recorder
   d) headphones

6. When should you replace batteries in your equipment?

   a) when they are 25 percent empty
   b) when they are 25 percent full
   c) once a week
   d) when they are totally drained

7. Which is the most common microphone for interviewing in the field?

   a) a dynamic cardioid handheld mic
   b) a condenser super-cardioid wireless mic
   c) a dynamic omnidirectional lavaliere mic
   d) a condenser cardioid windscreen mic

8. Which person is most likely to have his or her microphone attached to the headphone assembly?

   a) a news reporter doing an interview
   b) a guitar player
   c) a helicopter traffic reporter
   d) a circumaural public affairs producer

9. Which system for getting sound from a location back to the studio uses frequencies set aside by the FCC specifically for this purpose?

   a) ISDN
   b) Telex
   c) cell phone
   d) RPU

10. Which of the following refers to people's incoherent voices in the background of an audio recording?

   a) walla walla
   b) atmosphere sound
   c) flat sound
   d) room tone

11. What are "dead spots"?

   a) a recording that is flat
   b) places where stringers find news
   c) places from which a transmitted signal will not reach its destination
   d) places at clubs where disc jockeys can place their turntables

12. Which of these does the selective attention principle enable you to do?

   a) Concentrate on having a conversation with someone even though there are many other noises nearby.
   b) Keep your equipment from locking up in high humidity.
   c) Hire a security guard to keep people away from your recording equipment.
   d) Scoop other reporters because interviewees will be talking to you, not them.

13. Which of the following would be most important if you, as a news reporter, are covering a man barricaded in a house who occasionally shoots from a house window?

   a) suntan lotion
   b) energy bars
   c) a bulletproof vest
   d) fireproof clothing

14. What methods are producers using today to get the audio signal back to the studio from remote locations?

   a) Skype
   b) smart phones
   c) iPush
   d) All of the above.

15. Which of the following would you be most likely to use to adjust the volume for several inputs, such as an announcer's voice and background noise from a busy street?

   a) a channel
   b) a mixer
   c) a hum
   d) circumaural headphones

ANSWERS

If you answered A to any of the questions:

1a. No. The sportscaster would be in an announce booth. (Reread 10.2.)

2a. No. You will use the same amount of gas regardless. (Reread 10.10.)

3a. Absolutely not. You should never park in a handicapped spot unless you have a sticker to park there. (Reread 10.4.)

4a. Correct. House sound can lead to distortion.

5a. This is not the best answer. Some condenser microphones have batteries that will discharge, but in the field you generally use a dynamic mic. (Reread 10.6.)

6a. No. You are confusing empty and full. (Reread 10.6.)

7a. Yes. This is the best choice. A dynamic mic is rugged and both the cardioid pickup pattern and hand holding are best for an interview.

8a. Wrong. When you are doing an interview you need to place the mic close to the interviewee, which you can't do if it's attached to your headphones. (Reread 10.8.)

9a. Wrong. ISDN is a wired service that does not use the airwaves. (Reread 10.9.)

10a. Right. Walla walla is the correct answer.

11a. Wrong. These are two entirely different concepts. (Reread 10.9 and 10.12.)

12a. This is correct. Human beings, unlike microphones, can decide what is most important to hear.

13a. No. You might want suntan lotion if you are outside, but it is not the most important thing on the list. (Reread 10.11.)

14a. While it is a method to get the signal back to the station, there are other choices as well. (Reread 10.9.)

15a. Wrong. The sound might be on a channel, but that is not what you would use to adjust it. (Reread 10.5 and 10.7.)

If you answered B to any of the questions:

1b. Correct. Of all the options given, the traffic reporter would be most likely to use a helicopter.

2b. Correct. You may need to get somewhere quickly and you need the gas to get there.

3b. Wrong. This is a nonsensical answer. (Reread 10.4.)

4b. This is not the best answer. You are not really in charge of cueing up the sound. (Reread 10.7 and 10.8.)

5b. Incorrect. Memory cards don't use batteries. (Reread 10.6.)

6b. Yes. This is the correct answer.

7b. Incorrect. None of the parts of this answer are correct. (Reread 10.5.)

8b. Wrong. A guitar player usually needs a mic by the mouth and one at the sound hole. (Reread 10.8.)

9b. Wrong. Telex does not use those frequencies. (Reread 10.9.)

10b. Incorrect. This is not the purpose of atmosphere sound. (Reread 10.12.)

11b. Wrong. These two subjects are unrelated. (Reread 10.9 and 10.10.)

12b. Wrong. The selective attention principle has nothing to do with humidity. (Reread 10.3.)

13b. No. This is probably the least important thing. (Reread 10.11.)

14b. While it is a method to get the signal back to the station, there are other choices as well. (Reread 10.9.)

15b. This is correct. You would use a mixer.

If you answered C to any of the questions:

1c. Absolutely not. (Reread 10.2.)

2c. Incorrect. Gas and battery chargers are not related in this case. (Reread 10.10.)

3c. Wrong. This is a somewhat nonsensical answer. (Reread 10.4.)

4c. Wrong. Any type of recorder will work. (Reread 10.6 and 10.7.)

5c. Yes. You must carefully monitor your recorder's batteries.

6c. No. It is more dependent on their use than any particular time frame. (Reread 10.6.)

7c. No. One part of this answer is right, but the rest isn't. (Reread 10.5.)

8c. Right. Traffic reporter is the correct answer.

9c. No. Cell phones have their own general use frequencies. (Reread 10.9.)

10c. Incorrect. Flat sound would not refer just to people's voices. (Reread 10.12.)

11c. Yes. This is the correct answer.

12c. No. Security and selective attention principle have nothing to do with each other. (Reread 10.3.)

13c. Yes. This would be most important.

14c. While it is a method to get the signal back to the station, there are other choices as well. (Reread 10.9.)

15c. No. This is totally off base. (Reread 10.7.)

If you answered D to any of the questions:

1d. No. Maybe for publicity purposes, but this is certainly not the best answer. (Reread 10.2.)

2d. Wrong. Gas does not operate the scanner. (Reread 10.10.)

3d. Correct. This is by far the best answer.

4d. No. House sound assumes the house has a mixer. (Reread 10.7.)

5d. No. Headphones is not correct. (Reread 10.6 and 10.8.)

6d. No. You don't want to wait this long as they may keep you from recording something important. (Reread 10.6.)

7d. Incorrect. All parts of this answer are wrong. (Reread 10.5.)

8d. Incorrect. This answer doesn't even make sense. (Reread 10.8.)

9d. Yes. RPU is the correct answer.

10d. Wrong. There is a much better answer. (Reread 10.12.)

11d. No. These two concepts are not related. (Reread 10.9.)

12d. Wrong. Selective attention principle has nothing to do with scooping reporters. (Reread 10.3.)

13d. No. There is a better answer. This would only be needed if the house catches on fire. (Reread 10.11.)

14d. Correct. All of the choices are methods to get the signal back to the station.

15d. No, you might use these to listen to the sounds but not to adjust them. (Reread 10.7 and 10.8.)

Projects

Listen and plan for sounds.

Purpose

To make you aware of various sounds in a location environment and to help you figure out ways to keep sounds you want and eliminate sounds you do not want.

Notes

1. You do not need to use the exact locations mentioned for this project or to use the exact recording ideas. Feel free to improvise, but keep within the spirit of the assignment.
2. You do not need to go to the locations in the order given; you can mix them up.
3. You may want to use a recorder with a built-in mic. Generally, you record with the same mic you used for your main recording, but in this case you do not have a main recording.

How to Do the Project

1. Secure a recorder (and a mic if the recorder does not have one built in).
2. Go to a noisy outdoor location, such as a corner on the main street of your town. Stand there for about 10 minutes. Turn the recorder on and also make a list of all the sounds that you hear. Pretend that you are preparing a news story about the number of cars that are parked illegally. Consider the following questions:
   a. Which of the sounds that you heard would you like to keep and why?
   b. Which would you like to get rid of and how might you get rid of them?
   c. Are there any sounds you would like to add?
3. Go to a quiet outdoor location such as a residential street. Again turn your recorder on and also make a list of all the sounds you hear in 10 minutes. Pretend you are preparing a documentary about the history of this particular street. Consider the three questions listed in Steps 2a, b, and c.
4. Go to a noisy indoor location, such as a restaurant at dinner time. As with Steps 1 and 2, record and make a list of sounds you hear. Then pretend that you are preparing a commercial for the restaurant. Think through the questions in Steps 2a, b, and c.
5. Go to a quiet indoor location such as your bedroom and record and list the sounds you hear. Pretend you are preparing to record an argument between a mother and a son in this room and again think about Steps 2a, b, and c.
6. Listen to your recordings and note any ways you might change your answers to questions based on Steps 2a, b, and c.
7. Write a report about your experience. Include your lists of sounds and your answers (and revised answers) to the questions above. Give any other observations you might have by answering such questions as:
   a. Is it easier to change the nature of sound inside or outside? In a noisy location or a quiet one?
   b. Does the type of program you are producing (news, documentary, commercial, drama) affect the need to alter sounds?
   c. Did your ability to hear all the sounds improve as you went from the first location to the last location?
   d. Were there any sounds you heard on the recorded material that you did not note on the list you made?
8. Turn your paper in to your instructor to get credit for this project.

Record atmosphere sound, room tone, and walla walla.

Purpose

To give you practice recording these three types of ambient sounds and to familiarize yourself with location recording.

Notes

1. You may have to experiment and record some of these (particularly walla walla) several times until they sound authentic.
2. Try to do all of these within a small geographic area so that you do not need to spend a great deal of time traveling.
3. Make sure that you know how to check out and use your facility's portable audio equipment. Check to make sure it is working before you leave the facility and report any problems when you bring it back.
4. If you can't check out an omnidirectional mic, a cardioid will do, but omnis are better for gathering general sound.

How to Do the Project

1. Think of some sound that you feel would make a good atmosphere sound—a river, cars on a busy highway, pots and pans banging. Secure an omnidirectional mic and a recorder. Hold the mic near where the sound is generated and record it. Listen to it and decide whether or not it sounds appropriate. If not, experiment by holding the mic closer or farther away until you achieve the sound that you want.
2. Find a room that has at least some noise factors in it—a fan, air conditioning, electronic equipment. Hold your mic in the middle of the room and record the room tone. Play it back and see if you can discern sound.
3. Gather a few of your friends together and have them say "walla, walla" at different speeds and with different emphasis. They should definitely not say "walla, walla" as a chorus; each person should say it with different timing and quality. They will probably feel pretty silly, but record them and see how they sound. If you don't think their recording could be used as background talking noise, record them again.
4. Turn in all three recordings to your instructor so you get credit for this project.

11

SOUND PRODUCTION FOR THE VISUAL MEDIA

11.1 INTRODUCTION

Most of what has been covered in this book deals with audio-only material, such as a disc-jockeyed show or a radio newscast. Just as important though, audio is used in conjunction with visual media—a TV production, a music video, a talk show, an animated cartoon, a dramatic movie, a video game, a documentary, and so on.

To some extent, dealing with audio joined with visuals is no different than dealing with audio alone. Microphones used for a TV game show have the same directionality and construction options as those used for a radio interview. The audio board used to record a television cooking show has many of the same controls as one used by a disc jockey. If the video is shot outside the studio, wind noises will be a problem, just as they are for audio recording. However, there are also differences in the audio techniques needed when a visual is an element—differences that this chapter will address.

11.2 THE IMPORTANCE OF SOUND TO A VISUAL PRODUCTION

Audio adds a great deal to any visual production. In video production, for example, some have estimated that audio supplies 80 percent of the information and that the video image supplies only 20 percent. This does not mean that audio gets as much respect as the video image, however. On a video or film shoot, much more time is usually consumed by the needs of the picture than the needs of the sound. If an audio operator (aka the Production Soundmixer, as they are referred to on the set) encounters a problem recording sound, the director is often less sympathetic than that director would be to a cinematographer's problem. One audio operator who regularly works on television productions has confessed to a method he uses to handle this situation. If he needs time to fix an audio problem he will say, "I think I saw a shadow on the actor's face" or "Is it my imagination or is that light in the corner flickering?" While crew members are scurrying trying to find what might be a source of a shadow, he has time to rectify his audio problem.

The good news is that as audience members build home theaters with surround sound, they demand the best in audio. Their demands are trickling down to the production process and audio is gaining more respect. There are exciting, challenging jobs available for those who want to work with the sound of visual production. Bear in mind that for every one picture editor assigned to a production, there is an entire crew of sound editors to deal with dialogue, rerecording, sound effects, and music.

11.3 THE NEED TO ACCOMMODATE THE PICTURE

Audio recordists who are preparing material to be used with images need to take the picture into account. For example, for some forms of programming, such as television news or music shows, it is perfectly acceptable for the microphone to show in the picture. But for fictional forms (dramas, sitcoms), the presence of a mic will break the illusion, because in real life, people who are having an argument or embracing each other do not wear microphones.

Audio must also be recorded in such a way that it fits the needs of the overall program. If the material is fictional, then the sound must serve the needs of the story; if the scene is tense, the music and sound effects must be intense. To a lesser extent, the same is true for documentary productions. Documentary sound tracks often include music and sound effects. Very few documentaries use Foley (explained later in this chapter), but none the less all the sounds elements, including interviews, location, and ambient sounds, are mixed down from multiple tracks. New productions usually just include voice tracks and sounds for story elements.

For productions with complicated audio, such as movies, a **sound designer** is often hired to plan the overall strategy for sounds and to consider the purpose of sound in relation to the image. Sound designers are also being used in television broadcast dramas. There is a minimal, yet present, amount of sound design in some reality television, too.

Throughout the remainder of this chapter we will cover various types of sound common to production with the visual media, how these sounds are recorded, and how they are incorporated in the final product.

11.4 RECORDING SPEECH

Recording the spoken word is very important in any type of program. If the audience members can't understand the **dialogue** of a drama or the conversation of a talk show, they will not be able to follow the story or learn from the information. As with any form of recording, you must decide the directionality of the microphone(s) that you plan to use. **Cardioid** or Hyper-cardioid (aka short shotgun) mics are the workhorses of video and film recording, because they do not pick up sounds well from one side, the side where the video or film equipment usually is. For example, in a studio shoot, cooling fans might be heard through an **omnidirectional** mic. Also, as with audio recording, you must decide on the construction of the mic. Dynamic microphones are best for controlling

background noise (handheld reporting) or loud sound effects (natural compression).

The positioning of the mic is usually much more complex when an image is involved than when only audio is being recorded. For an interview situation, a television reporter might use a handheld mic in a fashion similar to how a radio reporter would hold it, but for dramas, sitcoms, and similar material, the microphone must not be obvious in the picture.

11.5 THE BOOMPOLE

The most common type of microphone holder used for visual media production is one that is hardly ever used for radio or other audio-only productions. It is called a **boompole** because it is a long pole with a mic at the end. The operator holds the boompole and moves it above the heads of the performers, pointing the end of the mic toward the actor's mouth. The best position is 1 to 3 feet above the performer, aimed at the nose of the talent, held at an angle 45 to 90 degrees overhead (see Figure 11.1(A)). It is particularly effective for recording dialogue, because the person holding the boompole can move the mic quickly to capture dialogue of different people (make note of the points raised in Figure 11.1(B)).

But the boompole has disadvantages. If the camera shot is at all wide, the mic is likely to show in the picture. Directors often like to use two cameras at the same time, one showing a long shot and another a close-up (see Figure 11.2).

FIGURE 11.1(A)   A boom operator, aiming at the nose of the talent, holding the boompole at an angle 45 to 90 degrees overhead. *(Image courtesy of Fred Ginsburg CAS PhD MBKS.)*

FIGURE 11.1(B) This picture shows the "incorrect" way to hold a boompole. Notice that the microphone is too far away from the talent, the boompole needs to be horizontal (not held like a flagpole), and both of the boomperson's arms should be overhead.

FIGURE 11.2 In this situation, two cameras are recording the same shot. The camera on the left has a wide shot and the camera on the right has a close-up. Therefore, the boomperson needs to hold the mic a little higher than would be proper if only the close-up camera were shooting. If the boom was placed from the camera axis, the mic could then be cued from side to side to cover both actors.

Therefore, audio operators have to be creative in the placement of the boom. Sometimes the solution is to place the boompole from below the frameline around the talent's waistline or thigh. Booming from the talent's feet is used for sound effects or footsteps. Some audio operators place a piece of white tape on the end of the mic. That way the camera operator can easily spot it if it dips into the picture, and the camera operator and audio operator, working cooperatively, can determine just how low the mic can come.

Another problem with the boompole is that it is likely to cast a shadow on the characters or the set. There usually is no way to prevent a shadow, but you can minimize it by positioning the boompole in such a way that the shadow falls onto the floor or onto a flowerpot or other piece of scenery where it will not be noticed. If that doesn't work, someone can stand by the microphone and look to see what light is causing the problem (see Figure 11.3). Moving that light slightly to the left or right can often redirect the shadow to a harmless area. If you can't move a light to solve a sound problem, you may be able to adjust the camera angle or actor position. Because a boom mic is often used to record several people talking, there is always the possibility that one person will talk more quietly than another. Usually this can be overcome by placing the mic closer to one person than to the other, but this involves skill and experience on the part of the boomperson.

PRODUCTION TIP 11A
Holding a Boompole

Hold the boom with the front supporting arm as vertical as possible, so that you are bench pressing the weight and not using the lateral support of your shoulder. Let the back arm control the aim (cue) of the mic. Holding your arms horizontally, as opposed to vertically, will tire you out very quickly. Placing a padded lighting stand under the supporting arm elbow can provide a resting place to take the strain off of your arm during a long take. If a crew member needs to "rescue" a tiring boomperson, they should place their palm under the supporting elbow, but never try to grab the pole itself.

11.6 THE LAVALIERE

As already mentioned in Chapter 4, "Microphones," another way to record speech is with a **lavaliere (lav)**, a small mic (see Figure 11.4) that can be clipped to a person's clothing. The best way to position it is 8 to 10 inches below the talent's mouth, attached to a tie or blouse, making sure that it is not in a position where it will pick up the rustle of clothing. Lav mics are often wireless so that they give the person wearing it more freedom (for example, the cooking show chef can move about the studio kitchen or the drama actor can run down the street), but they can also be totally

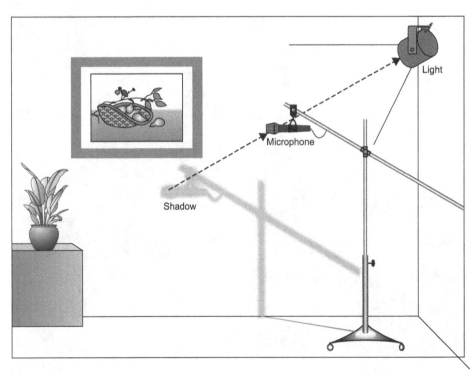

FIGURE 11.3 You can see what light is causing the microphone shadow if you stand in the shadow and look at the microphone. The offending light will be the one you see behind the mic.

FIGURE 11.4 A lavaliere microphone. *(Image courtesy of Bosch Communications Systems.)*

wired. In either case, they usually include a power supply/transmitter antenna that can be worn on a belt or waistband or buried in a set. If the mic is wireless, the antenna transmits the signal to a receiver that is then attached by wire to a recording device. If it is wired, a connection from the power supply goes to an audio board or recorder. If you are using condenser mics, they are connected to the board through their power supply (for example, an AA battery located in the mic shaft) or directly from the mic when using phantom power. Each person has his or her individual lav, so it is easy to record each person's speech on a different track of an audio recorder.

A disadvantage of lavalieres is that they are sometimes visible in a close-up. Also, the cables that are part of the lav assembly should be strung under the person's clothes because, even if it is all right for the mic to show, a cable draped down the front of a shirt or blouse looks sloppy. Depending on what the person is wearing, the stringing of the cables can be difficult. Also, because each person must be miked separately, if the people get too close together, there is the danger of **multiple-microphone interference** (see Chapter 4) or **phase cancellation**.

11.7 OTHER FORMS OF MICROPHONE POSITIONING

Some large-scale productions, such as musical reviews, use **perambulator booms**, which are movable platforms that can move the mic in, out, up, down, and various other directions. More modest **boom stands** consist of a long horizontal pipe with a standard thread for a

microphone on one end and a weight on the other end to balance the microphone. This pipe is attached to a vertical stand that is attached to wheels. The boom stand can be moved around to record different people in much the same way as the boompole. It is easier on the arms of its operator than a boompole, but not as maneuverable as a human body.

Mics can also be hidden on the set. Hidden mics are also referred to as "planted" or "plant" mics. Frequently microphones are placed in recesses of a news sets directly in front of where the anchors sit. For dramas, they can be hidden in flowerpots or other props. For a TV studio shoot, they can hang from the **grid** in the ceiling. Hidden locations free the talent of the need to deal with mics or even look at them, and as long as they pick up the sound well, they represent a valid way to mic. The main problem with them is that they do not move and, for most video or film productions, the performers move. The mic in the flowerpot might pick up the actress well while she is sitting on the sofa, but lose her sound as she walks to the door. Worse yet is hiding the mic in something that moves during the course of a shot—when the uncle picks up an umbrella to exit and the sound goes out the door with him.

There are times when **desk stand** mics are used for guests on public affairs TV shows, when **floor stands** are used for music videos, when hand mics are used by talk show hosts who converse with people in the audience, when **headsets** are used by TV sports announcers, and when **shotgun mics** are used to record distant sounds for a video documentary. The use of these devices has the same advantages and disadvantages as have already been discussed in Chapters 4 and 10.

A times it is important to "relieve" a tired boom person during a take. The trick is for the assisting person NOT to try and grab/support the (wavering) boompole, but rather to slip a supporting hand under the elbow of the tiring boomperson to take the strain off of his/her shoulder during a take. If the boom operator expects a long take, then you can tape a pad or pillow onto a lighting stand and use that as an elbow rest during the take, or during just part of the take. This way, the boomperson can begin the scene fully handheld and follow the actors, then settle onto the rest while the actors are stationery, and then go handheld again when they up and move (see Figure 11.5).

11.8 CONTINUITY AND PERSPECTIVE

Continuity and perspective are two concepts that play a larger role in visual productions than in audio-only productions. **Continuity** refers to shooting in such a way that there will be a consistent progression from one shot to another when the final product is assembled. On a movie set, continuity is almost an obsession, mainly because scenes are shot out of order. One of the crew positions is a **script supervisor** who has as a main part of his or her job

FIGURE 11.5   A boom pillow. *(Image courtesy of Fred Ginsburg CAS PhD MBKS.)*

making sure shots will work together. Is the actor wearing the same tie as he enters his car in a scene shot on Monday as the one he is wearing when he gets out of his car in a scene shot on Thursday? In a documentary about an author, was the author holding the book in her right hand both for the long shot and for the close-up that was shot after the interview?

Script supervisors are supposed to check for audio continuity also, but in reality, that job often falls to the audio recordist. If the air conditioning is operating during a long shot of a couple arguing, it should also be on during the close-ups of the couple, so that when the editor cuts from a long shot to a close-up, the background noise level will not change. When everyone breaks for lunch, you should note how far away the mic was from the performers so that when shooting starts again after lunch, the mic can be in the same position. To the extent possible, you should use the same mic for the same person throughout a production. Because each mic has its particular sound, switching mics can change the sound. Sometimes this is not possible, however. You may be using a boom for the bulk of the shooting but need to switch to a wireless lav for a long shot of a person running. In such circumstances, you should listen to recordings of various available mics and choose ones that sound fairly similar.

Perspective relates to distance. If a dog is barking in the background of a shot, the dog's sound should be different than if it is in the foreground of the shot. In this way, the viewers will feel closer to or farther from the dog and the sound will reinforce what they are seeing on the screen and

help give the two-dimensional frame more of a three-dimensional feel. The easiest way to achieve proper perspective is to move a mic farther away or closer to the sound. If the mic is on a boompole, it usually needs to be farther away for a long shot because it will show if it is right above the actor's head. Lav mics present a problem with perspective. They are always the same distance from the person's mouth so should not be used when perspective changes are needed. Hidden mics can be good for perspective if the picture relates to where the mic is placed. If the shot widens as the actress moves back from the hidden mic, perspective will be accurate. But if the camera follows the actress with a close-up while she is moving farther from the mic, the perspective will be incorrect.

Intimate or close-up perspective relates to hearing less background in the close-up and more background in the wider angle shot. Note that it does not refer to the close-up sounding louder than the medium shot or wide angle shot, as dialog volume should match. Dialog only gets recorded lower or louder if the DISTANCE changes, not the lens angle! Wide shot perspective for a lav can be achieved by mixing in some ambiance along with the dialog, usually from an open mic on the set.

11.9 THE RECORDING PROCEDURE

The simplest way to record sound for video is to plug the end of the microphone cable into the camera and record the sound on the tape or hard drive that is part of

FIGURE 11.6 To sync picture shot on film and sound recorded on a separate recorder so that they could ride together for editing purposes, filmmakers used (and still use) a clapper, such as the one shown here. At the beginning of each shot, a crew member holds the clapper in front of the film camera and claps the sticks together. In the editing suite, the frame where the clapper closes is lined up with the sound of it closing. Then the rest of the shot that follows is in sync. *(Image courtesy of iStockphoto, Tomasz Rymkiewicz, Image #5540076.)*

the camera assembly. This works well for a single sound, such as a news reporter doing a stand-up. But if the sound is more complex, involving a number of different voices or a number of different microphones, it is a good idea to have additional equipment that can control sound. If a program is being shot in a TV studio, the mics run through an audio board and the sound is adjusted just as it is for any production that involves studio audio. In the field, many audio recordists use a portable mixer so that they can manipulate sounds—for example, record a soft-spoken person at a high volume level on one track while recording a person with a louder voice at a lower level on another track.

The output from this mixer can be sent directly to the camera (either through cables or in a wireless manner) or it can be recorded separately on an audio recorder. Recording separately is referred to as **double-system sound** and recording onto the same source as the video is called **single-system sound**. Double-system recording was historically used for movies shot on film, because the sound couldn't be recorded on the film in an effective manner that allowed for editing. Sound and picture were recorded on separate pieces of equipment and later synched by lining up the picture and sound of a **clapper** (see Figure 11.6).

Video is edited electronically, so there is no technical reason to record sound separately. However, double-system sound recording is often used with video recording for a number of reasons. Not having an audio cable or a wireless receiver attached to the camera allows the camera operator more flexibility of movement. If sound is recorded using both single- and double-system recording, the editor has more flexibility once the project is in the editing suite. Tracks recorded direct to video are often limited to just 2, or 4 at the most. Double system recording allows us to record a live mix, as well as backups of

every mic, known as ISO's (short for "isolated"). Perhaps more important is that the camera operator is busy framing the picture and often doesn't have time (or the inclination) to make sure the sound is being recorded properly. A person who devotes his or her entire attention to audio recording will produce a better product. In fact, big-budget productions usually have at least three audio people—one to hold the boompole, one to operate the audio recording equipment, and the other to serve as a utility sound technician.

PRODUCTION TIP 11B
Recording with Your Eyes Shut

There is a great deal happening during a video or film shoot: people moving props, makeup artists powdering actors' foreheads, technicians rearranging lights. It is easy to get distracted. If you are in charge of recording audio, try to arrange to sit right on the edge of the set, just out of frame, so that you can SEE the actors and the mic placements. This allows you to anticipate audio problems BEFORE they happen, rather than just reacting to what is in the headphones and already recorded. Video monitors often only show what the camera sees, and not the interplay of other actors and mic positions. (see Figure 11.7). Wear high-quality headphones so that you can hear the sound you are recording distinctly and then close your eyes. This approach will allow you to concentrate on the sound and ignore the other hubbub. You will find it much easier to identify audio problems, such as extraneous noises that should not be part of the recording. A portable cart comes in handy to carry all of your gear (see Figure 11.8).

FIGURE 11.7 Shown here is production sound mixer, Richard Lightstone CAS, working on the ABC show *Lincoln Heights*. Note the use of the Yamaha 01V96 digital mixer and the Zaxcom Deva V 10-track digital recorder, appropriate for mixing and recording multiple actors on the set. *(Image courtesy of Omar Milano.)*

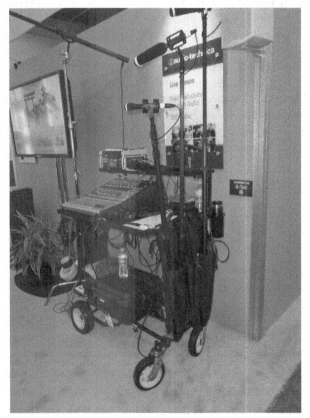

FIGURE 11.8 The RocknRoller® Multi-Cart® carries huge amounts of set equipment and folds small for storage. *(Image courtesy of Fred Ginsburg CAS PhD MBKS.)*

11.10 RECORDING SOUND EFFECTS

Many sound effects for visual productions are obtained from CDs or the Internet, in the same manner as for audio productions (see Chapter 1, "Production Planning"). But there is a different concept that must be considered: **sync**

sound versus **non-sync sound**. Synched sounds are those that are tied to the picture—when a dog that is onscreen barks, the audience expects to hear the bark and they expect it to coordinate with the picture. Barks that occur out of sync with the opening and closing of the dog's mouth are unacceptable. However, if the dog is offscreen, the bark does not need to be synched to the picture. In fact, you can use a standard sound effect, eliminating the need to record the sound at all.

Most commonly, onscreen synched sound effects are recorded during production; in fact, it is almost impossible not to record them. For that reason, they need to be carefully miked. If the person talking is petting the dog, the boom mic picking up the voice is adequate for picking up the barks. If the dog is by itself but onscreen, it may need to be separately miked—not necessarily an easy chore. You need to experiment with different mics and different positions to see what works well. You might be able to hide or hang the mic or use a boompole or shotgun mic. If the dog's barking occurs when the actor is not talking, you can, in a pinch, rely on sound effects of a dog barking that can be synched in postproduction. But if the man is talking while the dog is barking and they are therefore on the same track of audio, you need to make sure that you are picking up both sounds adequately. If the sound effect is non-sync, the process is much easier. You can place the mic wherever it sounds best, because there is no chance of it showing in the finalized production.

Video games have presented interesting challenges in terms of sound effects. For starters, the types of sounds used in video games often need to be "manufactured." They are not sounds that can be found in everyday life or on sound effects sources. Many of them are created by manipulating normal sounds in computers—raising the **pitch**, changing the **timbre**, stretching the vowels. Others are ingeniously constructed by such methods as running a razor over a bell or hitting a pillow with a wrench. Microphone selection and placement is varied until the proper effect is achieved. Even sounds based on real ones such as gunfire need to be manipulated, because they change as the dynamics of the game change. For example, if a player is too close to a cannon when it fires, this action might trigger an automatic change in loudness that brings about a realistic ringing in the player's ears. Sometimes sounds are "placed" on tires or windshields and triggered only when the player hits one of those areas. Everything needs to work together. If the player does hit the windshield, there needs to be the additional sound of glass falling to the ground. All of this organization takes a great deal of foresight and enables many creative recording sessions.

11.11 RECORDING AMBIENT SOUNDS

Ambient sounds, as discussed in Chapter 10, "Location Sound Recording," are general background noises, such as room tone, atmosphere sounds, and walla walla. They are just as important for visual programs as for audio productions. It is often harder to record **room tone** on a video or film shoot, however, because there are more crew members. Getting everyone to be quiet for 30 seconds so that you can record the ambience of the place where you are shooting can be a difficult task. Camera operators, grips, and lighting technicians are usually eager to pack up the equipment, but you should record the room tone with all the equipment still in place. Otherwise, the room will have more of a hollow sound. The best mic to use is the one that you used to record whatever was happening in the room—an interview, dialogue for a drama, and so on. Doing so will further help ensure a similar sound quality that can be used later in postproduction. (One trick for getting perfect room tone is to record 30 seconds at the beginning of take one, when everyone is primed and in position.)

Atmosphere sounds are usually recorded at a different time than the main portions of a production or they are obtained from prerecorded sources. As with audio only, these sounds, such as a bubbling brook or a noisy highway, add to the feel of the final product. With video or film productions, they must also coordinate with the picture. Although they do not need to be in sync with the picture, they must portray reality. If the highway in the picture shows a traffic jam, the atmosphere sound should not be that of traffic whizzing by. Most video cameras have a built-in microphone that should not be used in most video production circumstances, because it picks up noises close to the camera much better than it picks up dialogue from someone 5 feet away. However, for atmosphere sounds, the **camera mic** is often the best one to use, because you are trying to record general sounds of a particular area (see Figure 11.9). Of course, you don't really need a camera to record atmosphere sounds; you can just use an audio recorder, but video people usually record everything through a camera so that it is all in one place.

Walla walla (the sound of people talking in the background) is more likely to be needed for visual productions than audio only productions. Often scenes such as those in a bar are filmed with only the main characters talking and everyone in the background moving their mouths but not making any sounds. This gives a clean dialogue track to which other sounds can be added in postproduction. Walla walla is best recorded with an omnidirectional mic (which can be a camera mic) either by hiring people to say "walla walla" or by going to an actual location, such as a bar, and taping the sound of people talking (see Chapter 10). You can also use a combination of mics on a set from using a boom, which by nature adds more ambience to the voice, lavs, which give a good clean signal if done properly, and set mics. Each mic is given its own track and they are later mixed together.

11.12 RECORDING MUSIC

Music for a visual project is usually recorded after the project has been edited, so that the music can fit precisely where it is needed. Some music, such as opening theme

FIGURE 11.9 On this camera, the camera mic is mounted right above the lens. *(Image courtesy of iStockphoto, Oktay Ortakcioglu, Image #4892147.)*

music, can be recorded ahead of time, and music that is part of the plot, such as a rock band song that the leading actor and some of his buddies play, is recorded during the course of the main production, but generally music recording is one of the final stages.

Music can come from a variety of sources. There are many companies that supply stock music for a modest fee (see Chapter 1), but it is hard to find prerecorded music that fits the exact mood and timing required by most video productions.

A great deal of music for video or film productions is supplied by composers who work at their computers laying down various tracks of synthesized instruments (see Figure 11.10). They may never meet anyone involved with the production; in fact, a composer can live in France and compose music for a movie made in Hollywood. Often the director or someone aligned with the production makes a **timing sheet** (see Figure 11.11) for the composer. It lists the **time code** numbers for places where the director feels that music should be added, and the composer is, of course, given a time-coded copy of the movie to use as a guide. The composer records music of the right mood and length and sends it over high-quality phone lines or the Internet to the director, who may ask for some changes or accept it as it is and incorporate it within the movie. All the "recording" is undertaken in the composer's computer.

FIGURE 11.10 TV composer Ray Colcord (*Family Affair, Touched by an Angel, The Simpsons*) works out of a studio in his home, where he has a full array of software and hardware to aid his composing.

TIMING SHEET

Title: A Thrill a Minute
Director: Adrian Jones
Scene: 15

Code	Time	Description
00:48:40:12		Phil and Shorty enter the bar, look around and sit at a table.
00:49:10:00	00:00:00:00	Music starts as Mitzi moves toward the two men. The camera follows her as she sits and gives them the eye.
00:49:15:23	00:00:05:23	Cut to CU of Shorty.
00:49:17:21	00:00:07:21	Shorty: We didn't come in here looking for trouble. So if you're trouble, you can get lost.
00:49:24:03	00:00:14:03	Cut to CU of Mitzi. Mitzi: Actually, I have some information for you.
00:49:29:25	00:00:19:25	Cut to MS of Phil and Shorty looking at each other.
00:49:31:50	00:00:21:50	Phil: Well, I guess we might be interested in that. You got a private spot around here where we can talk?
00:50:40:40	00:00:30:40	Music swells then fades out as Phil, Shorty, and Mitzi head toward a hallway in the bar.

FIGURE 11.11 A timing sheet.

Other times, the composer might write a musical score designed to be played by an orchestra or band hired by the production company. The composer (often acting as the conductor) leads the group of musicians through the music. The portions of the production that require music are projected onto a screen in front of the composer so that he or she can make sure the recording fits the length and needs of the production. This is undertaken in a studio environment with technicians working at a large audio console to fade-in and mix various groups of instruments. The miking and recording process is similar to that for the musical recordings discussed in Chapter 10.

Smaller budgets (which encompass most non-broadcast video and film productions) often use free music or low-cost Internet sites, and other fee-based music libraries. Your school may have a music department program, where you can have original music composed.

When music is the mainstay of a program that is live or recorded as though it were live, a large audio crew is needed. *American Idol*, for example, has one group of audio technicians attending to the band and another attending to the singers. In addition, there are audio technicians handling the sound for the audience, the host, and the judges. All of these people need to work in close coordination with each other so that the sounds blend properly. They also must work cooperatively with the crew members handling the light cues, the camera work, and the scenery in order for the program to reach the audience as a seamless whole.

11.13 RECORDING ADR

ADR stands for **automated dialogue replacement** (sometimes also called **looping**). There really is nothing automated about it. It is a rather tedious process of rerecording dialogue that for some reason was recorded improperly during production. The reasons for needing ADR are usually legitimate—such as noises from a busy airstrip that were impossible to eliminate, or an actor who was too out of breath from physical activity to record useable dialogue.

If the director knows that the sound is going to need to be rerecorded, the audio crew does not need to try to record it properly. Instead they can record a **scratch track**—audio that is clear enough to be used as a guide during ADR, but not good enough for the final production. (One should always attempt to get the best image or audio signal that you possibly can. You may have to rethink what you are doing to get cleaner recorded audio, such as if there is a wind problem, maybe you can move a little to perhaps the other side of the building, etc.) During the ADR process, actors come to a soundproof room where they watch short segments (loops) of themselves on screen and listen through earphones to the scratch track for that loop. They rehearse lines until they feel they can say them in sync with their lips on the screen. Then they record the lines into a mic—usually a cardioid mic on a boom stand (see Figure 11.12). When they finish they go on to the next loop. A sound technician sits in a separate room by an audio console, playing back the loops and recording the new dialogue.

FIGURE 11.12 An ADR facility. Actors undertaking ADR look at the screen to sync the sound. They can stand or sit on the stool and that way they can move if they need to change perspective. Having a podium where they can place a script or notes also helps the process. *(Image courtesy of Sony Pictures Entertainment.)*

Of course, one problem with ADR is that it is recorded in pristine conditions, whereas the original audio (including parts of the dialogue that were perfectly fine) was recorded in a location with background noise. Here is where recorded room tone is a necessity. What is recorded in ADR will need to be mixed with room tone for the final product. If possible, try using the same mic to record ADR that was used to record the actor on the set. This can help in matching the recorded dialogue to that recorded on the set. ADR is expensive, so it is always preferable to record the sound properly on location, but sometimes looping is unavoidable.

A similar process to ADR is used for dubbing video productions into foreign languages or recording voices for an animated film. Although no faulty dialogue is being replaced, a technician plays short loops of video to actors who rehearse and then record the appropriate words.

11.14 RECORDING FOLEY

Most **Foley** work involves recording sounds that relate to actions in a movie or video production—for example, sounds like footsteps on concrete, the thud of a falling book, the clink of water glasses. **Foley walkers**, agile people who are often dancers, watch the video or film on a screen and perform the acts needed to provide synchronous sound for the various movements. The room where they record is equipped with various surfaces—sand, cement, tile, gravel—and various materials that they need to perform—high-heeled shoes, tree branches, water, utensils. They rustle cloth if a woman on screen walks close to the sofa, they don tennis shoes and run on cement if a man is jogging, they fall during fight scenes, scream in terror, and do whatever else is needed to enhance the sounds for the movie (see Figure 11.13).

A technician sits in an adjoining room recording all the sounds through an audio board onto a server or recorder. This technician or the Foley walkers make sure a microphone (usually cardioid) is properly placed to pick up the sound. If a Foley walker is trudging through sand, the mic will be by his or her feet. If he or she is pouring water from one pitcher to another, the mic will be at table height. Because the mic needs to be moved a great deal, it is usually mounted on a **boom arm**. Foley walkers often work in pairs with a man performing the acts related to men and a woman imitating women's movements, but sometimes one Foley walker undertakes all the actions. Foley is completed after the video or film is edited so that the Foley walkers are recording sounds as they will be needed for the final version of the project.

11.15 RECORDING VOICE-OVERS

Voice-overs (VO) (also called **narration**) are easier to record than ADR or Foley, because they do not involve synced sound. Depending on the type of production, they can explain complicated processes, indicate what a person is thinking, or comment on what is occurring in the picture.

FIGURE 11.13   This Foley walker is in a Foley pit moving branches in relation to what he sees on the TV screen. The mic on a boom stand is pointed toward the branches. *(Image courtesy of Henninger Media Services.)*

They are often recorded in soundproofed rooms with a setup similar to that of ADR, but they can be recorded from beginning to end without the looping. The person doing voice-over usually needs to watch the video so that the pace of reading matches what is on the screen. Sometimes, however, the voice-over is recorded before the production is put together and then it is edited to match the narration.

A cardioid mic attached to a desk stand or boom arm is generally used for voice-overs. Readers usually have scripts, and these are best displayed on a computer screen behind the mic. If the script is on paper, its pages should not be stapled together, because when they are turned they will make a shuffling noise. Rather readers should have loose script pages that they can lift off and gently place to the side.

11.16 POSTPRODUCTION CONSIDERATIONS

Before a video or film production is complete, the sounds need to be edited and mixed together. This can be a simple operation, such as making sure the musical interludes for an interview show are the correct length and do not overpower the speech. Or it can be a complex task that involves mixing dialogue, sound effects, ambient sounds, music, ADR, Foley, and voice-over. If there are many sound sources, they are often initially dealt with separately from the picture. While an editor is putting together the picture and dialogue, the sound designer or a **sound editor** is collecting and organizing the sound effects and ambient sounds and making sure that any ADR, Foley, or VO is under way.

Some of the sounds undergo **sweetening** at this point. Sweetening refers to making changes in the audio, usually to make everything sound better. For example, a door slam recorded during production might not sound realistic enough so the audio technician would sweeten it by adding some **reverb**. Audio technicians replace poorly recorded dialogue with ADR. They **filter** out an electrical hum accidentally recorded during a documentary interview. If there are continuity errors, they try to fix them, perhaps by adding walla walla to cover a gaff.

Once the editor has completed a rough cut of the picture, the sound editor can view it and create a **spotting sheet**, which indicates where in the production each effect should be placed (see Figure 11.14). The timing sheet for the composer also comes at this point.

Sound technicians place the various sound effects and other sounds on audio tracks, usually within a computer program (see Chapter 3, "Digital Audio Production"). Once the picture and dialogue are complete, the sounds (with the aid of time code) are positioned where they are needed—the gunshot when the trigger is pulled, the chirping birds throughout a rural scene. Sometimes, such as for a music video or a documentary with elaborate voice-over,

SPOTTING SHEET

Project: A Day in the Field   Page Number: 2
Spotter: Alicia Fong   Director: George Walters

Item	Sync/Non-Sync	In-Time	Out-Time	Description
20	NS	01:42:15:10	01:43:12:09	Airplanes landing
21	NS	01:45:26:29	01:46:15:12	Muddled voices
22	S	01:49:52:12		Suitcase hits floor
23	NS	01:49:52:12	01:53:05:05	Traffic noises
24	NS	01:54:03:27	01:56:22:17	Restaurant noises
25	S	01:56:13:25		Door slam
26	NS	02:03:14:23	02:04:03:10	Cell phone ringing
27	NS	02:10:52:12	02:11:01:24	Glass breaking
28	NS	02:11:27:03	02:12:04:19	Dog barking
29	S	02:12:46:18		Glass breaking
30	NS	02:15:33:16		Thunder
31	NS	02:16:12:15	02:19:22:04	Rain, wind, and thunder
32	S	02:21:45:23		Car backfire
33	NS	02:21:50:20	02:23:47:04	Children playing
34	NS	02:23:47:04	02:26:33:12	Traffic noises
35	NS	02:24:55:23		Thunder
36	NS	02:26:10:10		Thunder
37	S	02:30:12:14		Hit glass
38	NS	02:33:05:06	02:34:55:07	Dogs barking

FIGURE 11.14   A spotting sheet.

the picture is edited to the sound, so the audio track is built first. Usually, however, the sounds are matched to the picture. They are initially laid down without taking volume into account; they are simply placed where they are needed in the production.

11.17 FINAL MIX

The **final mix** is the time when the relative volumes are adjusted to fit the needs of the story. It can be undertaken within a computer program or in a room with a large screen, an audio board, and excellent speakers. Several technicians sit at the audio board, perhaps with one concentrating on dialogue, one on music, and one on sound effects. They watch the production in small increments, altering the volume of various elements so that the sounds have the proper **balance**—music does not overpower the sound of a car engine starting or one person's dialogue is not noticeably louder than another. They double-check perspective to make sure that sounds in the distance are fainter than sounds in the foreground.

They **fade-in** and **fade-out** music, **cross-fade** Foley sounds, **segue** sound effects, mix room tone with ADR, and so on. Sometimes they **upmix** dialogue or other sounds that were recorded in **mono** so that they become **stereo** or **surround sound**. They do this by using the **pan** control on the audio board to place the sound to the left or right of one of surround sound's 5.1 positions. They create the enveloping sound effects sounds, such as that of an airplane that sounds like it is flying behind you from left to right.

They usually have a **cue sheet** (see Figure 11.15) to work from so that they know what the director has in mind. The technicians must work cooperatively so that all of the sounds will work together harmoniously.

CUE SHEET

Title: The Next Episode   Sound Designer: Susan King
Director: Philip Acosta   Page: 3

Min.: Sec.	Track No. 1 Dialogue	Min.: Sec.	Track No. 3 Music A	Min.: Sec.	Track No. 4 Music B	Min.: Sec.	Track No. 6 Effects A	Min.: Sec.	Track No. 7 Effects B
		02:15	/ Fade-in Guitars			02:15	Wind blows		
02:31	Henry: Try to see							02:41	Tree branch falls
02:47	(last word) obvious								
02:49	Marcia: No, never again								
03:15	(last word) yesterday								
		03:21	Cross-fade\	03:21	/Cross-Fade Drums	03:21		03:21	Rain
03:25	Henry: Someday								
04:02	(last word) gun			04:03	Fade-out\				
						04:06	Gun shot		

FIGURE 11.15   A cue sheet.

If the final mix is completed using a computer program, one person does the whole thing. Under these circumstances, sweetening and the final mix can be one and the same—the audio operator alters and fixes sounds while mixing them, one small section at a time. If a production is simple (for example, adding opening and closing music and two sound effects), using a computer editing program is adequate. However, for complicated audio that needs extensive mixing and a large number of cross-fades, having several people work on sound together is faster and more likely to yield a better result.

If sound is going to be used on a computer for web-based material or video games, it is best to mix it within computer software. Video games present particular challenges when the games try to create different sounds depending on what the player is doing. For example, in first-person shooter games, if the player is holding the weapon steady, the audio focus may be on heartbeat and music intensity. If the player seems less confident, the sound track might consist of many distracting ambient noises.

When the final mix has been completed, it is placed in sync with the image. Where no more changes are expected in the editing of the images, other than color correction, it is called "picture locked."

11.18 CONCLUSION

Sound production used in conjunction with visual media is more complex than sound production for audio-only productions. It requires a knowledge of the principles related to all audio recording plus an ability to record and post-produce audio so that it enhances the overall production. Keep in mind that sound has specific requirements depending upon the visual production. For example, in video productions (and other media) there is a direct relationship to type of program. In news production, audio is the easiest and fastest to organize, while documentary productions can get complicated. TV (talk shows to talent shows) and TV soap operas (even though they are slowly going away) are produced in controlled broadcast studios, while dramas are quite often shot in the film style of a single camera. Those that can effectively manipulate audio in the service of the story are valued in video and film production.

Self-Study

QUESTIONS

1. What is the name of a sheet given to a composer that includes time code numbers of where the director thinks that music is needed?

   a) cue sheet
   b) spotting sheet
   c) timing sheet
   d) segue

2. Which of the following would be a perspective problem?

   a) A person is seen in a close-up, but the person's voice sounds like it is coming from a distance.
   b) The rumble of a train is heard when a man is talking to a woman but is not heard when the woman replies to the man.
   c) The sound cuts in and out because the microphones are too close together.
   d) The script supervisor neglected to note where the mic was positioned when the crew broke for lunch.

3. Why is a boompole effective for recording audio for video?

   a) It is also known as a fishpole.
   b) It is likely to cast a shadow.
   c) It needs to be held 6 feet in front of the talent.
   d) It can move above the heads of several performers.

4. What is a disadvantage of using a lavaliere mic to capture sound for a movie?

   a) It does not fit well on a desk stand and desk stands are the most common mic holders for movies.
   b) It might be visible in a close-up.
   c) It can be wireless.
   d) It is hard for the script supervisor to see.

5. Which of the following is a movable platform that can be used to move a microphone in various directions?

   a) grid
   b) shotgun
   c) table stand
   d) perambulator boom

6. Why are cardioid mics appropriate for recording dialogue for video?

   a) They generally don't pick up sounds from the area where the video equipment is operating.
   b) They are rugged.
   c) They are less likely to cast a shadow than omnidirectional mics.
   d) They are large microphones.

7. Which of the following would be an example of double-system sound?

   a) recording some material in sync and some out of sync
   b) recording to an audio recorder rather than the camera
   c) recording without using a mixer
   d) recording for a video game by using a clapper to ensure correct pitch

8. Which of the following is most likely to need various walking surfaces such as brick, tile, and cement?

    a) Foley
    b) ADR
    c) single-system sound
    d) VO

9. For which type of television show is there the most need for the microphone to be invisible?

    a) news report
    b) drama
    c) music video
    d) talk show

10. What is another name for "automated dialogue replacement"?

    a) ATM
    b) looping
    c) scratch track
    d) narration

11. Which of the following is least likely to be an ambient sound?

    a) atmosphere sound
    b) walla walla
    c) room tone
    d) music

12. Which of the following is most likely to involve filtering out a hum from an audio track?

    a) transferring audio to multitrack tape
    b) cross-fading
    c) sweetening
    d) recording with a boom stand

13. What is one way that dealing with audio joined with video is different from dealing with audio only?

    a) Wind noises are more of a problem for video, because it is usually shot in a studio.
    b) Cardioid microphones are never used for video shoots.
    c) Audio uses only digital audio consoles and video uses analog audio consoles.
    d) Microphone placement can be more difficult for video, because the microphone should not show in the picture.

14. What is the title of the person who plans the overall strategy of sounds for a production?

    a) sound designer
    b) script supervisor
    c) Foley walker
    d) composer

15. What might a microphone that is put in a vase of flowers that is on the set be called?

    a) floor mic
    b) omnidirectional mic
    c) hidden (or plant) mic
    d) flower mic

ANSWERS

If you answered A to any of the questions:

1a. Wrong. This is used for the final mix. (Reread 11.12 and 11.17.)
2a. You are right.
3a. A fishpole is for fishing. (Reread 11.5.)
4a. No. Desk stands aren't used often. (Reread 11.6 and 11.7.)
5a. No. This is not the correct term. (Reread 11.7.)
6a. Correct. Cardioids are less likely to pick up unwanted equipment noise.
7a. Wrong. Sync has nothing to do with double-system recording. (Reread 11.9 and 11.10.)
8a. This is the correct answer. These would be part of the Foley stage.
9a. No. In news reporting it is acceptable to hold a mic. (Reread 11.3.)
10a. Incorrect. This is where you get your money. (Reread 11.13.)
11a. No. Atmosphere sound is considered ambient sound. (Reread 11.11.)
12a. Wrong. Transferring to tape won't filter out unwanted sound. (Reread 11.16.)
13a. No. Wind would be the same for both types of productions and there wouldn't be wind in a studio. (Reread 11.1)
14a. Yes, sound designer is the correct answer.
15a. Wrong. This is not a name for a microphone. (Reread 11.7.)

If you answered B to any of the questions:

1b. Wrong. A spotting sheet is for other kinds of sounds. (Reread 11.12 and 11.16.)
2b. No. This would be a continuity problem. (Reread 11.8.)
3b. No. This would be a reason it is undesirable. (Reread 11.5.)
4b. Yes. This could be a disadvantage.
5b. Wrong. A shotgun isn't a platform. (Reread 11.7.)
6b. Not the best answer. Cardioids might be rugged and they might not depending on their structure. (Reread 11.4.)
7b. Yes. This is the correct answer for double-system sound.
8b. No. There is a better answer. (Reread 11.13 and 11.14.)
9b. Correct. You would not want a mic to be seen in a drama.
10b. Right. Looping is the correct answer.
11b. No. Walla walla is considered ambient sound. (Reread 11.11.)
12b. Incorrect. Cross-fading has nothing to do with filtering. (Reread 11.16 and 11.17.)
13b. Absolutely not. Cardioid mics are used for both, but are particularly useful for video shoots. (Reread 11.1 and 11.4.)
14b. No, the script supervisor deals with continuity. (Reread 11.3 and 11.8.)
15b. No. It might be omnidirectional, but pickup pattern doesn't have anything to do with its placement. (Reread 11.7.)

If you answered C to any of the questions:

1c. Yes. A timing sheet is the correct answer.
2c. This describes an entirely different phenomenon. (Reread 11.6 and 11.8.)
3c. No. That doesn't determine effectiveness. (Reread 11.5.)
4c. No. This is more likely to be an advantage. (Reread 11.6.)
5c. Incorrect. This is not a platform. (Reread 11.7.)
6c. Not the best answer. The type of mic is not what causes a shadow. (Reread 11.4 and 11.5.)
7c. Wrong. A mixer can be useful, but it's not the answer. (Reread 11.9.)
8c. Incorrect. This answer is unrelated. (Reread 11.9 and 11.14.)
9c. No. Singers often hold mics when they are performing. (Reread 11.3.)

10c. Wrong. Although a scratch track can be part of the process, it is not the other name. (Reread 11.13.)

11c. No. Room tone is considered to be ambient sound. (Reread 11.11.)

12c. Right. Sweetening is the correct answer.

13c. No. This wouldn't make any sense. (Reread 11.1)

14c. Wrong. This person only deals with Foley, not other aspects of sound. (Reread 11.3 and 11.14.)

15c. Correct. It could be called a hidden mic.

If you answered D to any of the questions:

1d. Wrong. This isn't even a sheet. (Reread 11.12 and 11.17.)

2d. Wrong. This could lead to a continuity problem but not a perspective problem. (Reread 11.8.)

3d. Yes. In this way, people can move.

4d. This is not the best answer. It may be a true statement but it is not a disadvantage. (Reread 11.6 and 11.8.)

5d. Yes. This is the correct term.

6d. Wrong. Size depends on construction. (Reread 11.4.)

7d. Very wrong. These elements have little to do with each other. (Reread 11.9 and 11.10.)

8d. Incorrect. (Reread 11.14 and 11.15.)

9d. No. A mic doesn't need to be invisible for a talk show. (Reread 11.3 and 11.7.)

10d. No. This isn't the right term. (Reread 11.13 and 11.15.)

11d. Yes. This is the correct answer.

12d. Wrong. A boom stand would not help. (Reread 11.17 and 11.16.)

13d. Yes, this is the correct answer.

14d. No. A composer only deals with music. (Reread 11.3 and 11.12.)

15d. No, there is no such thing. (Reread 11.7.)

Projects

PROJECT 1

Determine the importance of sound and picture.

Purpose

To help you decide whether you agree that audio provides more information than video.

Notes

1. Different genres of programs have different emphasis on video and audio. A talk show, for example, places a great deal of dependence on audio, and a western with many action scenes emphasizes video. Therefore, you may wish to use several shows for this project and mention the differences in your report.
2. If it is possible to record the programs at the same time as you are undertaking this project, you will then be able to look at them afterwards to see what was really happening.

How to Do the Project

1. Sit in front of your TV set and mute the sound. Watch for about 15 minutes and see how much of what is going on you understand.
2. Jot down your reactions—what you think you understand and what you don't and what degree of frustration you think someone might have if there really were no sound.
3. Turn the sound back on and sit with your back to the TV set so you can't see the picture. Listen for about 15 minutes and consider how much of what is going on you understand.
4. Jot down your reactions—what you think you understand and what you don't and what degree of frustration you think someone might have if there really were no picture.
5. Using your notes, write a report about your experience. If possible, try to figure out what percent of the material you understood when you only watched the picture and what percent you understood when you only listened to the sound.
6. Give your report to your instructor to receive credit for this project.

PROJECT 2

Record sound for a video project.

Purpose

To give you experience coordinating sound elements with the needs of the picture.

Notes

1. The website that accompanies this book contains some video material that you can use for this exercise, but you can also create something of your own.
2. If you are enrolled in a video production class, offer to be the sound designer for a group project and use that experience rather than what is on the CD-ROM.
3. Don't bite off more than you can chew. Be realistic about what types of sounds you can actually obtain. Don't spend a lot of money buying sound effects or traveling to exotic locations to record something. Use your ingenuity rather than your pocketbook.

How to Do the Project

1. View the video material on the website and plan the sounds you would like to add to it in terms of voice-over, sound effects, atmosphere sounds, music, and so on.
2. Gather all these sounds from CDs, from the Internet, or by recording them yourself.
3. Use whatever editing program you have access to for inputting the audio and video and editing it. You might want to reread some of the material in Chapter 3 that is related to editing.
4. Turn your project in to your instructor to receive credit for this project.

12

INTERNET RADIO AND OTHER DISTRIBUTION PLATFORMS

12.1 INTRODUCTION

Digital technology has changed audio production techniques as well as the very nature of how audio content is delivered. Radio stations, program syndication companies, and other content providers have found they must align themselves with the Internet and other "non-traditional" distribution outlets. For many, this has meant building **web pages** that tie programming to promoting contests, concerts, and other special events. Nearly all stations also **webcast**, or stream their broadcast signal over the Internet, broadening their audience from a local area to potentially the whole world. Radio stations and other content providers have also turned to "apps" (applications) that allow listener access to a wider variety of live and recorded content, including concerts and podcasts. As Internet access has spread from the desktop computer, to the laptop, to smartphones and other mobile platforms, audio content creation and delivery options have become nearly limitless.

College radio stations that have existed solely over-the-air on AM or FM can stream over the Internet and can reach international audiences. Many colleges that lack the resources or available broadcast frequencies in their area now have online radio stations and training programs. College radio webcasting provides a particular benefit to students who go away to college, because their friends and relatives back home can listen to their programs. Alumni also use Internet radio to stay informed about their school, and university sporting events are a popular draw for a variety of online listeners.

Anyone with the proper equipment can now program a "radio station" that can be heard around the world (see Production Tip 12A). A report by Edison Research showed that approximately 94 million people listened to online radio on a weekly basis in 2014 (up from 57 million in 2011), and that number is expected to rise. Established broadcasters and neophytes alike hope to attract these listeners to their programming and increase their advertising revenue.

This chapter looks at the techniques you should consider for creating an Internet radio station or production service, as well as several online and over-the-air radio distribution methods.

PRODUCTION TIP 12A
Ch-ch-ch-changes

I'd say I have had an interesting history in audio production and the broadcast industry. College gave me the tools and foundation I needed to pursue my interests in audio. Since graduating, I've worked in commercial radio and broadcast television, and now have my own side business producing voice overs, with some more detailed audio production sprinkled in.

A lot has changed over the past decade. Gone are the days of mailing demo cassettes (yes, cassettes) to production houses and potential clients. Gone are the days of rushing to FedEx to overnight a finished product on CD to your client. Gone are the days of installing expensive ISDN lines into your home studio to support remote voice over sessions. Gone are the days of literally walking from agency to agency to pitch your demo and promote yourself.

All of my recording, editing, and production is done in my home studio. This has played a huge part in obtaining work and delivering finished products to my clients. My website, easily created by anyone with basic web knowledge, has provided me with a solid online presence, where others can easily view and hear my demos and reach out to me directly. Online networking, searching out potential client leads, and finished product delivery all takes place in the digital realm. Online resources and products like LinkedIn, Google, Facebook, SquareSpace, and Dropbox have made this possible for me. What used to take days or weeks to turn around and get into a client's hands now takes mere hours. Sometimes even just minutes!

Independent production, digital presence, and internet delivery isn't the future of the audio production industry, it's the now.

FIGURE 12.1   Photo by Rob Fissel.

Rob Fissel *www.robfissel.com*

12.2 WEB PAGES

While many online listeners access programming through media players such as Apple's iTunes or Microsoft's Windows Media Player to download content to their iPods, MP3 players, or smartphones, chances are that at some point they will want to visit a station or program home page. The home page is still an important component of online broadcasting and streaming, because it provides detailed information about the station or program's content and links to additional material.

Most **home pages** contain at least one hyperlink that allows listeners to hear a station live. Some stations have developed their own proprietary players, which allow them to promote their brands and display revenue-generating advertising. In addition to live content, many station home pages also provide links to previously recorded **on-demand** content that can be listened to or downloaded at any time. For example, one link might lead to last night's basketball game, while another goes to the most recent edition of a documentary series, and another link contains an unedited version of an interview that was broadcast in an abridged form. In commercial radio, this recorded content sometimes requires a subscription fee for access, although many stations will provide material for free. Some stations direct listeners to places like Audible.com, where they will have to pay a monthly fee to access archived audio content. Public radio stations typically offer more free on-demand content than commercial stations, but may also require the user to listen to a short pitch for the station's latest fundraising campaign before the program begins.

A station's website may also include text and graphic information about contests, program schedules, disc jockeys, news stories, and other features. How web pages are organized is largely up to each individual station or their corporate owners, and is beyond the scope of this book. There are many software programs and online providers (Adobe® DreamWeaver®, and Wordpress.com,

for example) that can be used to build eye-catching web pages. This chapter focuses on what you need to consider in order to create audio content for online distribution.

12.3 OVERVIEW OF THE AUDIO PROCESS FOR STREAMING

For real-time **streaming** of a live broadcast to occur, the output of the on-air console needs to get to the **sound card** of a computer that contains encoding software. This **encoder** system translates the analog audio from the sound card into a digital streaming format and sends it to a **server**, which is a computer with much more storage space, faster processors, and more RAM. The server is connected to the Internet, and sends the audio data stream to the **player** software on a listener's computer, where it is decoded and played over the speakers (see Figure 12.2). The output of the audio console must be changed from analog to digital, because analog audio can't be streamed. It also must be converted back to analog on the listener's end, because the human ear can't hear digital frequencies.

Connecting the output of the audio console to the encoding computer with a sound card is fairly simple. As discussed in Chapter 5, "The Audio Console," most consoles have several built-in components or outputs, so at least one can be designated for the Internet while transmitting an on-air signal at the same time. The input to consumer-grade sound cards is usually a **miniphone connector**, while the input to professional-grade sound cards is either **RCA**, **XLR**, or a digital connection that can be made from a console with a digital output.

When it comes to providing previously recorded content that can be downloaded on-demand, the content needs to be created and stored as a **sound file**, in a format such as MP3, AIFF, WAV, or WMA. This is accomplished by using an audio editing software program such as Audacity, Pro Tools, or Adobe® Audition®. The file size and quality of the sound file must be considered carefully to insure that listeners can play it easily and that the server has the necessary bandwidth capacity to send it to a number of listeners at any given time. Although broadband connections are becoming more common, there are still many people in the United States, and many more abroad, who access

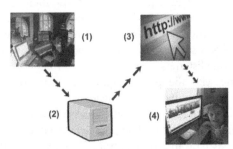

FIGURE 12.2   In the streaming process, audio from a studio/console (1) is encoded and sent to a server (2). The server sends the signal to the Internet (3), where it is downloaded and listened to through a computer with the appropriate software and speakers (4).

the Internet via low bandwidth, dial-up telephone connections of 56 kilobytes (Kb) per second or even less. These potential listeners will turn away if they try to access files that can't stream fast enough to be continuously heard, or take a very long time to download.

Different types of audio content require different settings to maximize their quality and minimize their size. For example, a sound file that contains only talk may be encoded at much lower quality than a music sound file, where a broader frequency range is desired. Encoding at a lower quality creates a much smaller file size, which is important when considering server space and bandwidth capabilities. Some online providers offer several different-sized files of the same program, so that people who are connecting via dial-up can choose a smaller file, while those connected through broadband can listen to a higher-quality (and larger) file.

Another consideration of online distribution is knowing there are two ways to listen to on-demand sound files—streaming and non-streaming. Streaming plays sound on a listener's computer soon *after* he or she clicks on the link to it. Streaming provides *near-instant* access to material, and playback may be interrupted if the speed of the listener's Internet connection can't consistently keep up with the amount of information being transmitted per second.

With non-streaming listening, the file is completely downloaded before it plays. This allows listeners with dial-up connections to listen to uninterrupted, high-quality audio files as long as they are patient enough to wait through potentially long download times.

Encoding software has utilized **variable bit rate (VBR)** processing more and more effectively over the last several years. Encoding software that uses VBR processing also automatically reduces the amount of data that makes up a file when there is little or no signal to digitize (such as silence or a single, constant tone). VBR processing automatically increases the amount of data that can be used for a sound file when a more complex sound is being produced (such as multi-instrumental, polyphonic music from a band or orchestra). The process of creating VBR content, however, takes time and can consume significant amounts of computer resources. As such, it is appropriate to use only for on-demand files, and not live streaming. In live streaming a **constant bit rate (CBR)** is used. CBR depends on the availability and capability of bandwidth processing, and unfortunately can result in more complex sounds being washed out if required bandwidth goes beyond what is available. Both Internet and satellite radio use CBR for live streaming.

12.4 ENCODERS

The encoding sound card in a computer is used to translate an analog signal into a digital format. The signal is sampled and then compressed so it can be sent efficiently over the Internet and received by people using DSL, satellite, cable,

or dial-up **modems** that can operate anywhere from 28 Kb to 12 megabytes (Mb) per second or even higher.

Many online content providers like to have their Internet files be CD-quality. As discussed in chapter 3, CD quality audio is produced and encoded using a **sampling rate** of 44.1 kHz, or 44,100 times per second. Additionally, each sample usually has a **bit depth** of 16 bits. One way to look at bit depth is its relationship to **signal-to-noise (S/N) ratio** (see Chapter 2, "The Studio Environment"). Generally speaking, the greater the number of bits, the higher the S/N ratio, and the better the signal quality. Each bit provides approximately 6 **decibels (dB)** of signal to noise, so 8-bit audio is about 48 dB of signal to 1 dB of noise; 16-bit audio provides a very respectable 96 dB to 1 dB of noise. However, a 44.1 kHz, 16-bit signal uses a lot of bandwidth; roughly 1.4 Mb per stereo second (to determine the amount of necessary bandwidth, multiply the sample rate and the bit depth together, and then multiply that number by two if the file is in stereo). Additionally, when data is sent over the Internet, it needs **overhead** (more bandwidth) so that the packets it is sent in arrive at their destination properly. What this means is that in order to receive pure CD-quality sound, a listener might actually need an Internet connection that operates at approximately 1.7 Mb per second. The good news is that many people have modems that operate from 1.5 to 12 Mb per second, so streaming live audio or downloading audio files is not as much of a problem as it used to be. As mentioned previously, however, there are still many homes in the United States and abroad that rely on dial-up modems that connect to the Internet at *56 to 300 kilobytes* per second (Kbps). It could take as long as 2 hours to download a 3-minute CD-quality song using a 56 Kbps modem. In order to insure that all listeners are able to stream and download content, the bandwidth used by an audio signal being sent over the Internet must be reduced as much as possible.

There are several ways to handle this reduction. One solution is to send a monaural signal instead of stereo, which cuts the amount of bandwidth needed in half. A more common method is to employ **compression**, which is used by most encoding systems and software. Compression degrades the quality of the signal somewhat, but it gets the required bandwidth down to about one-tenth of what is needed for CD-quality sound. Most compression systems use some form of **perceptual coding** that is based on characteristics of the human ear. For example, there are certain frequencies that humans can't hear (such as those below 20 Hz or those above 20,000 Hz). With perpetual coding, these frequencies are eliminated from the sound file. There are also frequencies that human ears hear better than others. If two sounds are playing at the same time, we tend to hear the louder one better than the softer one. Perpetual coding reduces or eliminates softer sounds in a sound file, thus reducing the size of a file even more. By taking human hearing into account, a great deal of what is in a normal sound signal can be eliminated without much damage to its quality. Some compression systems also deal with silence.

They take out the silent places within the audio and insert a code that places the silence back in when the sound is **decompressed** before it gets to the listener's audio speakers.

There are many encoding programs on the market, such as the LAME MP3 encoder for Audacity, and Radionomy's SHOUTcast. Most have the capability to encode for both live streams and on-demand content and have various settings that allow for limits on file size, audio degradation, and the amount of time needed to encode. However, keep in mind that there are always trade-offs. The concept of having to choose between "cheap," "fast," and "good" applies to audio encoding. If you have the time and resources to allow your computer to do several passes at compressing a file, the result can be cheap and good, especially in regard to the amount of bandwidth used. Just going with cheap and fast, however, can result in a poorer quality file that may not be appreciated by listeners.

12.5 SERVERS

Computers that are used as **servers** must run software that is specially designed to send multiple audio data streams to the Internet simultaneously. Unlike over-the-air radio where one signal can reach an infinite number of listeners, online distribution must provide a separate audio stream from the server for each listener. This can use up large amounts of bandwidth, especially if the audio is encoded at high quality and file size. Some online providers find they must therefore limit the number of listeners who can tune in at any one time, or time out due to inactivity. One way around this is to incorporate multiple servers with high bandwidth connections to the Internet, but that can be prohibitively expensive. Another solution is to contract out to local **Internet service providers (ISPs)** that have multiple servers already in place. This can be troublesome as well however, since many ISPs charge higher rates for the amount of bandwidth consumed in a given period of time. Most colleges and universities have on-campus servers that can help stream their radio station(s) online.

12.6 PLAYBACK SOFTWARE AND APPS

Streaming player software resides on a listener's computer or mobile device. It receives the audio data stream from the server and decodes and decompresses it into analog sound. However, most streaming audio does not stay on the listener's hard drive; it streams in and is then discarded by the listening device.

Inside the listening device, the initial compressed audio data is placed into a **buffer area**, similar to a small storage tank. Once a certain amount of data has partially filled the "tank," the sound file begins to play. Incoming streamed audio continues to be placed near the top of the buffer while the playback audio is being drained from the bottom.

FIGURE 12.3 Wi-Fi Internet radios, such as the IRC6000 Mondo from Grace Digital, allow listeners to hear static-free programming of Internet radio stations and podcasts as well as music from sources such as Sirius/XM, Pandora, and Live365. Listeners can tune into online international broadcasts as well. *(Image courtesy of Grace Digital Inc., copyright 2012. Reproduced with permission of Grace Digital Inc. from www.gracedigital audio.com.)*

The filling of the buffer is slightly faster than the draining, so there is always enough in reserve to continue playing the audio if the input stream is momentarily interrupted.

Playback software on most listening devices is usually available for free or at a low price. Users simply download it, or it comes bundled with store-bought devices. The majority of audio players used on computers are stand-alone programs such as Apple's iTunes and Microsoft's Windows Media Player. On portable devices and smartphones, there are also a variety of listening apps that can be downloaded, such as Pandora, Spotify, and TuneIn Radio. While many of these apps are free to download and use, some will offer commercial-free versions for a small price. The majority of playback software, whether on a computer or mobile device, is created using Adobe®'s Flash Player®, which is the same player used for most of today's Internet video and animation applications. Therefore, In order to use practically any audio player, the Flash Player plug-in is required on your device.

Several companies have developed stand-alone radios that are designed to receive Internet radio stations, podcasts, and other audio streams via Wi-Fi without a computer or mobile device, such as the Grace Digital radio shown in Figure 12.3.

12.7 SOFTWARE OPTIONS

Sometimes a company that creates and provides encoder and server software to a station is the same company that creates and provides playback software to the listener, and sometimes this arrangement can present a problem. Listeners may need one brand of software to listen to one station and another to listen to a different station. Although both brands of software can be downloaded free or for a minimal cost, the two may compete to be the default software on the consumer's computer, and some playback apps may not provide access to as many radio stations as others. It isn't uncommon today to have several brands of playback programs or apps on one computer or mobile device, just to make sure the listener has access to as much content as possible.

In addition, playback software is always being updated by the manufacturer. The major manufacturers in the market want to keep providing better, more dependable products in order to attract more listeners from their competitors. Updated encoding and playback software is often very appealing to content providers, in that it often provides better-quality sound with smaller file sizes, and allows more simultaneous outgoing streams. However, if listeners have not upgraded their playback software (or don't know how to), they will only be frustrated by this jump to the cutting edge. It is just as important for content providers to inform their listeners of any software upgrades as it is for them to provide programming.

As of this writing, there are two major playback software providers for desktop and laptop computers: iTunes from Apple (www.apple.com), and Microsoft's Windows Media Player (www.microsoft.com). Other computer playback software providers include:

Real Media Player from Real Networks
www.real.com
Radionomy's Winamp
www.winamp.com
Spotify
www.spotify.com
Pandora
www.pandora.com
Google Play Music
www.play.google.com
Last FM
www.last.fm
iHeartRadio
www.iheart.com
TuneIn
www.tunein.com

Some of these sites also offer listening apps that can be downloaded to smartphones and tablets. Keep in mind that while most of these services are free to listeners, some charge monthly subscription fees. Real Media Player was the first major online audio player to enter the market in 1995, and is still a popular option for listeners today. It dominated the streaming content market in the mid- to late 1990s, but since then players bundled with operating systems, such as iTunes and Windows Media Player have captured the bulk of market share.

The various types of proprietary audio file formats can cause headaches for listeners. While the MP3 format is universal and will play back on any software platform, its lack of security combined with the advanced compression techniques of several other formats do not allow it to be widely-used for streaming. Additionally, MP3 is audio only, and with more and more applications combining video and animation with audio, today the MP3 format is used more for on-demand file storage and playback of content. Perhaps the only mainstream audio content providers that still use the MP3 format are Amazon's music download service (www.amazon.com) and Google's Music Play service (www.play.google.com); however, both rely on a proprietary software package to download and process music on a customer's computer. Downloaded content from Amazon or Google can be automatically stored and played back through iTunes or Windows Media Player.

Microsoft's proprietary WMA (Windows Media Audio) format and the MP4 format used by Apple are both more capable for streaming audio, video, and animation. To confuse things, however, the WMA format can only be played on Apple computers after a third party plug-in has been installed. Likewise, the MP4 format will not play through Microsoft's Windows Media Player unless a plug-in has been installed, and to top it off, Real Media has its own file format (RM), which can be played on both Apple and Windows machines, but not without the proper plug-in application. When creating your own content and posting it online, it is very important to let your listeners know about these limitations and the various plug-ins available. Doing so goes a long way toward alleviating any possible frustration on the listening end.

More recently, many radio stations have bypassed providing their own streams and have instead opted to make their content available through a centralized provider such as Tune In, Last FM, or iHeartRadio. Although stations and other providers may have to pay fees to host their streams or pre-recorded material on these sites, some may find it worth the cost to let a third party handle the technical requirements associated with streaming and providing content online.

PRODUCTION TIP 12B
Internet Audience and On-Air Talent Interaction

The Internet is a very flexible tool for exchanging information and communication, and can be used to expand any business, including radio. Allowing listeners to feel that they are a part of a radio station or show is a good way to assure they keep coming back; however, it can backfire as well, so care must be taken in the introduction of such technologies and add-ons.

The Internet and a good webmaster can offer countless cheap and easy ways to invite listeners to get involved in a station: iPad and Android apps, social media links to Facebook and Twitter, and email and instant message clubs are just a few. However, it is important for users of these services to feel that they are active, so they are encouraged to continue using them. If you set up a station Facebook page for instance, or a text club with little or no activity, they will look dead to visitors. It is better to start small and then branch out to new options or threads after the site is already populated and activity is consistent.

Another concern is that interactive social media outlets run the risk of being "trolled" by mischievous or malicious individuals who may post incendiary content. It is a good idea to have a moderator for these outlets, who can review comments for unwanted content before they are posted or remove them shortly after. However, if you plan to moderate your sites, make sure that you have the personnel to do so and make sure you notify your fans and followers of your commenting policies.

12.8 ON-DEMAND FILES AND PODCASTING

As mentioned previously, one of the advantages of Internet radio is that programs can be placed on a server for listeners to access any time. Listeners can use this benefit to hear programming at times that are convenient for them, and also determine exactly what they want to listen to.

When recorded material is to be placed in an on-demand mode, special care should be taken to make a high-quality recording because these files also go through encoding, compression, and other processes. The more useful audio information the system has to analyze, the better. Try to record with as little background noise as possible, with high-quality equipment.

Most sound-file names will end with file extensions such as WAV, AVI, RM, WMA, AIFF, or MP3. For recording material that will be made available for downloading, it is best to produce and save your files in the MP3 format.

MP3 is an audio file format commonly used on the Internet and supported by a large number of applications. MP3 is short for **MPEG-2, Audio Layer 3. MPEG** stands for **Moving Picture Experts Group**, which is the industry group that developed compression systems for video data. This particular MPEG system encompasses a subsystem for audio compression that is referred to as Layer 3, and is based on the "perceptual coding" method discussed earlier. Practically every streaming program or app supports this format in addition to their own proprietary format(s).

Podcasting is another form of sending digital files (mostly MP3) over the Internet, but is different from regular, on-demand downloadable files, in that it relies on **RSS (Really Simple Syndication)** feeds with the audio file in the enclosure tag for playback on a computer or mobile device. "Podcast" is a term that refers to both the audio content and the method of delivery. The term originated by combining "iPod" and "broadcasting," but has come to be short for a more generic, and personal, on-demand form of listening. Podcasts cover almost any topic, including music shows, verbal journals, political essays, lectures,

or talk shows. Podcasts can be syndicated by a producer or subscribed to by a listener, and new content can be downloaded automatically through an aggregator that is capable of reading RSS feeds. RSS feeds can also be programmed to automatically let listeners know about new and updated podcasts. Apple's iTunes is one popular source for finding and subscribing to podcasts and offers several tutorials on developing and posting podcasts on the Web.

The basic process for creating a podcast is not difficult. First, you have to create the audio file or content of your program. Regardless of what audio format is used to create the files, they will need to be converted into MP3 format before they can be podcast. Next, you need to create an RSS feed file, which provides keywords that tell listeners about the content of your program, and allows aggregators or readers to find and access your podcast. While there are templates and tutorials on the Internet to manually write an RSS file, there are also programs such as PodcastGenerator (www.podcastgen.sourceforge.net) that will simplify the process. Next upload both the audio file and RSS file to your website or to the site of a firm that specializes in hosting podcasts, such as blogtalkradio.com or SoundCloud. com. Make sure that your MP3 audio file and the RSS file are located in the same directory. You also want to validate your RSS feed file, and can do so by going to feedvalidator.org and typing in the URL of the RSS file. Finally, you need to publicize your podcast on your site and list it in various podcast directories. Listeners who subscribe to your RSS feed will receive notifications of your podcast and many will then download it and listen at a time that is convenient for them.

12.9 BUILDING A HOME STUDIO FOR INTERNET AUDIO PRODUCTION

Building a mini production studio in your own home is not difficult at all. In fact, with just a computer, a microphone and appropriate audio editing software, you could record productions for Internet radio stations, podcasting, and other Internet applications. Add a few additional pieces of audio equipment and your home studio becomes even more "professional" and offers more flexibility in use. However, before assembling any equipment, consider where you will locate the studio. Although it's not likely that you can dedicate a whole room to a studio and turn it into an acoustically correct studio environment, consider finding a quieter spot—perhaps a corner of a room away from windows and doors that can allow exterior noise to leak in. Also, avoid any room that might have an air conditioner or other background noise source that occurs on a regular basis.

The computer should have a fairly fast CPU of at least 2 GHz, a minimum hard drive size of 500 GB, and at least 4 GB of RAM. Any upgrading beyond these specifications will make your system work even better. Your computer

FIGURE 12.4 M-Audio's FastTrackPro is a 4-input, 24-bit/96 kHz USB audio recording interface designed to let you record with professional results. *(Image courtesy of M-Audio.)*

should also have a CD/DVD drive and a good sound card. Instead of a computer, you could consider using a multitrack digital workstation, but in most cases, a computer will be less expensive and easier to work with, and you may already have the equipment in your home.

Although many computers include an adequate sound card, you should consider investing in one of better quality or consider using an external audio interface. For one thing, most computer sound cards will have tiny ⅛-inch input and output jacks, which are difficult to work with; better-quality cards offer ¼-inch or XLR I/O. Lynx, Digital Audio Labs, Creative Labs, and Digigram offer sound cards with digital and analog inputs, as well as other desirable features that make them popular for home studios. Another approach is to use an external interface, such as an M-Audio USB recording interface (see Figure 12.4) or a small audio mixer from Mackie or Behringer. Often this setup will give you more inputs, better mixing control, and direct connectivity to your computer.

The microphone used in your system should be a general-purpose microphone, like the Shure SM57, which can be found for about $100. You may also consider a USB microphone, such as those described in chapter 4, designed to plug directly into your computer. Remember, if you plan to do interviews, record group discussions or a band, you will need more than one microphone to get good, high-quality recordings.

A home recording studio also needs monitor speakers or headphones. You can get by with the speakers that came with your computer, but better-quality speakers will allow you to judge the quality of your productions better. If you are using an external interface or mixer, it probably has an output to drive a pair of speakers. Of course, headphones will also work, but again, look for some that are fairly good quality.

The final piece of "equipment" needed for a home studio is the editing/recording software. Audacity can be downloaded from the Internet at no charge, and other programs such as Adobe® Audition®, Avid Pro Tools, or Sony Sound Forge that offer multitrack capability, sound processing, and additional editing features, are fairly inexpensive. Put these pieces of equipment together and you'll be ready to record in the privacy of your own home studio.

12.10 COPYRIGHT

Copyright in relation to online content and distribution has become a major concern. Because it is so easy to copy digital material, most copyright laws (written in the 1970s) either do not cover the concept or are ineffective in stemming the tide of illegal copying. An Internet radio station or producer must remain aware of the latest in copyright legislation and enforcement or they can find themselves in major legal trouble.

There are three agencies that license music to broadcast radio stations—ASCAP (www.ascap.com), BMI (www.bmi.com), and SESAC (www.sesac.com). These organizations have been in business for many years and have negotiated license agreements that enable radio stations to play music without having to negotiate for each song. The agencies collect a standard fee from each station and then require stations to submit music logs of what they play. They then distribute the money collected to composers and music publishers based on an approximation of how often each selection was played. There are three agencies, because each of them handles different types of music, and a station that subscribes to all three can feel confident that it is playing music without violating copyright laws both on-air and online. Each organization charges a percentage fee based on a combination of the amount of revenue generated by your site, the number of times you played specific songs through your site, and the estimated number of online listeners. Additional fees apply when your site turns a profit, and special arrangements are in place for stations broadcasting religious music. Additionally, ASCAP and BMI offer discounts on license fees if your station is affiliated with live365.com. If you are starting an Internet radio station and you are planning to use pre-recorded music of any kind, you should make sure you contact these agencies.

Another copyright licensor to consult is the Recording Industry Association of America (www.riaa.org), which regularly updates information about copyright procedures. This site can help determine what licenses you need and how to obtain them.

Licensing requirements are also administered by SoundExchange (www.soundexchange.com), an independent, nonprofit performance rights organization designated by the U.S. Copyright Office to collect and distribute digital performance royalties for recording artists and copyright owners when their recordings are performed on digital cable, satellite television music channels, or Internet and satellite radio.

12.11 INTERNET RADIO STATION LISTING SITES

Once you have established an Internet radio station, you want people to know about it. With all the material on the Internet, it can be hard for anyone to find your station. To help solve this problem, there are Internet sites that list and link radio stations by format type, geographic area, and various other parameters. Some charge for listing and some do not. You would be well advised to contact as many of these as possible to see whether they will list your station.

Several of the more popular station listing sites include:

- Streamfinder (www.streamfinder.com)
- Live 365 (www.live365.com)
- RadioTower (www.radiotower.com)
- Shoutcast (www.shoutcast.com)
- Live-Radio (www.live-radio.net)
- Radio Station World (www.radiostationworld.com)
- iTunes Radio (itunes.apple.com/radio)

While many computer audio players such as iTunes and Windows Media Player have built-in radio station listing services, this feature can be both good and bad for radio stations. It can be good in that it allows listeners to easily find their station; it can be bad in that it only provides a link to the audio stream, and not to the website, where additional advertising, appeals for support, and branding can occur.

Establishing and operating an Internet radio station can be a rewarding activity and a great learning experience for anyone interested in a career in either traditional radio or new media technologies.

12.12 OTHER DISTRIBUTION MEANS

New distribution techniques, including the Internet, are not unusual in radio programming. As each new form has taken hold, the old ones change, and sometimes fade from popularity. Always keep in mind, however, that regardless of how the material is distributed, content must be produced in a way that meets the entertainment and information needs of the public.

AM broadcasting started in the 1920s and was the dominant form of content distribution for many years. FM was developed during the 1930s but did not come to fruition until the 1960s, and it later overtook AM as the favored listening form. When cable TV went through a growth spurt in the 1980s, it added audio-only services to its menu, and although many cable subscribers use the service, it has never achieved the popularity level of AM or FM radio. Recently, the development and implementation of **digital audio broadcasting** via satellite and traditional terrestrial methods have competed with each other for listeners and advertising dollars.

12.13 SATELLITE RADIO

Satellite radio (as its name implies) beams programming directly from a satellite to a home, automobile, or portable receiver. Signals are distributed nationally with no need for local stations. Programming is up-linked from several earth-based studios to satellites designed to transmit the signal back to earth and provide coast-to-coast coverage. If the satellite signal is interrupted, ground repeaters are used to beam the signal to listeners. Because satellite radio is a digital service, its signals are higher quality than standard analog AM and FM broadcasting.

In 1997, the FCC granted Sirius Satellite Radio and XM Satellite Radio the right to broadcast via satellite. In 2008, the FCC allowed these two companies to merge into a single satellite radio service (Sirius/XM Radio). Consumers pay a monthly fee for access to hundreds of channels of music, news/talk, sports, and entertainment programming. Many channels are narrowly focused to particular interests such as *Talk for Truckers*, *Krishna Das Yoga Radio*, *The Howard Stern Show*, *Mad Dog Sports Radio*, and others. Some of the channels are commercial free (usually the music channels), and others carry a handful of ads. Some of the stations on Sirius/XM, such as Los Angeles' KIIS-FM, are simultaneous broadcasts of terrestrial broadcasts with all commercial breaks included. Some programs, such as *The Howard Stern Show*, that were once available only on traditional radio stations have moved to satellite radio. Satellite radio uses different frequencies than AM or FM broadcasting, so anyone wanting to listen needs a specific receiver with their subscription. Sirius/XM has increased its popularity by increasing its availability. In addition to striking deals with auto manufacturers to provide satellite receivers as standard equipment in most new cars, listeners can also access Sirius/XM via satellite TV, cable TV, and the Internet.

12.14 CABLE AND SATELLITE TV AUDIO

Many cable and satellite TV systems provide digital CD-quality audio services that are piped into the home along with TV service. In some instances, the cable and satellite providers charge extra for these services, but in some the services are part of the basic package. On some cable systems the audio is used as background for text channels listing local events or program schedules, and some satellite companies provide audio services to commercial establishments such as doctors' offices or department stores. Many colleges operate public access cable channels and place student-produced audio on the channel, often behind text-based announcements.

12.15 OVER-THE-AIR BROADCASTING

The most common way that radio programming is delivered and listened to is by local broadcast stations. Audio goes from the station's studio to its **transmitter**, where it is **modulated**.

This means the signal is superimposed onto a **carrier wave** that is created by using a station's specific frequency. Some stations operate in the **AM** band, where the frequencies are 535 to 1,705 kilohertz (kHz), and others are in the **FM** band of 88 to 108 megahertz (mHz). These frequencies are much higher than sound frequencies and can travel much farther than sound. Once the sounds have been modulated, they are sent out of the station antenna into the air. The carrier wave is picked up by a radio receiver that demodulates the signal and turns it back into sound that can be heard.

Technically, the difference between AM and FM involves how the on-air signal is modulated. AM stands for **amplitude modulation** and involves a method that varies the amplitude, or height, of the carrier wave in order to transmit the sound. FM stands for **frequency modulation** and is a process that changes the frequency of the carrier wave in order to transmit the sound. Figure 12.5 shows a representation of the two types of modulation.

Most student-operated radio stations are in the FM band. This has nothing to do with modulation or bandwidth, but more with politics. Many AM stations were operated by educational institutions in the 1920s and 1930s, however, as radio's popularity grew and frequencies were in short demand, universities lost out to commercial interests. When FM was established, educators convinced the FCC to reserve some of the frequencies (88.1 to 91.9) for non-commercial use. Today, many of these frequencies are licensed to colleges and universities that use them for public radio programming or as student stations.

12.16 HD RADIO

In 2002, mainly to help terrestrial radio compete with satellite radio, the FCC approved **in-band, on-channel (IBOC)** technology to allow AM and FM radio stations to transmit a digital signal at the same time and the same frequency as their analog signals. The sole developer of the technology is iBiquity Digital Corporation. **HD Radio**, as IBOC is commonly known, provides CD-quality FM radio and AM radio that rivals the quality of FM. HD Radio signals can't be received through a standard AM/FM radio, so listeners must use a specific radio. An important thing to remember is that HD Radio sets can tune in to analog AM and FM signals as well as digital signals.

HD Radio technology offers the ability to multicast, so in addition to a digital main channel, two secondary channels can provide different programming. In 2015, more than 3,400 HD Radio channels were in operation in the United States, up from 2,100 in 2012.

Figure 12.6 shows how HD Radio works. (1) An analog signal is sandwiched with digital information that can include text (such as artist and song information and more). (2) The digital signal layer is compressed using iBiquity's HDC compression technology. (3) The combined signal is transmitted on the station's regular frequency (88.9 FM or 1030 AM for instance). (4) Multipath distortion (radio interference commonly found with AM and FM broadcasting) can occur when part of a signal bounces off an object and arrives at the receiver at a

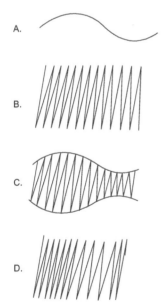

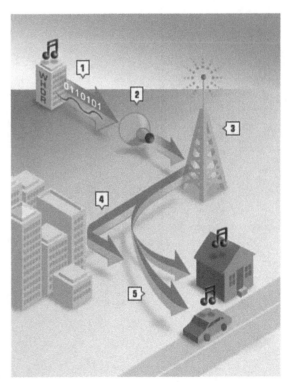

FIGURE 12.5 (A) An electrical wave that represents original sound. (B) The carrier wave of a radio station that is of a much higher frequency. (C) How the carrier wave is changed for amplitude modulation. Note that the height of the carrier wave is varied to take into account the characteristic of the sound wave, while the frequency stays the same. (D) How the carrier wave is altered for frequency modulation. Note that the frequency of each wave is varied and related to the original sound wave, while the height (amplitude) remains the same.

FIGURE 12.6 How HD Radio works. *(Image courtesy of iBiquity Digital Corporation.)*

different time than the main signal. (5) HD Radio receivers pick up the digital signal while older, analog radios that can't receive HD Radio signals are still able to receive the regular AM or FM signal. HD Radio receivers are designed to sort through multipath interference and reduce static, hiss, pops, and fading.

12.17 CONCLUSION

This chapter presents just a glimpse into the wealth of outlets available for student audio work. In addition to stations operated by colleges and universities, there are Internet sites, cable services, and broadcast stations that accept individual programs produced by students. Anyone wishing to get experience and exposure while a student can certainly find a distribution outlet for quality audio work, and with more distribution outlets than ever before, employment prospects after graduation are better than they used to be. Additionally, with the cost of computing technology and software continuing to go down, setting up and producing content at home has never been easier. The biggest challenges are keeping up with the ever-changing methods of delivery and production, technically maintaining your online platforms, and promoting your work so that listeners know about them. Students must do what they can to stay in "the know" with constantly developing changes in digital radio and audio technology.

Self-Study

QUESTIONS

1. How has the Internet been important to college radio?

   a) Stations can have an international audience.
   b) Alumni can keep in touch by listening to the college radio station over the Internet.
   c) Colleges that could not have a station because of lack of available frequencies can now have one.
   d) All of the above.

2. Which of the following defines on-demand programming?

   a) programming that is streamed live over the Internet
   b) files of programmed material that can be accessed over the Internet
   c) the icon that listeners click on to receive live streaming
   d) a home page as opposed to a web page

3. Which of the following aspects of Internet radio is on the listener's computer, tablet, or smartphone?

   a) encoder
   b) server
   c) player
   d) overhead

4. Which type of sound file allows listeners with low-bandwidth connections to the Internet to still listen to high-quality, on-demand sound files?

   a) streaming
   b) nonstreaming
   c) FM
   d) Flash

5. Someone with a modem operating at 56 Kb per second can easily receive uncompressed, analog audio signals from the Internet.

   a) true
   b) false

6. As an audio signal requires more and more bandwidth, which of the following will occur?

   a) the faster the system will operate
   b) the lower the S/N ratio will be
   c) the greater the frequency of the sound will be
   d) the higher the quality of the sound will be

7. Approximately how much does the S/N ratio improve with each additional bit when you are dealing with bit depth?

   a) 44.1 kHz
   b) 48 dBS
   c) 6 dB
   d) 1,411,200 bits

8. Which is true about perceptual coding?

   a) It takes into account what sounds the human ear can't hear.
   b) It involves sending a mono signal instead of a stereo one.

c) It is a compression system that increases the amount of bandwidth a signal uses.

d) It places more silence in a signal so it can be easily decompressed.

9. Servers that provide Internet radio stations do not need to have as much bandwidth as those that receive the signal.

a) true

b) false

10. What is one difference between Internet radio and over-the-air radio?

a) Internet radio stations do not need to pay license fees to play music.

b) Internet radio covers a smaller geographic area than over-the-air radio, enabling only people who are close to its server to receive its signal.

c) Each listener is served individually for Internet radio, but an over-the-air signal can be picked up by anyone who has a radio within the service area.

d) Internet radio was developed before over-the-air radio.

11. Streamed audio stays permanently on a listener's computer in a large buffer area that acts like a storage tank.

a) true

b) false

12. Which two audio streaming formats are currently used by the most stations?

a) QuickTime and Real

b) Windows Media and QuickTime

c) Real and Winamp

d) Real and Windows Media

13. Which of the following software audio players has a simple interface and few compatibility issues?

a) Winamp

b) Windows Media Player

c) QuickTime

d) iTunes

14. What is the audio compression system based on the one developed by the Moving Picture Experts Group known as?

a) WAV

b) MP3

c) QuickTime

d) AVI

15. Which of the following is a music licensing agency?

a) ISP

b) RIAA

c) ASCAP

d) MPEG

16. Which of the following delivers CD-quality radio content, such as commercials, to radio stations around the country?

a) digital audio workstation

b) digital distribution network

c) digital audio card

d) digital audio editor

17. Which of the following was developed the earliest?

    a) AM radio
    b) FM radio
    c) cable radio
    d) satellite radio

18. Which mobile form of radio requires a subscription and a special receiver?

    a) cable radio
    b) Internet radio
    c) FM radio
    d) satellite radio

19. Which company was one of two designated by the FCC to establish a satellite radio service?

    a) Apple
    b) Ibiquity
    c) Sirius
    d) Microsoft

20. What is one difference between AM and FM?

    a) AM stations have greater bandwidth than FM stations.
    b) AM stations modulate by varying the height of the carrier wave, whereas FM stations modulate by varying the frequency of the carrier wave.
    c) FM stations are more subject to static than AM stations.
    d) More college radio stations are on the AM band than are on the FM band.

21. Analog audio can be streamed online.

    a) true
    b) false

22. VBR streaming takes time and can consume a lot of computer resources. Therefore, it is more appropriate to use for:

    a) streaming live audio
    b) downloading on-demand files
    c) both A and B
    d) none of the above

23. Producers of online audio content should not concern themselves with informing their listeners of any software player updates or upgrades.

    a) true
    b) false

24. Encoding at a lower quality creates a _____ file size, which is important when considering server space and bandwidth capabilities.

    a) larger
    b) smaller
    c) comparable
    d) incompatible

25. HD Radio signals can be received using a standard AM/FM radio.

    a) true
    b) false

If you answered A to any of the questions:

1a. This is basically correct, but there is a better answer. (Reread 12.1.)

2a. No. By definition, on-demand is not streamed live. (Reread 12.2.)

3a. No. An encoder would be used by the station. (Reread 12.3, 12.4, and 12.6.)

4a. Wrong. A quality streaming sound file will not play easily on a computer with low bandwidth. (Reread 12.3.)

5a. Wrong. (Reread 12.4.)

6a. No. It would be just the opposite. (Reread 12.4.)

7a. No. This relates to sampling, not bit depth. (Reread 12.4.)

8a. You are right. Perceptual coding takes into account what a human can hear.

9a. No. It is the opposite. (Reread 12.4 and 12.5.)

10a. Wrong. They must both pay. (Reread 12.5 and 12.10.)

11a. No, not on the listener's computer. (Reread 12.6.)

12a. No. These are not the top two. (Reread 12.7.)

13a. This is the best response. Nullsoft's Winamp is a bare-bones player and one with few compatibility issues.

14a. No. This is a type of sound file, but not a compressed one. (Reread 12.8.)

15a. No. (Reread 12.5 and 12.10.)

16a. No. A DAW could be utilized to produce a commercial, but it is not used to send spots around the country. (Reread 12.14.)

17a. Yes, in the 1920s.

18a. No. The cable TV system is not mobile. (Reread 12.13 and 12.15.)

19a. Wrong. This is not a satellite radio company. (Reread 12.7 and 12.13.)

20a. Wrong. It is the other way around. (Reread 12.16.)

21a. Incorrect. Analog audio must be converted to digital before it can be streamed online (Reread 12.3)

22a. Wrong. Streaming live audio uses CBR, not VBR (Reread 12.3)

23a. No. It is a good idea to inform listeners of software upgrades (Reread 12.7)

24a. Incorrect. Encoding at lower quality does not create larger files (Reread 12.2)

25a. Incorrect. You need a special receiver to listen to HD Radio (Reread 12.16)

If you answered B to any of the questions:

1b. This is basically correct, but there is a better answer. (Reread 12.1.)

2b. Correct. A listener can click on a file on a web page.

3b. No. A server would be used by the station. (Reread 12.3, 12.5, and 12.6.)

4b. Correct. Nonstreaming files (on-demand) can be downloaded and then listened to.

5b. Right. Data must be compressed or else the information would take forever to download.

6b. Wrong. This has more to do with bit depth than bandwidth. (Reread 12.4.)

7b. No. This number was used in reference to 8-bit sound. (Reread 12.4.)

8b. No. That is a form of bandwidth reduction, and it is not related to perceptual coding. (Reread 12.4.)

9b. Yes. You are correct.

10b. Wrong. It would be the opposite. (Reread 12.5 and 12.16.)

11b. Correct. It does not stay on the listener's computer.

12b. No. These are not the top two. (Reread 12.7.)

13b. No. This is considered a more complicated player and has compatibility problems. (Reread 12.7.)

14b. Yes. This stands for MPEG-2, Layer 3.

15b. No. This organization has been involved with copyright disputes, but it doesn't license music. (Reread 12.10.)

16b. Yes. This is the correct response.

17b. No. There is a better answer. (Reread 12.12.)

18b. No. Internet radio sometimes requires a subscription, but does not require a special receiver. (Reread 12.13.)

19b. Wrong. This is not a satellite radio company. (Reread 12.7 and 12.16.)

20b. Correct.

21b. Correct. Analog audio can't be streamed online.

22b. Correct. It is best to use VBR for downloading pre-recorded files.

23b. Yes. Keep your listeners happy!

24b. Correct. Encoding at lower quality creates smaller-size files.

25b. Correct. HD Radio signals can't be listened to on a standard AM/FM radio.

If you answered C to any of the questions:

1c. This is basically correct, but there is a better answer. (Reread 12.1.)

2c. No. The program is not an icon, and they would not click it to stream. (Reread 12.2.)

3c. Yes. The listener's device has the player.

4c. No. FM is not a form of a digital sound file. (Reread 12.3 and 12.16.)

6c. Wrong. Bandwidth and frequency are not related here. (Reread 12.4.)

7c. Yes. This answer is the correct choice.

8c. No. This is wrong for a lot of reasons, one of which is that compression does not increase bandwidth. (Reread 12.4.)

10c. Right. Internet radio requires a separate stream for each listener.

12c. No. Winamp is not a file format. (Reread 12.7.)

13c. No. This is considered a more complicated player and has compatibility problems. (Reread 12.7.)

14c. No. This is Apple based, not MPEG. (Reread 12.7 and 12.8.)

15c. Correct. ASCAP is one of the music licensing agencies.

16c. No. A digital audio card is part of a desktop radio system. (Reread 12.14.)

17c. Wrong. This didn't come along until much later than some of the other choices. (Reread 12.12.)

18c. No. FM radio, though mobile, does not require a subscription. (Reread 12.13 and 12.16.)

19c. Yes. XM Satellite Radio was the other company.

20c. No. It is the other way around. (Reread 12.16.)

22c. Incorrect. It is not possible to download pre-recorded files that are streamed live. (Reread 12.3.)

24c. No. This answer does not make sense. (Reread 12.2.)

If you answered D to any of the questions:

1d. Yes, this is the best answer. All of the answers apply.

2d. No. On-demand programming could be placed on a home page or web page, but it is not either of them. (Reread 12.2.)

3d. Wrong. Overhead is not a relevant term here. (Reread 12.3, 12.4, and 12.6.)

4d. Wrong. Although Flash is capable of transmitting sound, it does not allow users to download files onto their computers for later use. (Reread 12.2 and 12.3.)

6d. Correct. More bandwidth equals higher quality.

7d. No. This is acutally the number of bits needed for a minute of stereo audio. (Reread 12.4.)

8d. No. This answer does not relate to perceptual coding. (Reread 12.4.)

10d. No. (Reread 12.5 and 12.12.)

12d. Correct. Real and Windows Media are the top two.

13d. Wrong. iTunes is considered more complex with its numerous options and has compatibility issues. (Reread 12.7.)

14d. No. This is a type of sound file, but not the right one. (Reread 12.8.)

15d. Wrong. This is an organization that develops compression standardization. (Reread 12.8 and 12.10.)

16d. No. A digital audio editor could be used to produce a commercial, but it's not used to send spots around the country. (Reread 12.14.)

17d. No. (Reread 12.12.)

18d. Correct. Satellite radio is a subscription service.

19d. No. Microsoft, though it seems to have its fingers in most businesses, is not directly in charge of a satellite radio company. (Reread 12.13.)

20d. Wrong. That is not the band where most college radio stations are located. (Reread 12.16.)

22d. No. There is a correct answer. (Reread 12.3)

24d. Wrong. This answer does not make sense (Reread 12.2)

Projects

PROJECT 1

Report on the differences and similarities among six radio station websites.

Purpose

To give you a feel for how stations are currently utilizing the Internet.

Notes

1. Try to select different types of stations—for example, Internet only, both broadcast and Internet, AM stations, and FM stations.
2. Consider both the audio and visual elements of the websites as you do your analysis.

How to Do the Project

1. Find six different stations that have websites on the Internet. Feel free to use the Internet sites that list radio stations, as mentioned in Section 12.11.
2. Go to one of the websites and click on a link for pre-recorded sound files or live streaming. If necessary, download any player software.
3. Make notes about the appearance of the site and about the type of audio material that is available. Look at the various pages on the site and see the types of materials that are included. You might want to consider such facets as: how much material is on the home page; whether the site contains live streaming and on-demand files and, if so, what types of on-demand files are included; the color scheme of the pages; the organizational principles around which the pages are presented; the extent to which on-air personalities are featured; the level of interactivity with listeners (social media, text clubs, email clubs, promotions, song requests, etc.) and so on.
4. Do the same for five additional sites.
5. Write a report that compares and contrasts the radio station presentations. See if you can find trends or recurring features. For example, do music stations tend to use the Internet in different ways than talk stations? Are there things that Internet-only stations include that broadcast stations do not?
6. Put your name on your report and label it "Website Analysis." Give the report to your instructor to receive credit for this project.

PROJECT 2

Tour a broadcast radio station transmitting facility.

Purpose

To enable you to see a broadcast transmitter firsthand and better understand the broadcast process.

Notes

1. Although you may have toured a radio station before, you may not have seen the transmitter facility, because many broadcast studios are physically separated from their transmitter site.
2. Try to arrange to talk with an engineer, or the person who can give you the most information.
3. Make sure that before you go you have some ideas about what you want to find out.
4. Your instructor may have set up a tour for the entire class, or the engineer at your college station may offer a tour of your own facilities. Either case is fine for completing the project.

How to Do the Project

1. Select a station. (If the instructor has arranged a station tour for the whole class, skip ahead to Step 4.)
2. Call someone at the station, tell him or her that you would like to see the station's transmitter site and talk with someone about it so you can write a report for an audio production class, and ask if you may visit.
3. If the answer is yes, set a date; if not, call a different station.
4. Think of things you want to find out for your report. For example:
    a. How does the line out of the audio console get to the transmitter?
    b. Is there any signal processing equipment at the transmitter site?
    c. Does the station use amplitude modulation or frequency modulation, and exactly where is this modulation equipment located?
    d. What is the physical layout of the transmitter site?
    e. How does the signal get from the transmitter to an antenna?
5. Go to the station transmitter site. Tour to the extent the station personnel will let you and ask as many questions as you can.
6. Write a report about what you found out.
7. Put your name and "Transmitter Site Report" on your report, and hand it in to your instructor to get credit for this project.

PROJECT 3

See what you have and what you need in order to build your own audio recording and editing facility in your home or dorm room.

Purpose

To show how easy it can be to set up your own recording facility and to discover what you might need (and how much it might cost) to make one.

Notes

1. Consult Chapter 12 for some suggestions about equipment and software.
2. Take note of the equipment used in your production studio(s) at school and determine if you need similar equipment.

How to Do the Project

1. Take an inventory of the digital computing equipment that you own, that could be used for recording and editing. What do you have?
2. Take note of the equipment you would need to create your own recording and editing facilities. What do you need?
3. Where can you find the equipment you need to create your own facilities? How much would it cost?
4. Create an inventory of what you have and what you would need to create a recording facility. Be sure to include where you would find the equipment you need, and how much it would cost.
5. Write a short report on what you discovered, put your name and "Home Recording Inventory" on your report, and hand it in to your instructor to receive credit for this project.

GLOSSARY

A-B miking A method of stereo miking where one microphone feeds the right channel and another microphone feeds the left channel. Also known as "spaced pair miking" or "split pair miking."

Absorption The process of sound going into the walls, ceilings, and floors of a studio.

Absorption coefficient The proportion of sound that a material can absorb in relation to the sound it will reflect back. A coefficient of 1.00 means that all sound is absorbed in the material.

Acoustics The science of how sound behaves in an enclosed space.

Acoustic suspension A speaker enclosure design that consists of a tightly sealed box that prevents rear sounds from disrupting main speaker sounds.

Actuality A voice report from a person in the news rather than from the reporter.

Actual malice As relates to libel, something that was known to be harmful and incorrect but which was reported anyway.

Adaptive transform acoustic coding (ATRAC) A data compression system used for Mini discs.

ADR See *Automated dialogue replacement.*

AES/EBU A digital connector standard set by the Audio Engineering Society and the European Broadcasting Union.

Aliased The shifting of frequencies that are not within the range of hearing into the audible range.

AM See *Amplitude modulation.*

Ambient sounds Background noises that are desirable for an audio or video production and that are often recorded and added to the production.

Amplifier A piece of equipment that boosts the strength of a signal, often as it goes from one audio element to another.

Amplify To make louder.

Amplitude The strength or height of a sound or radio wave.

Amplitude modulation A form of radio transmission in which the amplitude (height) of a carrier wave is varied according to the characteristics of the sound signal being broadcast.

AMS See *Automatic music sensor.*

Analog A recording, circuit, or piece of equipment that produces an output that varies as a continuous function of the input, resulting in degradation of the signal as material is copied from one source to another.

Announce booth See *Performance studio.*

Anti-aliasing Filtering the input signal during the digital process to prevent the creation of unwanted frequencies, usually ones not within the audible range.

ATM Asynchronous transfer mode; a high-speed telephone system capable of transmitting audio.

Atmosphere sounds Recorded sounds that give a sense of location or a particular feeling.

ATRAC See *Adaptive transform acoustic coding.*

Attack The time it takes an initial sound to build up to full volume.

Audio board See *Audio console.*

Audio card A connection between a computer-based audio workstation and other audio equipment; also known as a "sound card."

Audio chain The route through various pieces of equipment that sound takes in order to be broadcast or recorded.

Audio console The piece of equipment that mixes, amplifies, and routes sound; also known as a "mixer" and "control board."

Audio routing switcher A type of patch panel that allows audio inputs to be switched to various outputs electronically.

Audio signal A sound signal that has been processed into an electromagnetic form.

Audio tape recorder A device that rearranges particles on magnetic tape in order to store sound.

Audition An output channel of an audio console.

Automated dialogue replacement (ADR) Rerecording dialogue in a studio situation that for some reason was not recorded properly during production; also known as "looping."

Automatic music sensor (AMS) A button on a digital audio tape recorder that allows the operator to skip forward or backward to the start of a new song.

Aux See *Auxiliary*.

Auxiliary An output channel of an audio console.

Backing layer The back side of audio tape—the side that does not have a magnetic coating.

Balance Recording or mixing sound in such a way that the proper elements are emphasized and one sound does not drown out another that is more important.

Balance control A knob on stereo input channels used to determine how much sound goes to the right channel and how much sound goes to the left channel.

Balanced cable A cable with three wires—plus, minus, and ground.

Band cut filter See *Band reject filter*.

Band pass filter A filter that cuts all frequencies outside a specified range.

Band reject filter A filter that allows all frequencies to pass except a specified frequency range.

Bass reflex A speaker enclosure design that has a vented port to allow rear sounds to reinforce main speaker sounds. Also known as a "vented box."

Bass roll-off switch A switch that turns down bass frequencies to counter the proximity effect.

Bidirectional Picking up sound from two directions; usually refers to a microphone pickup pattern.

Binary A number system that uses two digits: 1 and 0.

Bit depth Number of data bits used to encode a digital sample.

Blast filter See *pop filter*.

BNC connector A bayonet type connector with a twist lock that is primarily used for analog and serial digital video production that can also be used for digital audio.

Boom A microphone stand that sits on top of a base or is held by a person in a fishpole style.

Boom arm A microphone stand for use in the radio studio, consisting of metal rods designed somewhat like a human arm; one end goes into a base that can be mounted on a counter near the audio console, and the other end supports the microphone.

Boompole A type of microphone holder that consists of a pole held by an operator.

Boom stand A stand that can be placed away from an announcer; usually consists of one vertical pipe with a horizontal pipe at the top of it.

Bouncing A multitrack recording technique that combines two or more tracks and transfers them to a vacant track; also known as "ping-ponging."

Boundary microphone See *Pressure zone microphone*.

Branding The act of defining a station so listeners know what to expect when they tune in. See *imaging*.

Broadband A high-speed data transmission system that can accommodate audio and video and connect to the Internet faster than dial-up telephones.

Buffer area The area in computer memory that temporarily stores downloaded data before it is played back.

Bumper A prerecorded audio element that consists of voice-over music that is used as a transition between different forms of content.

Cable Wire that carries audio signals.

Cable radio Music services offered by cable TV systems.

Camera microphone A microphone that is built into a camera or permanently mounted on it.

Cannon connector See *XLR connector*.

Capacitance The ability of a piece of electronic equipment to hold or store an electrical charge.

Capacitor microphone See *condenser microphone*.

Capstan A metal shaft that controls the speed of a tape recorder.

Cardioid Picking up sound in a heart-shaped pattern; usually refers to a microphone pickup pattern.

Carrier wave A radio wave that is constant in amplitude or frequency but can be modulated by some other audio signal; used to deliver a radio broadcast signal.

Cartridge A device that converts the vibrations from the turntable stylus into variations in voltage; also, the endless-loop tape container used in a tape recorder.

Cartridge recorder A tape recorder that uses tape that is in an endless loop.

Cassette A plastic case containing ⅛-inch audio tape.

Cassette tape recorder A tape recorder that records and plays back ⅛-inch tape housed in a plastic case.

CBR See *Constant bit rate*.

CD See *Compact disc*.

CD player See *Compact disc player*.

CD-R See *CD recorder*.

CD recorder A type of CD machine that can record as well as play back CDs.

CD-RW (CD-rewritable) A CD format that can be recorded on more than once.

CF See *CompactFlash*.

Channel The route an audio signal follows; also, a grouping of controls on an audio console associated with one input.

Chorusing A multitrack overdubbing technique in which an announcer reads the same script on several different tracks to give a "chorus" effect.

Circumaural Refers to covering the outer part of the ear. See *Closed-cushion headphones*.

Clapper A wooden palette with a hinged piece of wood across the top that is used to sync sound, because the sound of wood hitting wood can be matched up with the picture of the two hitting each other.

Closed-cushion headphones A ring-shaped muff that rests on the head, not the ear, through which a person can hear sound.

Close-proximity monitoring See *Near-field monitoring*.

Coding Assigning a 16-bit binary "word" to the values measured during quantizing.

Cold ending A natural, full-volume ending of music or a song.

Combo The working procedure in which the radio announcer is also the equipment operator.

Compact disc (CD) A round, shiny disc onto which sound is recorded digitally so that it can be read by a laser.

Compact disc (CD) player The piece of equipment that uses a laser to play back CDs.

CompactFlash (CF) One of the many formats of portable digital memory upon which large quantities of broadcast-quality audio can be recorded.

Compander Signal processing equipment that compresses dynamic range during recording and expands it during playback.

Compression A sound wave characteristic that occurs when the air molecules are pushed close together; also, a system for encoding digital data bits so fewer can be placed on a recording medium yet still represent the original data.

Compressor A volume control usually associated with the transmitter that boosts signals that are too soft and lowers signals that are too loud.

Condenser microphone A microphone that uses a capacitor, usually powered by a battery, to respond to sound. Also known as "capacitor microphone."

Connector adapters Freestanding connector parts that allow one connector form to be changed to another.

Connector Metal device to attach one piece of audio equipment to another.

Constant bit rate (CBR) A technology used for live streaming, wherein data—complex or simple—has the same bit rate.

Continuity Consistent and unobtrusive progression from one shot to another without distractions in terms of audio or video elements that are present in one shot and absent in the other.

Control board See *Audio console*.

Convergence The amalgamation of various media, such as radio and television, so that they share common traits.

Copy marking A means of using a system of graphic symbols whereby supplemental notations and punctuation are added to the script to assist in the interpretations of written copy.

Copyright The exclusive right to publish or sell certain literary, dramatic, musical, or artistic works.

Countdown A type of music show in which the most popular songs are played in reverse order of their popularity, for example, a Top 10 countdown would start with the tenth most popular song and end with the most popular.

Cross-fade To bring up one sound and take down another in such a way that both are heard for a short period of time.

Crossover An electronic device that sends low frequencies to the speaker woofer and high frequencies to the tweeter.

Cross-talk The picking up on a tape track of the signal from another track.

Cue To preview an input (such as a CD or audio tape) before it goes over the air; also, to set up an audio source at the point where the input is to start.

Cue sheet List used during the final mix that indicates when each sound of a production should be brought in and taken out.

Cue talent A signal given to talent that means, "You're on"; it is given by pointing the index finger at the talent.

Cue wheel Part of a CD player that allows the operator to find the exact starting point of the music.

Cut A hand signal given to talent at the end of a production; it is given by "slicing your throat" with your index finger.

Cut 'n' splice A form of editing used during the analog era wherein audio tape was cut with a razor blade and put back together with splicing tape.

DASH A linear reel-to-reel recording standard that uses a stationary recording head, and is capable of recording up to 48 tracks of audio on a ¼-inch or ½-inch tape.

DAT See *Digital audio tape*.

Data compression In audio production, a digital transduction process that eliminates frequencies above and below the threshold of hearing. This is done in order to create smaller file sizes for streaming, downloading, and storage.

DAW See *Digital audio workstation*.

dB See *Decibel*.

dbx® A noise-reduction system that compresses both loud and soft parts of a signal during recording and then expands them during playback.

Dead air A long pause on a broadcast when no sound is heard.

Dead spot A place from which a remote unit cannot send a signal to a studio or other destination, usually because buildings or hills are in the way of a line-of-sight transmission.

Dead studio A studio with very little echo or reverberation, caused by a great deal of absorption of the sound.

Decay The time it takes a sound to go from full volume to sustain level.

Decibel (dB) A measurement to indicate the loudness of sound.

Decompress The process of restoring an audio signal that has been encoded with some type of data compression system.

De-esser A processor that gets rid of sibilant sounds without affecting other parts of the signal.

Demographics Information related to specific statistics about a population such as sex, age, occupation, and education.

Depth The apparent placement of a microphone to receive sound between the front and back planes of a recording environment. It provides the aural image of foreground and background space.

Desk stand A microphone stand for a person in a seated position.

Dialogue Words that are spoken by a person in a radio or television production.

Diffusion   Breaking up sound reflections by using irregular room surfaces.

Digital   A recording, circuit, or piece of equipment in which the output varies in discrete on/off steps in such a way that it can be reproduced without degradation of the signal.

Digital audio broadcasting (DAB)   A transmission process that compresses and digitizes a radio station's analog signal.

Digital audio tape (DAT)   High-quality cassette tape that can be dubbed many times without degradation because of the sampling process of its recording method.

Digital audio workstation (DAW)   A computer-based system that can create, store, edit, mix, and send out sound in a variety of ways, all within one basic unit.

Digital cart recorder   A piece of equipment that operates similarly to an analog cart machine but stores sound on a computer disk.

Digital delay   A unit that holds a signal temporarily and then allows it to leave the unit.

Digital distribution network   A network that links ad agencies, production houses, or record companies with radio stations to deliver CD-quality audio via PC-based servers and phone lines.

Digital radio   In-band, on-channel radio that AM and FM stations use to provide high-quality digital sound.

Digital reverb   A unit that produces reverberation electronically.

Digital signal processor   A type of electronic audio card used for computer editing.

Digital versatile disc (DVD)   A data storage format that has the capacity to hold a feature-length movie on a CD–styled medium. It can also be used for music and computer data.

Direct sound   Sound that goes straight from a source to a microphone.

Disc jockey (DJ)   A person who introduces and plays music for a radio station. The term arose because the person plays recorded discs and "rides gain" on the audio board.

Distortion   A blurring of sound caused by overamplification or other inaccurate reproduction of sound.

DJ   See *Disc jockey*.

Dolby   A noise-reduction system that raises the volume of the program signal most likely to be affected by noise during production, then lowers it again during playback so that the noise seems lower in relation to the program level.

Double-system sound   A recording method wherein sound is recorded on one piece of equipment and picture on another.

Dovetailing   A multitrack overdubbing technique in which a single announcer appears to have a dialogue with himself or herself by recording different parts of a script on different tracks.

Dramatic pause   A little extra silence before making an important point.

Driver   A loudspeaker component that receives and reproduces frequencies from the crossover. See *Tweeter* and *Woofer*.

Drop-out   A flaking off of oxide coating from audio tape so that the total signal is not recorded.

DSP   See *Digital signal processor*.

DSP audio card   A necessary component in order to use standard computers to edit audio.

Dubbing   Electronically copying material from one tape to another.

Duration   The time during which a sound builds up, remains at full volume, and dies out.

DVD   See *Digital versatile disc*.

Dynamic microphone   A microphone that consists of a diaphragm, a magnet, and coils. It is extremely rugged and has good frequency response, so it is used often in audio recording. Also known as "moving-coil microphone" and "pressure microphone."

Dynamic range   The volume changes from loud to soft within a series of sounds; also, the amount of volume change a piece of equipment can handle effectively.

Dynamic speaker   A speaker with a cone attached to a voice coil. Electrical current in the voice coil creates a magnetic force that moves the cone. Also known as an "electromagnetic speaker."

Earbud   A headphone that fits in the ear.

Echo   Sound that bounces off one surface.

Edit decision list   A feature of audio editing software that documents the edits made during a production session and allows the user to display previous edits and access them at any time.

Electret microphone   A type of condenser microphone with a permanently charged capacitor.

Electromagnetic speaker   See *Dynamic speaker*.

Electrostatic headphones   Headphones that require external amplification.

Electrostatic loudspeaker   A speaker design that generates sound using a thin plastic membrane (usually coated in graphite) suspended between two electrostatically charged plates.

Encoder   In streaming audio, a computer or software that converts the analog signal into a digital format.

"Engineer-assist"   The working procedure in which the radio announcer is assisted in the operation of the equipment by an engineer.

Envelope   The stages that sound goes through during its duration from full volume to silence.

EQ   See *Equalization*.

Equalization (EQ)   The adjustment of the amplification given to various frequencies such as high frequencies or low frequencies.

Equalize   To adjust the amplification of the various frequencies of an audio sound.

Equalizer   The unit that adjusts the amount of amplification given to particular frequencies such as high or low frequencies.

Equal loudness principle   Sounds that are equally loud will not be perceived as being equally loud if their pitch is different.

Ergonomics Design factors within a facility that reduce operator fatigue and discomfort.

Etherne An international standard for local area network transmission technology.

Expander Signal processing equipment that increases the dynamic range of sound, for example, making low-volume passages even lower in volume.

Fade To gradually increase or decrease the volume of music to or from silence.

Fade-in To bring sound up from silence to full volume or the desired level.

Fade-out To take sound from full volume to silence.

Fader Part of an audio console that moves up and down to control volume. Also known as a "slider."

Feedback A howling noise created when the output of a sound (usually from a speaker) is returned to the input (usually a microphone).

Filter A unit of softare setting of an audio editing program that cuts or reduces a specific frequency range in an audio signal.

Filtering The process of eliminating or reducing specific frequencies in an audio signal.

Final mix The last sound work in an audio or video project where the volume levels of the various tracks are set and made to coordinate with the picture.

Firewire A digital connection that enables audio and video signals to go from one piece of equipment to another; often referred to as "1394," "1394b," "400," and "800."

Flanger A device that electronically combines an original signal with a slightly delayed one.

Flat A method of recording wherein everything is recorded at approximately the same volume.

Flat frequency response The quality of a frequency curve wherein all frequencies are produced equally well.

Floor stand A microphone stand for a person in a standing position.

FM See *Frequency modulation*.

FM microphone Another name for a "wireless microphone."

Foley To record sounds, such as footsteps and branches moving, in sync with the picture.

Foley walker A person who creates sounds within a facility where sounds are recorded in sync with the picture.

Frequency The number of cycles a sound wave or radio wave completes in one second.

Frequency modulation (FM) A form of radio transmission in which the frequency (wavelength) of a carrier wave is varied according to the characteristics of the sound signal being broadcast.

Frequency response The accurate reproduction of both high and low frequencies that a piece of audio equipment reproduces.

Full-track A recording method that uses the whole tape for one monophonic signal.

Fundamental A basic tone and frequency that each sound has.

Gain control A knob or fader that makes sound louder or softer.

Gain trim Controls on an audio board that are used to fine-tune the volume of each input.

Geosynchronous satellite A satellite positioned approximately 22,233 miles above the earth that travels the same orbit as the earth thereby appearing to hang indefinitely above a certain earth position.

Give mic level A signal given to talent to tell him or her to talk into the microphone so the audio engineer can set controls properly. It is given by "chattering" one hand, with the palm down and the thumb under the second and third fingers.

Graphic equalizer An equalizer that divides frequency responses into bands that can then be raised or lowered in volume.

Grease pencil A crayon-like substance used to mark edit points on an analog tape.

Grid In a television studio, pipes close to the ceiling from which lights are hung.

Half-track mono The recording of two separate mono signals on a tape—one going to the left and one going to the right.

Hand signals A method of communication that radio production people use when a live microphone prohibits talk or when they are in separate rooms.

Hard drive A large storage medium either built into a computer or offered in a portable mode.

Hardwiring Connecting equipment in a fairly permanent manner, usually by soldering.

Harmonics Exact frequency multiples of a fundamental tone.

Harmonizer An effects unit that raises or lowers the pitch of a sound to create a desired effect.

HD Radio A name for in-band, on-channel digital radio that AM and FM stations use to provide high-quality digital signals.

Head An electromagnet that rearranges iron particles on tape; also, the beginning of an audio tape.

Headphone jack The female connector input on an audio console or other piece of equipment that is used for headphones.

Headphones Tiny speakers encased in something that can be placed in, or close to, the ear.

Headset A combination headphone and microphone in one unit, usually used by sportscasters and helicopter traffic reporters.

Hear-through cushion headphones See *Open-air headphones*.

Hemispherical A microphone pickup pattern that captures sound well within a 180-degree angle.

Hertz (Hz) A measurement of frequency based on cycles of sound waves per second.

Hiss A high-frequency noise problem inherent in the recording process.

Hoax A deceptive trick, often done just for the sake of mischief.

Home page   The initial or index page for an Internet website.

House sound   Sounds generated and mixed by technicians at a venue such as an auditorium that can then be picked up by someone else plugging into the final output of the venue equipment, usually so that they can record the sound.

Hum   A low-frequency noise problem caused by leaking of the 60-cycle AC power current into the audio signal.

Hyper-cardioid   Picking up sound well from the front, but not the sides; usually refers to a microphone pickup pattern.

Hz   See *Hertz*.

IBOC   See *In-band, on-channel*.

Idler arm   A tension part of a reel-to-reel tape recorder that will stop the recorder if the tape breaks.

Imaging   The apparent space between speakers and how sounds are heard within the plane of the speakers; the act of defining a station so listeners know what to expect when they tune in.

Impedance   The total opposition a circuit offers to the flow of alternating current.

In-band, on-channel (IBOC)   A method of digital audio broadcasting used by terrestrial radio stations.

Indecency   Language that, in context, depicts or describes, in terms patently offensive. as measured by contemporary community standards for the broadcast medium, sexual or excretory activities or organs.

Indirect sound   See *Reflected sound*.

In phase   A combination of two sound waves such that their crests and troughs exactly align.

Input selectors   Switches that are used to choose microphone or line positions on an audio board.

Insert edit   See *Punch in*.

Internet service provider (ISP)   A business or organization that provides access to the Internet.

In the mud   Operating volume consistently below 20 percent on the VU meter.

Invasion of privacy   The act of not leaving someone alone who wants to be left alone, especially a celebrity.

ISDN   Integrated Services Digital Network; a type of digital phone line for audio and data transmissions.

Jacks   Female connectors.

Jewel box   A plastic case for a CD.

Jingle   A short audio piece that includes the singing of such elements as the station call letters or slogan.

kHz   See *Kilohertz*.

Kilobyte (kB)   A unit of computer memory or data storage capacity that is approximately 1,000 bytes.

Kilohertz (kHz)   1,000 cycles per second.

Laser   An acronym for "light amplification by simulated emission of radiation"; a narrow, intense beam in a CD that reads encoded audio data.

Laser diode   A semiconductor with positive and negative electrons that converts an electrical input into an optical output.

Latency   The short amount of time required to convert analog audio into digital audio, add a digital effect to audio, or move audio from one location to another.

Lavaliere microphone (lav)   A small microphone that can be attached unobtrusively to a person's clothing.

Leader tape   Plastic tape that does not contain iron particles to record; it is used primarily before and after the recording tape so that the tape can be threaded.

LEDE   See *Live end/dead end*.

LEO   See *Low-earth orbit satellite*.

Libel   Broadcasting or printing something about a person that is both harmful and false.

Life cycle   Components of direct and indirect sound.

Limiter   A compressor with a large compression ratio that won't allow a signal to increase beyond a specified point.

Line level   An input that has already been preamplified.

Line of sight   A method of transmitting a signal wherein the transmitter and the receiver are lined up in such a way that there could be a visual communication between them.

Liner   A sentence or two that a disc jockey says over the opening of a song or between songs.

Liner notes   Information found on the back of vinyl albums or inside CDs about the songs, artists, writers, and so forth; often used by announcers to provide ad-lib information.

Live bouncing   A multitrack recording technique that combines two or more tracks plus a live recording and transfers them to a vacant track.

Live end/dead end (LEDE)   A studio where one end of the studio absorbs sound and the other end reflects sound.

Live studio   A studio with a hard, brilliant sound caused by a great deal of reverberation.

Long play (LP)   A record that can hold at least an hour of music.

Loop   To record something such as a sound effect over and over in order to make it longer.

Looping   See *Automated dialogue replacement*.

Low cut filter   A filter that eliminates all frequencies below a certain point.

Low-earth orbit (LEO) satellite   A satellite positioned approximately 200 to 400 miles above the earth that is often used for satellite phones.

Low pass filter   A filter that allows all frequencies below a certain point to go through unaffected.

LP   See *Long play*.

Magnetic layer   The part of the tape that contains the iron oxide coating.

Magneto-optical design   A recordable CD that records on a magnetic alloy and uses laser light to play back.

Master fader   The control that determines the volume of the signal being sent from the audio console.

MD   See *Mini disc*.

Megabyte (MB) A unit of computer memory or data storage capacity that is approximately one million bytes.

Memory card A chip that can retain information without a power source and can be recorded on and then erased and recorded on again.

Microphone A transducer that changes sound energy into electrical energy.

Microphone level An input that has not been preamplified.

MIDI See *Musical instrument digital interface.*

Mid-side miking (M-S miking) A method of stereo miking where three microphones are arranged in an upside-down T pattern.

Mini See *Miniphone connector.*

MiniDisc (MD) A 2½-inch computer-type disk that can hold 74 minutes of digital music.

Minidrama A short episode or story line such as is often found in commercials.

Miniphone connector A small connector with a sleeve and a tip.

Mixer See *Audio console.*

Modem A device used for communication of computer data over standard telephone lines that converts digital signals to analog and vice versa.

Modulated A radio wave whose frequency or amplitude has been changed according to the characteristics of another audio signal to broadcast that signal.

Mono See *Monoaural.*

Monaural One channel of sound coming from one direction.

Monitor amplifier A piece of equipment that raises the volume level of sound going to a speaker.

Monitor speaker A piece of equipment from which sound can be heard.

Motor The part of a turntable that makes the platter turn.

Moving-coil microphone Another name for a "dynamic microphone."

MPEG-2, Audio Layer 3 (MP3) A data compression algorithm that relies on perceptual coding, and is the most common format used for digital audio production and streaming.

MP3 See *MPEG-2, Audio Layer 3.*

M-S miking See *Mid-side miking.*

Multidirectional microphone A microphone that has switchable internal elements that allow it to employ more than one pickup pattern.

Multiplay A type of CD player that can hold up to 200 CDs and access material on them according to a prescribed pattern.

Multiple-microphone interference Uneven frequency response caused when microphones that are too close together are fed into the same mixer.

Multitrack recorder A machine or computer software program that can record four or more tracks, all going in the same direction.

Musical instrument digital interface (MIDI) A communication system that allows musical instruments, other electronic gear, and computers to communicate with and control each other.

Music bed Background music used in commercial production to convey the tone or mood of the commercial.

Music library A CD or Internet site that has numerous selections of music that are copyright cleared.

Mute switch A control on an audio console that prevents the audio signal from going through a channel; similar to an on/off button.

Narration A non-sync voice track that accompanies a picture.

Near-field monitoring Placement of monitor speakers on a counter on each side of an audio console so that they are extremely close to the announcer. Also known as "close proximity monitoring."

Needle See *Stylus.*

Netcast To "broadcast" radio programming or similar material, using audio streaming on the Internet.

Noise Unwanted sound in electronic equipment.

Noise gate A signal processing device that reduces noise by suddenly turning way down any audio signal below a set threshold point.

Noise reduction Methods of eliminating unwanted sound from a signal.

Nondirectional Another word for "omnidirectional"; usually refers to a microphone pickup pattern.

Non-sync sound Audio that does not need to match the picture, either because it is a general sound or it is offscreen.

Notch filter A filter that eliminates a narrow range of frequencies or one individual frequency.

Obscenity Material that contains the depiction of sexual acts in an offensive manner; appeals to prurient interests of the average person; and lacks serious artistic, literary, political, or scientific value.

Octave An interval between one musical note and another whereby the change in pitch is caused by doubling or halving the original frequency.

Off-axis Sound picked up away from the front of a microphone on a 360-degree axis.

Omnidirectional Picking up sound from all directions; usually refers to a microphone pickup pattern.

On-air light A signal that comes on to indicate that a live microphone is on in the studio.

On-air studio The studio from where programming is broadcast.

On-axis Sound picked up at the front of a microphone, or zero degrees on a 360-degree access.

On-demand Programming that can be accessed at any time.

On-mic When a sound has good presence because it is recorded directly into the microphone as opposed to at the side or back of it.

On/off switch   A control to stop or start an electronic device.

Open-air headphones   Headphones with a porous muff that produce little or no pressure on the ear. They are more susceptible to feedback if their volume is too high when used near a microphone.

Out of phase   A phenomenon that occurs when the sound wave from one microphone or speaker is up and the sound wave from a second microphone or speaker is down; the combined result is diminished or canceled sound.

Output selectors   Buttons that determine where a sound goes as it leaves the audio console.

Overdubbing   Adding new tracks to something that is already recorded; usually a multitrack recording technique.

Overhead   Additional bandwidth that is available if necessary for audio streaming on the Internet.

Overmodulated   Too loud for the equipment to handle.

Overtones   Pitches that are not exact frequency multiples of a fundamental tone.

Pan   To use controls to send part of a stereo signal to the left channel and part to the right channel.

Pan knob   The part of an audio board that controls how much sound goes to the right channel of a stereo system and how much goes to the left channel; also known as "pan pot."

Pan pot   See *Pan knob*.

Parabolic microphone   A microphone housed in a large concave "bowl" that is used to enhance and record ambient and background sound.

Parametric equalizer   An equalizer that can control the center frequency and the bandwidth of frequencies selected that will have their volumes raised or lowered.

Patch bay   See *Patch panel*.

Patching   Connecting equipment together through the use of jacks and plugs.

Patch panel   A board that contains jacks that can be used to make connections with plugs. Also known as a "patch bay."

Payola   Accepting money under the table from a record producer in return for playing a particular piece of music on the air.

Peaking in the red   Modulating a signal so that it reads above 100 percent on the VU meter.

Pegging the meter   Operating sound so loudly that the needle of the VU meter hits the metal peg beyond the red area; also known as "pinning the needle."

Penetration   Sound going through a surface and being transmitted into the space on the other side of the surface.

Perambulator boom   A large, three-wheeled, movable platform that holds a microphone operator and a microphone in such a way that the microphone can move in various directions to capture sound.

Perceptual coding   Data compression based on characteristics of human hearing.

Performance release   A form that gives a producer the right to include words and actions from a person in a media program.

Performance studio   A studio used primarily by actors or musicians that has microphones but no other production equipment.

Personal audio editor   A compact portable digital editor that records onto a hard drive.

Perspective   The spatial relationship of sound; for example, sounds that are supposed to be distant should sound distant.

Phantom power   Power that comes from a recorder or an audio board through a microphone cable to a condenser microphone.

Phase   The up and down position of one sound or radio wave in relation to another.

Phase cancellation   See *Out of phase*.

Phone connector   A connector with a sleeve and a tip.

Phono connector   See *RCA connector*.

Photodiode   The part of a CD player that provides the data signal that will be converted to an audio signal.

Pickup pattern   The area around a microphone where it "hears" best.

Pinch roller   A rubber wheel that holds tape against the capstan.

Pin connector   See *RCA connector*.

Ping-ponging   See *Bouncing*.

Pinning the needle   See *Pegging the meter*.

Pitch   Highness or lowness of a sound determined by how fast its sound wave goes up or down.

Planar-magnetic speaker   A loudspeaker that uses a thin film membrane with an imprinted voice coil suspended between two magnets. It produces high-frequency sound with extraordinary precision, but low frequency sounds are less defined. For this reason, the planar magnetic speaker is usually used only as a tweeter.

Plastic base   The middle part of audio tape, usually made of polyester.

Platter   The part of the turntable on which the record rests.

Play-by-play   A term designating sports broadcasting from the scene.

Player   In relation to audio streaming, the software or computer used to decode and play back audio on the listener's computer.

Plugola   Accepting money under the table for promoting a business or service on the air.

Plug   A male connector.

Podcasting   Transmitting material so it can be viewed or heard on portable devices such as iPods.

Polar pattern   A two-dimensional drawing of a microphone's pickup pattern.

Pop filter   A ball-shaped accessory placed over the microphone to reduce plosive sounds.

Pot   See *Potentiometer*.

Potentiometer (Pot)   A round knob that controls volume.

Preamplifier   A section of an audio console that increases the voltage of an input source.

Preamplification   The initial stage at which volume is boosted.

Pressure microphone Another name for a "dynamic microphone."

Pressure zone microphone (PZM) A flat microphone that, when set on a table or other flat surface, uses that surface to collect the sound waves and therefore can pick up audio levels from a fairly widespread area. Also known as a "boundary microphone," "plate microphone," or "surface-mount microphone."

Print-through The bleeding through of the magnetic signal of one layer of tape to an adjacent layer of tape.

Prism system The part of a CD player that directs the laser to the disc surface.

Production studio The place where material for radio is produced before it is aired.

Program An output channel of an audio console.

Promo An announcement that is usually 30 or 60 seconds long that promotes an upcoming station event.

Proximity effect A boosting of bass frequencies as a sound source gets closer to a microphone.

PSA See *Public service announcement.*

Public figures Well-known people—they have to prove actual malice in a libel suit.

Public service announcement (PSA) A commercial-like announcement, usually 30 or 60 seconds long, that extols a charity or other nonprofit organization that does not pay for the airtime.

Punch in To edit by recording over one section of a track but leaving what was before and after the edited section intact.

PZM See *Pressure zone microphone.*

Quantizing Determining how many levels or values each sample will be broken down into; the standard for most digital recording is 65,536 quantizing levels (16-bit).

Quarter-inch phone See *Phone connector.*

Quarter-track stereo The recording of two stereo signals on one tape in which two signals go to the left and two go to the right.

Radio microphone Another name for a "wireless microphone."

Rarefaction A sound wave characteristic that occurs when the air molecules are pulled apart.

Rate The number of words spoken in a given time period.

Razor blade Used to cut analog tape in order to edit it.

RCA connector A connector with an outer sleeve and a center shaft; also known as a "pin connector" and a "phono connector."

R-DAT See *Rotary head digital audio tape.*

Reel-to-reel tape recorder A tape recorder that uses open reels of tape placed on a supply reel and a take-up reel.

Reflected sound Sound consisting of echo and reverberation that bounces back to the original source.

Region In digital audio editing, a common designation for a section of audio that is to be edited or saved for later use.

Regulated phase microphone A microphone that consists of a wire coil impressed into the surface of a circular diaphragm that is suspended within a magnetic structure.

Reinforced sound Sound that causes objects to vibrate at the same frequency as the original sound.

Release The time it takes a sound to die out from a sustained level to silence.

Remote pickup unit (RPU) A piece of equipment used to send a location feed back to a destination such as a radio station; it operates by using frequencies assigned by the FCC for electronic news transmission.

Remote start switches Buttons that enable a piece of equipment to be operated from a distance.

Reverb See *Reverberation.*

Reverberation (Reverb) Sound that bounces off two or more surfaces.

Reverb ring The time it takes for a sound to go from full volume to silence.

Reverb route The path a sound takes from a source to a reflective surface and back again.

Reverb time See *Reverb ring.*

Revolutions per minute (RPM) The number of times a record on a turntable makes a complete rotation within a minute.

RF microphone Another name for a "wireless microphone."

Ribbon microphone A microphone that consists of a metallic ribbon, a magnet, and a coil. Because it is bulky, heavy, and fragile, it is rarely used in radio anymore.

Riding levels See *Riding the gain.*

Riding the gain Adjusting volume during production. Also known as "riding levels."

Rip and read To read news copy from the wire service machine with very little editing.

Room tone The general ambient noise of a particular location that is often recorded so that it can be mixed with sound during postproduction.

Rotary head digital audio tape (R-DAT) Another name for "digital audio tape."

RPM See *Revolutions per minute.*

RPU See *Remote pickup unit.*

Really Simple Syndication (RSS) A web feed format used to notify users of updated material posted on a website or blog.

Rundown sheet A list of segments that will be included in a program and the approximate time each will run.

SACD Super audio compact disc; a higher-quality CD format.

Sampling Reducing a continuous signal to a discrete signal by taking readings from the original sound source to convert to binary data. A common example is the conversion of a sound wave (continuous signal) to a sequence of samples (discrete-time signal).

Sampling rate The number of times per second that a reading of the sound source is taken in order to convert it to binary data.

Satellite phone   A special telephone that transmits to satellites in low-earth orbit, allowing audio to be transmitted from distant outlying areas.

Satellite radio   Beaming radio programming from a satellite directly to home or automobile receivers.

Scanner   A piece of equipment that can be used to listen to frequencies used by police and fire departments.

Scratch track   Dialogue that is recorded during the shooting of a video production as a temporary placeholder with the knowledge that it will not be used but will serve as a guide for automated dialogue replacement and in the editing process.

Script   A written guideline from which to produce a media-related program.

Script supervisor   The person whose job it is to make sure that there are no potential continuity problems while a production is being shot.

SD   See *Secure digital*.

Sealed box   See *Acoustic suspension*.

Secure digital (SD)   A form of flash media that is used for recording, storage, and editing with some digital handheld audio recorders.

Segue   To cut from one sound at full volume to another sound at full volume.

Selective attention principle   The ability of humans to ignore some sounds in their surroundings and concentrate on sounds they want to hear.

Sel sync   Selective synchronization; a tape recorder feature that makes a record head act as a play head.

Sensitivity   The ability of a microphone to efficiently create an output level.

Server   A powerful master computer used to store and distribute files on demand.

Shellac   A combination of a resin excreted by a beetle and alcohol that was once used to make records.

Shock jock   A disc jockey or someone who intentionally makes off-the-wall or titillating comments with the intent of shocking audience members.

Shock mount   A microphone holder that isolates the microphone from mechanical vibrations.

Shotgun microphone   A highly directional microphone that consists of a microphone capsule at one end of a tube or barrel that is aimed toward the sound source.

Signal processing   Manipulating elements of sound, such as frequency response and dynamic range, so that the resulting sound is different from the original sound.

Signal-to-noise ratio (S/N)   The relationship of desired sound to inherent, unwanted electronic sound. The higher the ratio, the purer the sound.

Single-system sound   A recording method wherein both sound and picture are recorded on the same recording medium.

Slapback echo   A unique echo effect created by dubbing one audio track over another without synchronization.

Slider   See *Fader*.

Slip cueing   Preparing a record to play by having the turntable motor on and holding the edge of the record until it should be played.

Slogan   A short, pithy group of words used to describe a radio station.

Slug   A part of a news script that gives the title or general content of a story.

SMPTE time code   An electronic language developed for video by the Society of Motion Picture and Television Engineers that identifies each picture frame.

S/N   See *Signal-to-noise ratio*.

Solid-state recorder   An audio recorder that uses CompactFlash or other solid-state storage devices as its recording medium.

Solo switch   A button that allows one particular audio board sound to be heard on the monitor.

Sound bite   A short audio statement in which a person gives information or comments on some topic.

Sound card   See *Audio card*.

Sound designer   A person who develops the overall approach to sound for a video production.

Sound editor   The person who spots for and acquires the various sounds needed to accompany a video production.

Sound file   A segment of audio recorded on a hard disk; usually associated with digital audio editors.

Sound lock   A studio design in which a small area is located outside both the control room and performance area that captures sound, not allowing it to pass through.

Soundproofing   Methods of keeping wanted sound in the studio and unwanted sound out of it.

Sound signal   A noise that has not been processed into an electromagnetic form.

Source/tape switch   A switch that allows someone to monitor either the input or the output of a tape recorder.

Spaced-pair miking   See *A-B miking*.

S/PDIF   A Sony–Philips Interface Format standard for RCA connectors.

Speaker   A transducer that converts electrical energy into sound energy.

Speaker enclosure   A sealed assembly that houses a loudspeaker's components, including crossover(s), driver(s), and the power supply.

Speaker level   An input that has been amplified several times in order to drive a speaker.

Speed selector switch   On a turntable, the control that determines whether the record plays at 33⅓ RPM or 45 RPM.

Splicing block   The device that held audio tape during analog editing.

Splicing tape   Special tape used for holding together audio tape in the analog editing process.

Split-pair miking   See *A-B miking*.

Spotting sheet   A paper that indicates what sounds are needed and when for a video production.

Stacking   A multitrack overdubbing technique in which an announcer "sings harmony" to a previously recorded track.

Standby A signal given to talent just prior to going on-air by holding one hand above the head with the palm forward.

Standing wave A combination of a sound wave going in one direction and its reflected wave going in the opposite direction.

Station ID A short over-the-air identification of a station usually giving its call letters and perhaps its slogan; the FCC requires station identification at least once an hour.

Stereo Sound recording and reproduction that uses two channels, one coming from the right and the other from the left, to imitate live sound as closely as possible.

Stereo microphone A microphone that incorporates small, multiple sound-generating elements as part of a single microphone housing that can record sound in such a way that when it is played back, it sounds like it is coming from two areas.

Stereo synthesizer A device that inputs a monophonic audio signal and simulates a stereo output signal.

Stinger Individual, short, sharp sound effects designed to capture immediate attention.

Stop set A series of commercials played over the air at one time.

Streaming audio Using the Internet to transfer audio data from one computer to another so that it can be heard in real time.

Stringer A person who is paid for news stories he or she gathers if a news organization uses them.

Stylus A small strip of metal with a diamond tip that is used to convert the vibrations generated by the playback of a record into sound.

Super-cardioid Picking up sound well from the front but not the sides; usually refers to a microphone pickup pattern.

Supply reel The reel on the lefthand side of a reel-to-reel or cassette tape recorder that holds the tape before it is recorded or played.

Supra-aural headphones See *Open-air headphones.*

Surface-mount microphone See *Pressure zone microphone.*

Surround sound A multichannel audio format that refers to five full-bandwidth channels (right, left, center, right rear, left rear) and one limited-bandwidth channel (bass subwoofer).

Sustain The amount of time a sound holds its volume.

Sustain ending Music or songs that end with the last notes held for a period of time, then gradually fadeout.

Sweeper A recorded element usually consisting of voice, music, and/or sound effects that creates a transition between songs or commercials and music.

Sweetening The act of improving sound during the post-production stage of audio or video production.

Sync sound Audio that must match the video, such as spoken words that need to match lip movement.

Take-up reel The reel on the right-hand side of a reel-to-reel or cassette tape recorder that holds the tape after it is recorded or played.

Talk-back switch A simple intercom on an audio console that allows the operator to talk with someone in another studio.

Tape guide A stationary pin that leads tape through the transport system of a reel-to-reel recorder.

Tape transport The part of a tape recorder that moves the tape from the supply reel to the take-up reel.

Target audience The group of people, usually defined by demographics, that a radio station particularly wants to attract.

Teaser A short segment, usually broadcast before a commercial break, to keep the listeners tuned to the station.

Telephone coupler See *Telephone interface.*

Telephone interface A piece of equipment that connects telephone lines to broadcast equipment.

Telex A location to station transmission system based on telegraph technology but adapted for short wave radio.

Tension arm A moveable guide for tape on a reel-to-reel recorder.

Three-pin connector See *XLR connector.*

Three-way speaker system A monitor speaker that divides sound not just to a woofer and tweeter, but also to another driver such as a midrange.

Threshold of hearing The softest sound the human ear can hear, noted as 0 decibels.

Threshold of pain The loudness level at which the ear begins to hurt, usually about 120 decibels.

Timbre The distinctive quality of tone that each voice or musical instrument has, composed of the fundamental tone, plus overtones and harmonics.

Time code A system of marking frames of audio and video that indicates hours, minutes, seconds, and frames and that can be used to place audio in sync with video because they both have the same time code.

Time line The part of an editing program that shows a graphical display of the sound so that it can be edited.

Timing sheet A list given to a music composer that gives time code numbers where music should be written for a video project.

Tone The quality of a particular sound.

Tone arm The device that holds the turntable cartridge and stylus.

Tone control A control that increases the volume of the high frequencies or the low frequencies.

Tone generator An element in an audio board or other piece of equipment that produces a tone that can be set to 100 percent to calibrate equipment.

Toslink A connector developed by Toshiba that is used for digital audio.

Track sheet A format for keeping notes of what material is recorded on what tracks of a multitrack recording.

Transducer A device that converts one form of energy into another.

Transmitter Equipment used to broadcast a radio signal.

Transport buttons Mechanism, such as play and fast forward, that move audio through a system, be it a tape recorder, an editing system, or another form of audio.

Tray   The area where the CD sits so that it can spin and be read by the laser. Also known as a "well."

Trim control   See *Gain trim*.

Trimmed   Cut down or edited, usually only a small amount.

Tube microphone   A condenser microphone that incorporates a vacuum tube.

Turntable   A device for spinning a record and converting its vibrations into electrical energy.

Tweeter   The part of a speaker that produces high frequencies.

Two-column script   A script that has video elements in the left-hand column and audio elements in the right-hand column.

Two-track stereo   The recording of two tracks on one tape, both going in the same direction to produce stereo sound.

Two-way speaker system   A speaker that has a woofer, a tweeter, and a crossover to send the sound to each.

Ultra-cardioid   Picking up sound well from the front, but not the back or sides; usually refers to a microphone pickup pattern.

Unbalanced cable   Cable with two wires, of which one is positive and the other is combined negative and ground.

Undermodulated   An audio signal or recording in which the volume of the signal is too low, causing noise to be more noticeable.

Unidirectional   Picking up sound from one direction; usually refers to a microphone pickup pattern.

Unity-gain   A device whose circuits do not make the output signal louder or softer than the input signal.

Upmix   To take a monaural sound and make it into stereo or surround sound.

USB   Universal Serial Bus; a connector used to link audio (and other things) to computers.

User interface   A device, such as a keyboard or mouse, that allows a person to interact with a computer.

Utility   See *Auxiliary*.

Variable bit rate (VBR)   A compression technology used with on-demand files that automatically reduces the amount of data where there is little or no signal, such as during silence.

Variable resistor   A device that controls the amount of signal that gets through the audio console, and thereby controls the volume.

VBR   See *Variable bit rate*.

Vented-box   See *Bass reflex*.

Vinyl   A durable synthetic resin used to make records.

Voice doubling   A multitrack overdubbing technique in which an announcer reads the same script on two different tracks to give a double voice effect.

Voice coil   An internal speaker component consisting of a cylinder wrapped in a coil of wire that is suspended between two magnets.

Voice-over (VO)   Speech over something else, such as music.

Voicer   A voice report in the news from the reporter.

Voicer-Actuality (V/O)   A report in the news containing an actuality with a "wrap-around" voicer from the reporter.

Voice tracking   The practice of a radio announcer recording various vocal elements that will later be incorporated into a regular on-air shift, either locally or in a different city.

Volume   Loudness.

Volume control   See *Gain control*.

VU (volume unit) meter   A unit that gives a visual indication of the level of volume.

Walking over   Talking over the vocal portion of a song, such as when an announcer is introducing a record; normally, an announcer talks over only the instrumental portion.

Walla walla   A background sound created by recording people's voices in a manner in which you cannot hear the actual words they are saying.

Waveform   The shape of an electromagnetic wave.

Wavelength   The distance between two crests of a radio or sound wave.

Web cast   The process of streaming a broadcast over the Internet.

Web page   A screen that can be called up on the Internet using a computer browser.

Well   See *Tray*.

Wild track   Background noises that are desirable for an audio or video production and that are often recorded and added to the production; they are also called ambient sounds and the name "wild track" comes from the fact that they are recorded "wild," as in separate from the video or main audio.

Windscreen   A ball-shaped accessory placed over the microphone to reduce unwanted sound coming from wind or breezes.

Wireless headphones   A device to hear sound that transmits an RF or IR audio signal from the source to the headphones.

Wireless microphone   A microphone that does not need a cable, because it consists of a small transmitter and receiver. Also known as a "radio microphone," "FM microphone," or "RF microphone."

Woofer   The part of the speaker that produces low frequencies.

Wow   Slow variations in sound speed.

XLR connector   A connector with three prongs. Also known as "cannon."

X-Y miking   A method of stereo miking in which two microphones are placed like crossed swords. Also known as "cross-pair miking."

APPENDIX

ANALOG AND DIGITAL AUDIO EQUIPMENT

A.1 INTRODUCTION

Analog technology was used in audio production for almost a century. Most analog technology has been replaced by **digital** equipment, although even some forms of digital production equipment seem antiquated today. Most analog equipment, and even some digital equipment, has lost much of its importance, and as such has been relegated to the appendix. It is unlikely that you will ever work in an analog-dominated environment, but many facilities still use analog equipment for specialized purposes. In this appendix, we will discuss many of these pieces of equipment, and give you the information you will need to know if you ever operate it.

Microphones and speakers will not be discussed in this appendix. Although each can contain digital elements, they are still basically analog and probably always will be. The human voice box and the human ear operate in an analog fashion, and barring some major genetic breakthrough, people are not going to spew forth 0s and 1s from their mouths or hear them with their ears. At the beginning and end of the audio process, analog will remain with us. Therefore, what was discussed in Chapters 4 and 7 about microphones and speakers is still relevant in covering the analog aspects of these pieces of equipment.

Even if you have no contact with the analog equipment discussed in this chapter, you may find it interesting. Knowledge of history is always advantageous. Some of the capabilities and quirks of modern digital equipment are there because people wanted to keep the best of the analog characteristics and make it easy for people who were experienced in analog to make the transition to digital. You may wind up marveling at what audio practitioners were able to do in the past, given some of the limitations of the equipment.

A.2 TURNTABLES

The **turntable** (see Figure A.1) was the first piece of analog equipment to succumb to digital technology when the **compact disc (CD)** player became popular in the early 1980s. Because a turntable cannot record, its functions relate only to playing back sound material. The two basic functions are to spin a record at the precise speed at which it was originally recorded, and to convert the variations in the grooves of the record into electrical energy. The main parts of most professional turntables include a platter, an on/off switch, a motor, a speed selector switch, a tone arm, a cartridge/stylus, and a preamplifier.

The **platter** is a metal plate about 12 inches across covered by a felt or rubber mat, which spins the record that is placed on top of it. The **on/off switch** turns on the **motor**, which turns the platter. The **speed selector switch** is needed because during the reign of turntables several types and sizes of records were developed. The first were 10-inch discs invented around 1900 that operated at 78 **revolutions per minute (RPM)**. They had very large grooves and held only 3 minutes of music per side. They were also made of **shellac** and were thick and heavy, and broke easily. In the late 1940s, two different formats emerged that were made of less breakable **vinyl**, and had smaller grooves. One was 7 inches in diameter, played at 45 RPM, which held about 5 minutes of music per side. The other was the **long-play (LP)** record, which was 12 inches wide, operated at 33⅓ RPM, and could hold about 30 minutes per side. As a result, turntables were developed with a speed selector switch that controlled the speed of the motor.

The **tone arm** is usually a metal tube attached to a pivot assembly near the back of the turntable. Its purpose is to house the **cartridge** and **stylus** and allow them to move freely across the record as it is played. The stylus

FIGURE A.1 A turntable showing the on/off switch, speed selector buttons, platter, tone arm with the cartridge and stylus at the end, and volume control. *(Image courtesy of Denon Electronics.)*

(sometimes referred to as the **needle**) is a small, highly compliant strip of metal with a diamond end that sits on the record groove. It picks up the vibrations from the record and sends them to the cartridge where they are converted into variations in voltage and then sent to the **preamplifier** to increase the level of the signal. From there, the signal can be sent to another piece of equipment, such as an amplifier or an audio console. Some consumer turntables had a volume control to adjust the loudness of the sound as it was produced.

A.3 TURNTABLE USE

Today, turntables are used mainly as performance instruments, an application that early audio technicians never envisioned. Club disc jockeys and other live performers use turntables to create unusual sounds that they incorporate within their performances. Turntables are also used to play vinyl recordings that are in someone's record collection and were never remastered for CD. Many new turntables have digital USB outputs to facilitate transferring vinyl discs to some form of digital media.

If you are ever working with a turntable, you should handle the tone arm gently. Most tone arms have small handles by the cartridge/stylus assembly that you should use to pick up the arm and place it on the record. Do not touch the stylus with your fingers. If it's necessary to remove dust, do so by blowing lightly on it, or use a fine-hair brush. If you are using a turntable made after 1960, it probably has only a two-speed speed selector switch of 45 or 33⅓ RPM, because by then, those formats had replaced the 78s. One other thing to watch for is that 45 RPM records have a larger center hole than the 33s, and turntables usually include a separate adapter to fill the hole. If that adapter has been misplaced (and many of them have), you will not be able to play 45 RPM records.

Care of the records is also very important. Dust can be a big problem, because it can fall into a record's groove and cause permanent popping on playback. Static electricity is produced by playing a record and this compounds the problem by attracting more dust. Use a good-quality record cleaner before playing records to help minimize dust problems. Unlike CDs, records can be wiped clean with a cloth and record-cleaning fluid, using a circular motion following the record grooves. Records should be handled by their edges to avoid getting fingerprints on the surface. Unless they're being played, keep records in their paper or plastic inner sleeve and cardboard jacket and store them in a vertical position to prevent them from warping.

Many production facilities still have a stack of vinyl LP records around, such as an old sound effects library or a production music bed library, and some radio stations even still have a collection of music on records. If you're looking to store your vinyl for the long term, do it the right way. Try to use a conventional wooden enclosure designed for record storage that is appropriate for the size of your collection. The most important point about vinyl storage is to keep the records absolutely vertical. If you do, the only weight at the edge of the record will be the record itself. If you allow the records to lean to one side or the other, then the weight will be unequal and it could cause the disc to warp. Also avoid storing records in either very dry locations, which will cause the paper record jackets to crumble, or very damp locations, which will cause mildew. A temperature between 65 and 70 degrees Fahrenheit with relative humidity between 30 and 40 percent offers the best environment for storing vinyl.

If you are going to be playing a record on a turntable or using a turntable as a musical instrument, you will need to use its ability to **cue**. When radio disc jockeys used turntables, their cueing process was fairly complicated by today's standards. Because a turntable's motor does not get up to speed instantly when it is turned on, the record has to start turning slightly before the stylus hits the groove. If the motor isn't up to speed when the music starts, the record produces a **wow** sound—a change in pitch caused by slow variations in the playback speed. (You can hear this sound on the book's accompanying website.) This was an undesirable sound on early radio, but is often used by today's club disc jockeys to create unique audio effects such as mixing and scratching.

There are several techniques for cueing records. For conventional use, you place the stylus on the outer groove and then rotate the turntable platter clockwise until the first sound is heard. Then, if you do not want the "wow," place the speed selector switch to "neutral" and back-track the platter (turn it counter-clockwise) about one quarter of a turn. You start the turntable just before you want the music to play. If you want a "wow" or a scratching sound you can start a record at any point and also rotate it while it is playing.

Another way to cue a record is known as **slip cueing**; hold the edge of the record with your finger, using enough force to keep it from spinning when the on/off switch is turned to "on" (the turntable platter will be spinning below the record). Release the record when the actual sound is to begin. You can attempt to slip cue only if the turntable platter has a felt mat; a rubber mat does not allow the platter to continue to spin as you hold the record edge. Both methods of cueing records take practice and skill and can produce interesting effects.

A.4 REEL-TO-REEL AUDIO TAPE RECORDERS

Prior to the advent of digital recorders, analog **audio tape recorders** were the workhorses of the production room. The **reel-to-reel tape recorder** first surfaced in the 1940s and was the most-used recorder for several decades (see Figure A.4). All tape recorders rearrange oxidized particles on magnetic tape so that sound impulses can be stored on the tape and played back later. The physical makeup of the tape consists of three basic layers: a **plastic base** sandwiched between a **backing layer** and a **magnetic layer** (see Figure A.2). The top layer is composed of tiny slivers of magnetic oxides that are capable of storing an electromagnetic signal. Another type of tape, called **leader tape**, is made of colored paper or plastic and is not capable of recording because it does not have a magnetic layer. Its main use is at the beginning and end of a tape so that it takes on the wear and tear of threading the tape through the recorder, and not the magnetic tape. It can also be written on, so it can be used to identify what is on the tape. Reel-to-reel recorders use magnetic tape and leader tape that is ¼-inch wide, but other types of recorders use different widths.

The rearranging of particles on the magnetic tape is done by the recorder's **heads**, which are actually small electromagnets. When a tape is blank, the magnetic particles on the tape are scattered about in a random pattern (see Figure A.3). The heads electromagnetically arrange these particles to represent sounds. Usually, professional-quality recorders have three separate heads: erase, record, and play (in that order). The erase head is always before the record head so that old material can be erased and new material recorded at the same time. With the play head behind the record head, it's possible on some decks to monitor what's just been recorded. When the machine is just in play mode, the erase and record heads are disengaged.

An important part of a reel-to-reel tape recorder is its **tape transport**. It is the part of the recorder that is involved with the actual motion of the audio tape. The audio tape is threaded manually from left to right; the left reel is known as the **supply reel** or feed reel (the reel that has audio tape on it as you begin to use the recorder) and the right reel is the **take-up reel**, which starts out empty. Behind each reel (inside the tape recorder) are motors that help drive the tape from one reel to the other and help maintain proper

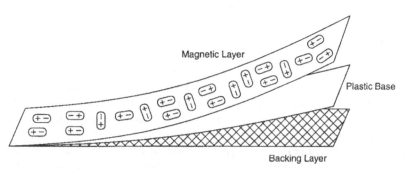

FIGURE A.2 The basic "layers" of audio tape are the magnetic layer, the plastic base, and the backing layer.

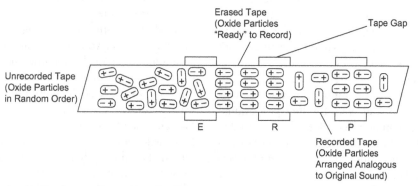

FIGURE A.3 Tape recorder heads align the metallic particles of audio tape in a pattern analogous to the original sound.

FIGURE A.4 The top part of this reel-to-reel recorder includes the supply and take-up reel, tension arm, tape guides, heads under the box with "Tascam" on it, capstan, pinch roller, and idler arm. The bottom section contains VU meters, pots, controls for playing, pausing, and so on, and various function switches to route the sound. *(Image courtesy of Tascam.)*

tape tension and speed. The standard reel sizes used in audio production for many years were 5-inch, 7-inch, and 10½-inch reels.

The audio tape is kept in line with various **tape guides** and **tension arms** so that it properly passes the heads, which are under the little box in the middle of the recorder. Sometimes the tape guides are stationary pins that provide a track that is just wide enough for the tape to pass through. The tension arms are generally more flexible. As the audio tape threads through them, they provide some spring, or tension, against the tape. One of these tension arms is known as the **idler arm**; if the tape breaks, this arm drops down into an "off" position, and the reel-to-reel recorder stops running. The heart of the tape transport is the **capstan** and **pinch roller**. Normally located just to the right of the tape heads, the capstan is a metal shaft, and the pinch roller is a rubber wheel. The audio tape must pass between these two components or it will not move. When the recorder is running, the pinch roller holds the tape against the revolving capstan, while the capstan controls the speed of the tape as it passes the heads.

Other components of a reel-to-reel recorder are its controls. Typically, recorders have buttons for rewind, fast-forward, play, stop, pause, and record. Often professional recorders also have a cue button that allows the tape to stay in contact with the tape heads during rewind and fast forward or even in the stop position, so you can find a certain spot on the tape. The electronics of the audio recorder include record-level and play-level potentiometers, VU

meters, and a **source/tape switch**. The record-level pots adjust the volume or level of the incoming sound signal while play-level pots control the volume of the sound signal as it's being played from the audio tape. The signal you see on the audio tape recorder VU meter is dependent on where the source/tape switch is set. In the "source" position, the VU meter shows the volume of the incoming signal while in the "tape" position the VU meter shows the output level of the reproduced signal at the playback head. Reel-to-reel recorders have counters that show minutes and seconds, and they also have speed selection switches because they are able to record and play back at varying speeds (mostly 7½ inches per second (IPS) and 15 IPS). The faster the tape speed, the higher quality the recording, but the sooner you will run out of tape.

A.5 REEL-TO-REEL RECORDER USE

Today, reel-to-reel recorders are used primarily to play back archived materials. Many important programs were stored on reel-to-reel tapes, especially the 10½-inch reels, because they could hold a great deal of information. Today the content of those reels, like the content of vinyl records, is being transferred to digital media. One aspect of the 10½-inch reels is that they had a larger hole in the middle than other size reels and therefore needed special hubs to hold them on the supply and take-up reels. These hubs, like the 45 RPM adapters, may have strayed from their recorders.

If you are playing something back on a reel-to-reel recorder and it sounds groggy or too fast, you probably need to change the speed control. If the sound is not consistent or fades in and out, you may have dirty heads. Normal use of a tape recorder leaves some oxide material on or near the heads. It's good production practice to clean the tape heads gently with cotton swabs and head cleaner or denatured alcohol before you begin any production work. When playing a tape, if there is one dominant sound but bits of other sounds that creep in, this probably means there is **cross-talk**—the signal from the dominant track is picking up sound from an adjacent track. You can check to make sure that the tape is properly aligned as it passes the heads, but the problem might not be correctable.

Another reason that you might hear garbled sound is that the recorder you are listening to has a different track configuration than the recorder used to originally record the material. Although all reel-to-reel recorders used tape that was ¼-inch wide, they used the tape space in different ways. There were **full-track** recorders that used the whole ¼-inch space to record one mono signal. There were also **half-track mono** recorders, in which one signal could be recorded on the top half of the tape and when the tape was turned over, another signal was recorded on the other part of the tape going the opposite direction. When stereo came along, two signals were recorded on the tape, both going the same way, a configuration known as **two-track stereo**. Later on, reel-to-reel recorders were configured to record two stereo signals or a total of four tracks (**quarter-track stereo**) with each occupying one quarter of the tape: two going one way and two going the other way.

This was all well and good, except that tapes recorded on one type of machine might sound incoherent on another. For example, a tape recorded on a quarter-track stereo machine can be played on a full-track mono recorder. All the tracks will be heard; however, there will be recorded material going both forward and backward, and the resulting sound will be a garble. If you run into this problem when you are trying to listen to a tape, there is little you can do except try to find a different, compatible reel-to-reel recorder.

Print-through is another problem, especially with older tapes. It is the transfer of the magnetic signal on one layer of tape to the magnetic signal on the next layer of tape, either above it or below it on the reel. Visualize a jelly sandwich stacked on top of a peanut butter sandwich. If the jelly soaks through the bottom piece of bread and onto the peanut butter, print-through has occurred. It's most audible when one of the tape layers contains a very loud sound and the adjacent layer contains a soft sound.

Perhaps the biggest problem with audio tape is signal loss due to drop-out. **Drop-out** is a defect in the oxide coating that prevents the signal at that point from being recorded. Drop-out can be a problem that occurred during the manufacturing of the tape, but it can also be caused by flaking of the oxide coating due to heavy use or improper storage. The results of these problems, such as incorrect speed, cross-talk, tracking incompatibility, print-through, and drop-out, can be heard on the web site that accompanies this book. To some extent these problems can be overcome with digital technologies. For example, sound can be artificially created to fill in where drop-out has occurred. Although important material can be restored that way, the process is tedious and expensive.

PRODUCTION TIP A.A
Sel Sync

If you are using a reel-to-reel recorder, you may need **sel sync** (selective synchronization). As already mentioned, the erase head is first followed by the record head, followed by the play head. Because of this head arrangement, you can't easily record one track in synchronization with a previously recorded track. For example, if you record one voice on one track and want to record another voice on another track on the same tape, you run into the following problem; as the previously recorded voice is playing, the sound signal is coming from the play head, but the second voice is recording at the record head. Because of the small distance between these two heads, you hear the previously recorded material a split second before you can record the second voice and therefore the two tracks will be out of sync when played back. To overcome this problem, some recorders have sel sync, which makes the record head also act as the play head. Now you're hearing the previously recorded material at the same time as you're recording the new material, so there is no time difference between them, and you can easily synchronize the two recordings.

A.6 CASSETTE TAPE RECORDERS

The **cassette tape recorder** found its way into broadcast facilities in the 1960s mainly because of its portability and ease of use, but also because a professional-quality cassette recorder could offer high-quality recordings. There are units, such as the one shown in Figure A.5, that mount into a rack in the studio, and there are also small portable cassette recorders that reporters and others can take into the field.

One of the reasons the cassette recorder became popular in a portable configuration is that the tape it uses is much

FIGURE A.5 A cassette recorder that mounts in a studio rack. *(Image courtesy of Tascam.)*

narrower (⅛-inch wide) compared to the ¼-inch reel-to-reel tape. The tape also moves at a little less than 2 inches per second, so not as much tape is needed to obtain a quality recording. This tape is housed in a plastic case (see Figure A.6) that slips into the recorder, which is a definite time saver, because no one needs to thread it manually.

A.7 CASSETTE RECORDER USE

Of all the analog tape equipment, the cassette is the one most likely to still be used in a production facility. If you do use one to record, you need to be aware of several issues. For starters, portable recorders have a built-in microphone, but you should avoid using this microphone for broadcast work, because it often picks up internal noise, such as the tape recorder motor. Use a good-quality microphone like those mentioned in Chapter 4.

Also, the tape in the cassette housing has a short leader tape attached at each end, and both ends of the tape are permanently attached to the reels. When recording onto cassettes, it's important to remember the leader tape,

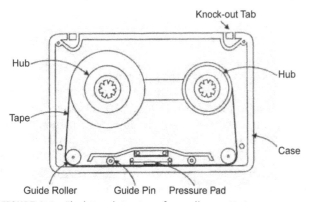

FIGURE A.6 The internal structure of an audio cassette tape.

because if you're at the very beginning of the cassette, the actual recording will not begin for a few seconds, until you are past it.

Another feature of the cassette is the knock-out tabs on the top edge of the cassette shell. There are two of these little plastic tabs (one for each side of the cassette), which allow the recorder to go into record mode. If someone has previously recorded on the tape and wants to be sure to save what was recorded, the tabs will be knocked out and you won't be able to record over the cassette. To solve this problem, put a small piece of cellophane tape over where the tab was, and it will record just as if the tab were still there.

You may also run into track problems, although they are not as confusing or severe as those related to reel-to-reel recorders. There are two basic cassette tape-recording methods. One is half-track mono and the other is quarter-track stereo. However, unlike reel-to-reel configurations, the tracks are laid out in such a way that the mono configuration and stereo configuration are compatible. In other words, you can hear a cassette tape recorded using half-track mono on a stereo recorder and you can hear a cassette recorded in stereo on a mono recorder, but of course it will be heard as mono, not stereo.

A.8 CARTRIDGE TAPE RECORDERS

The **cartridge recorder** (shown in Figure A.7) became predominant in radio facilities during the 1960s and was used to play anything from songs, to commercials, to public service announcements, to station promos, and jingles. The audio tape cartridge is constructed as a plastic container with a continuous loop of tape inside. The tape pulls from the inside and winds on the outside of the spool.

Cartridge tapes (commonly referred to as "carts") have an inaudible cue tone placed just in front of the recorded

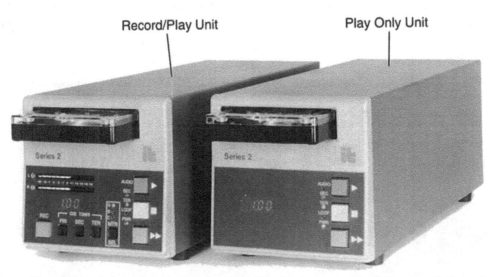

FIGURE A.7 The cartridge recorder and player once played a major role in delivering the commercials, PSAs, promos, jingles, and music heard on the radio. *(Image courtesy of International Tapetronics Corporation.)*

information so that when music or a spot is played, the tone signals the machine to stop the tape before it repeats itself. This automatic re-cueing also allowed several different spots to be put on one cartridge and played one at a time without fear of playing another before stopping the machine. As cart machines became more sophisticated, they were able to put secondary and tertiary cue tones on the cartridge that could be used to indicate the end of a spot, to start another cart machine, and to activate other programming features often found in automated situations.

A.9 CARTRIDGE RECORDER USE

You are not likely to use analog cart machines, as they have been replaced by a variety of digital equipment and software programs replicating the cart machine. Most carts did not hold more than 5 minutes of tape, so they were not used for archival purposes. However, if you do need to operate one, they are simple. Some are play only, and because they cue themselves and stop themselves, all you really need to do is push the PLAY button. Decks that record have a record button and a meter for setting levels.

A.10 TAPE-BASED DIGITAL RECORDERS

Although most of the tape formats discussed so far have been retired after decades of use, some more recent digital tape formats have also found themselves on the fringe of common use. The decline in their use has nothing to do with their technology, however, but more with their applicability and functionality in the studio.

One of these digital tape formats is the **DASH** system, which stands for **Digital Audio Stationary Head**. DASH is a reel-to-reel standard for digital recorders, like the Sony PCM-3202 or Mitsubishi X-86. Because of their high costs, multitrack reel recorders like these are used mainly in high-end recording studios for music and film production.

Another digital tape-based format that found some broadcast application but has faded from regular use is **DAT** (**digital audio tape**). The DAT recorder, also known as **R-DAT** (**rotary head digital audio tape**), is based on VCR and CD technology. The DAT system records using rotating heads, putting digital data on a ⅛-inch tape in a series of diagonal tracks similar to VCR recording. DAT recording is 16-bit with sampling rates of 32, 44.1, or 48 kHz.

The DAT cassette tape is designed similarly to a VCR tape and consists of two small tape reels encased in a plastic housing about the size of a pack of playing cards (see Figure A.8). DAT, like CDs, have several controls for selecting specific songs on a tape. An **AMS (automatic music sensor)** button allows the operator to skip forward or backward to the start of a recorded track.

Although DAT recorders have the same superior sound quality associated with all digital equipment—exceptional

FIGURE A.8 The DAT tape cassette is similar in design to a VCR tape. *(Image courtesy of TDK Electronics Corporation.)*

FIGURE A.9 A splicing block. *(Image courtesy of Xedit Corporation.)*

frequency response and S/N ratio; wide dynamic range; and virtually no wow, flutter, hiss, hum, or distortion—being tape-based became a disadvantage because audio tape was subject to breaking, deterioration, and manufacturing defects. Today, if a DAT system is used in an audio production facility or radio station, it is generally used for archival and storage purposes.

A.11 ANALOG TAPE EDITING TOOLS

The tools used to edit tape during most of the analog days were not electronic—they consisted of a grease pencil, a splicing block, a razor blade, and splicing tape. The editing process was called "**cut 'n' splice**" and has all but been replaced by computer editing. But if you have a need to undertake cut 'n' splice editing, you will use a white or yellow **grease pencil** to physically mark the points where you're going to cut on the back or unrecorded side of the tape. You do this utilizing a reel-to-reel recorder, where the tape is totally exposed, because it is *extremely* difficult and practically impossible to cut 'n' splice with a cassette or cartridge tape. The **splicing block** is a small metal block (see Figure A.9) with a channel wide enough to hold the audio tape and two grooves to guide the razor blade when cutting it. The grooves are usually at 45-degree and 90-degree angles to the audio tape. For almost all production work, you'll use the diagonal cut, because it provides more surface area for contact with the splicing tape at the point of the edit and it provides a smoother sound transition. Any standard single-edged **razor blade** will work for cutting audio tape. **Splicing tape** is commercially available, although a bit hard to find these days, and is specially designed so that its adhesive material does not soak through the audio tape and gum up the heads of the tape recorder. Do *not* use cellophane tape to do editing work. Splicing tape is slightly narrower than audio tape so that any excess adhesive material does not protrude beyond the edge of the audio tape.

Audio was sometimes edited using a **dubbing** method, and some facilities still use this method for digital as well as analog editing. It requires the use of two tape recorders. You simply dub or copy from one to the other with the "master" tape recorder in the play mode, and the "slave" tape recorder in the record mode. You cannot do the precise types of editing that computer editing allows, but through careful manipulation you can transfer the material you want and eliminate material you don't want. Usually dubbing produces a glitch at the edit point that can range from barely noticeable to terrible, depending on the tape recorders used. Dubbing works very well if you simply want to make duplicate copies of any existing tape.

A.12 MAKING EDITS

If you are editing with the cut 'n' splice method, you will find that it involves a two-step process: marking the edit points and then making the cut. Because audio tape passes through the recorder from left to right, sounds are recorded on the tape in the same manner. For example, the phrase "editing is really a two-step process" would be recorded on audio tape in this manner: "ssecorp pets-owt a yllaer si gnitide"—the rightmost word ("editing") would be recorded first. If you want to edit out the word "really" in this phrase, you would make two edit points, one on each side of the word. To do this, you find the playback head on the reel-to-reel recorder then play the tape, stopping it right before the word "really." By rocking the tape reels back and forth in the stop position and cue mode, you can hear (with some practice) the beginning and end of the word "really." Make your mark at those two places in line with the middle of the play head, being careful not to get grease pencil on the head because it will cause head clog. If you are making more edits, you can proceed to mark them, too.

Once you have marked the edit points, it's time to perform the actual cut 'n' splice (see Figure A.10). Normal splicing technique follows these steps:

A. *Position the tape at the first edit mark in the splicing block.* You can do this by placing the splicing block below the recorder play head and pulling the tape down to it. The unrecorded side of the tape should be facing up in the splicing block, and the edit mark should be at the 45-degree cutting groove.
B. *Cut the tape at your first mark.* A simple slicing motion with the blade through the groove should cut the tape

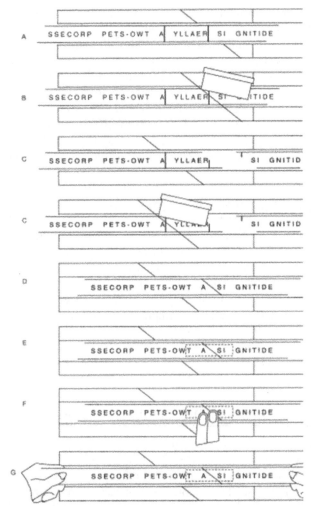

FIGURE A.10 To edit audio tape: (A) position tape at first edit mark; (B) cut tape at first mark; (C) position and cut tape at second edit mark; (D) butt tape ends together; (E) apply splicing tape; (F) smooth out splicing tape; (G) remove audio tape from splicing block.

cleanly. Be careful! Razor blades are sharp and will cut your fingers as easily as they cut audio tape.
C. *Repeat Steps A and B at the second edit mark.* Remove the unwanted piece of audio tape, but don't discard it yet. It's good production practice to hang on to cut-out tape until after you're sure the splice has been accomplished as you want it. It's possible (although difficult) to splice the cut-out piece of tape back in and try the splice again if you've made a mistake.
D. *Butt the remaining tape ends together.* Move both pieces of the tape slightly left or right so that you don't butt them together directly over the cutting groove.
E. *Apply the splicing tape on the edit.* About ¾ inch of tape is the proper amount. The splicing tape should be centered at the edit. Because the splicing tape is narrower than the audio tape, it should not protrude over either edge of the audio tape.

FIGURE A.11   An analog audio console. *(Image Courtesy of Toft Audio Designs.)*

F. *Smooth out the splicing tape.* Be sure to get air bubbles out from under the splicing tape for a strong bond. Rubbing your fingernail over the splice will usually take care of this.

G. *Remove the audio tape from the splicing block.* Never do this by grasping one end and lifting; the edges of the channel in the splicer can damage the audio tape. The proper procedure is to grasp both ends of the tape just beyond the splicing block, apply slight pressure to the tape by pulling your hands in opposite directions, and lift straight up. The tape will pop right out of the block, and you will have completed your splice.

It's best not to fast forward or rewind the tape until you've played it, in order to help secure the splice. Manually wind the tape on the feed reel until you're past the splice, then thread the tape on your recorder, and listen to the edited tape. If it came out as you wanted, you can discard the unwanted tape section. Sometimes you may find it necessary to shave a piece of the edit by splicing off one edge of the tape. If you've made a good edit mark, however, you'll rarely have to do this.

Editors often encounter problems with their first few splices. These are usually overcome with practice and experience. One of the most common problems with splicing-tape manipulation is simply using too much; a piece of splicing tape that's too long is difficult to position properly on the audio tape and makes the tape too stiff at the edit, which prevents proper contact with the tape recorder heads. On the other hand, a piece of splicing tape that's too short may not hold the audio tape together during normal use. Other problems arise when the splicing tape is put on crooked. A portion of the splicing tape will hang over the edges of the audio tape, making it impossible for the tape to glide through the transport properly. Another problem is leaving a gap as you butt the two tape ends together. Obviously, a gap at the edit point will be heard as an interruption of sound or too long a pause. On the other hand, if you overlap the two tape ends as you butt them together, the splice won't occur where you thought it would. By now, you may be glad you were not editing tape during the analog era.

A.13 ANALOG AUDIO CONSOLES

It is not uncommon to still find analog **audio consoles** in a production facility, such as the one shown in Figure A.11. They are built so that they can handle digital inputs (such as CD players) and output to digital sources (such as a computer), but the electronics of the console itself are analog. Some people feel that going through an analog process gives audio a mellower, pleasing sound.

A.14 ANALOG AUDIO CONSOLE USE

The functions of an analog audio console, whether one from the past or one that is new, are essentially the same as those discussed in Chapter 5 ("The Audio Console"), for digital consoles. They have **input selectors** and **output selectors**, **VU meters**, **headphone jacks**, and so on. This is due in part to the fact that digital consoles were made to imitate analog consoles so that the operators switching from analog to digital would not have a steep learning curve. In addition, the functions that you need from a mixer (the ability to adjust **volume**, **equalize**, **pan**, and so on) are the same for the digital realm as for analog.

The main difference is the interface and the size. With a digital console, you may call up and execute functions by operating a mouse or touching a screen, and you can place only those functions that you need on a computer screen. With an analog board, you physically manipulate levers, switches, buttons, and knobs. If these are numerous, then the console is going to be very large, whereas the digital console, regardless of the number of functions, can be the size of a computer screen.

A.15 THE MD RECORDER/PLAYER

Mini Disc (MD) recorders and players saw moderate success in the audio production studio, but lately have seen reduced use. Originally developed by Sony as a digital replacement for the cassette, an MD deck can be a handheld, tabletop, or rack-mounted system (see Figure A.12).

FIGURE A.12 The recorder in a rack mounted configuration (and combined with a CD player) can still be found in some audio production facilities. *(Image courtesy of Tascam®.)*

FIGURE A.13 A portable MD allows recording and editing in the field, as well as all the other features of the format. *(Image courtesy of HHB and Sennheiser Electronic Corporation.)*

Employing a small disc, the MD can still hold up to 80 minutes of music because of its data compression scheme. The MD is *not* actually CD quality, but is still an extremely high-quality audio medium. The MD also features a "shock absorber" system that uses a memory buffer to store music that continues to play for a few seconds if the player mistracks, until the pickup can return to its correct position.

The portable MD recorder, an example of which is shown in Figure A.13, is still a serious competitor for other recorders when it comes to news gathering or any production situation requiring field recording. The stand-alone unit offers recording, playback, and editing tracks on the move.

A.16 THE MINIDISC

The MiniDisc is permanently sealed in a plastic cartridge (see Figure A.14), so it can be handled without worrying about dust, dirt, or fingerprints. The actual disc is much smaller than a CD—about 2½ inches in diameter—but otherwise shares many of the CD's characteristics, including its storage of digital data in a spiral of microscopic pits. The MD, like a computer disk, has a sliding protective shutter that automatically opens when the disc is put into a recorder or player.

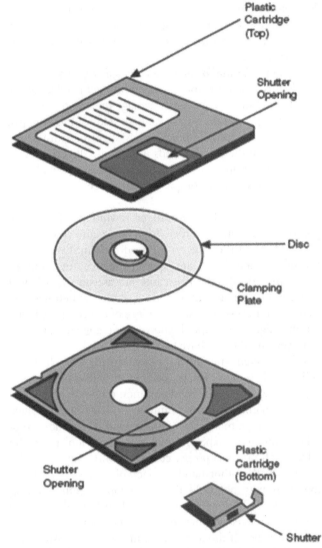

FIGURE A.14 The MD is housed in a plastic case to keep it free from scratches, dust, and fingerprints.

Recordable MDs employ a **magneto-optical design**. A magnetically active layer of the disc is heated (by a laser) to the point where its magnetic orientation can be changed by a magnetic recording head. As the heated spot moves out of the laser beam, it rapidly cools, "freezing" its magnetic polarity according to the data applied to the recording head. During playback, the laser senses variations in reflected light in relation to the magnetic orientation. New material is recorded directly over the old with the format. In order to record large amounts of digital data in a small disc-recording medium, the MD recording process uses a data **compression** system developed by Sony known as **ATRAC (adaptive transform acoustic coding)** to provide digital-quality sound.

The popularity of MD technology has waned recently, especially with the growth of flash media technology for field recording. The MD was never really a true competitor for studio playback technology, so the advancements in field production technology took away the only distinct advantage it had. Additionally, MD technology is proprietary-based, and some radio stations and production facilities found its cost prohibited its use, especially when compared to flash media and other storage and playback options.

A.17 CONCLUSION

In addition to MDs and DAT, other digital recording and production formats have mostly gone by the wayside. Floppy disks, Zip drives, and removable, IDE (integrated drive electronics) computer drives have all had their moments in the world of audio production. Over time, however, each has faded from regular use as technology has improved.

For years, turntables and tape recorders were the primary workhorses in most radio production studios and cut 'n' splice was the only way to edit audio tape. With the advent of digital equipment and the quality and convenience it offers, the use of older analog equipment and production techniques have all but disappeared. Analog had an illustrious past and deserves occasional use and a revered position in the history of sound. As time goes by, further advances in technology, as well as how producers use it, will determine the obsolescence of both analog and digital audio production technology.

INDEX

Locators in **bold** refer to diagrams, figures and tables

A-B stereo miking 70–1, **71**
absorption **24**, 24–5
absorption coefficient 25
accessories, microphone 74–6
accessories, studio: see studio
  accessories
acoustic panels **25**
acoustics 21, **24–5**, 124–6, **125**
actual malice (libel) 11
actuality 164
A/D (analog-to-digital) converter
  42, 65–6
adaptive transform acoustic
  coding (ATRAC) 253
Adobe Audition 47–54, **48–50**,
  144, 214
ADR (automated dialogue
  replacement) 203–5, **202**
advertisements 4, 21; see also
  commercials
AES/EBU (Audio Engineering
  Society/European
  Broadcasting Union) 131
aesthetics, studio 26–7
age, target audience 3
agencies, media 5, 219
AIFF file format 111
air, dead 6, 159, 177
air personality 1, **24**, 162
Akustia 65
alerts, news 2
aliasing 41
all-digital facility **45**
all-news station **164**
AM (amplitude modulation) 221,
  221
ambient sounds 185, 199–200
American Society of Composers
  Authors and Publishers
  (ASCAP) 219

amplification 91
amplifier, monitor 93, **94**, 127
amplitude **30**, **42**
amplitude modulation (AM) 221,
  221
AMS (automatic music sensor)
  249
analog 115, 243
  editing 249–51, **250**
  recording 29, 147
analog-to-digital (A/D) converter
  42, 65–6
anchorperson **5**, 7
announce booth 21
announcers 6, 7–8, 77, 143, **181**
announcing 161–2, 163–4
antenna **22**
anti-aliasing 41
antiphase, in 30
anti-static measures 27
AP (Associated Press) 5
arm, tone 243–4
arms, idler/tension 245–6
ASCAP (American Society of
  Composers Authors and
  Publishers) 219
assignment of tracks 54
Associated Press (AP) 5
Asynchronous Transfer Mode
  (ATM) 182
atmosphere 27
atmospheric sounds 185, 201
ATRAC (adaptive transform
  acoustic coding) 253
attack **31**
attention, selective 176
AUD (audition) output 95–6
Audacity 43, 214
audience
  response 3, 213

target 3, 161
  importance of 1, 193–4
  production 1
audio
audio chain **22**
audio cards, DSP (digital signal
  processing) **44**
audio connectors **130**, 214
audio consoles **22**, 89–91, **89–95**,
  98, **214**, **251**
  equalizers and pan pots 96–7
  functions 91
  input 91–3
  monitoring 93–4
  other features 97–8
  output 91–3, 95–6
  VU meters and cue 94–5;
    see also digital audio
    production
Audio Engineering Society/
  European Broadcasting
  Union (AES/EBU) 131
audio equipment 243, 253
  analog tape editing 249–51
  cartridge tape recorders 248–9
  cassette tape recorders 247–8
  reel-to-reel tape recorders
    245–7
  tape-based digital recorders 249
  turntables 243–5
audio filters 146
audio recorders **22**, 180, **181**
audio routing switcher 129, **129**
audio signal 29
audio source **93**
audio streaming **214**, 214–15, 218
AudioVAULT 91
audition (AUD) output 95–6
automated dialogue replacement
  (ADR) 203–5, **202**

automatic music sensor (AMS)
  249
auxiliary (AUX) output 95–6
Avid Pro Tools 144, 214

B, Dolby 147
background noise: see noise
backplate (mic) **67**
bag, survival 184, 184–5
balance control 97
balanced cable 132
balanced wiring scheme **132**
band pass/reject filter 146
bandwidth 146, 214, 215
barefoot headphones 128
bass 31, 69–70, **125**
bass reflex **125**
bass roll off 69–70
bay, patch 129
Bayonet Neill-Concelman (BNC)
  connector 131
bed, music 159
bidirectional pickup pattern **68**
binary 41
binoculars 185
bit depth **42**, 215
bit rate, constant/variable (CBR/
  VBR) 215
blackbox signal processing 144
blast filters 74–5, **75**
block, splicing **249**
BMI (Broadcast Music Inc.) 219
BNC (Bayonet Neill-Concelman)
  connector 131
board operator 1
boards 89, 93
boompole **194**, 194–5
booms 23, 75–7, **76**, 197, **197**
boomy (sound quality) 125
boosting (frequency) 26

booth, announce 21
boots 185
bouncing 51
boundary microphones 73–4
brainstorming **5**
branding 160–1
bribery 11
brilliance 25
broadband wireless transmission **180**
Broadcast Electronic 91
broadcast monitors 125
Broadcast Music Inc. (BMI) 219
broadcasting 1, 22–3, 47, 51, 220–1
buffer area 216
bulletproof vest 184
bumper 160
buttons
   AMS 249
   PGM/AUD/AUX/UTL 96
   PLAY/RECORD/STOP 48
   PREVIEW 146
   SELECT/RAISE/LOWER 144

C, Dolby 147
cable television 220
cables **132**
cables, balanced/unbalanced 132
call-in talk show 165
camera microphone **201**
cancellation, phase 197
cancelling (frequency) 26
Cannon connector 130–1, **130**
capacitance 66
capacitor microphones 66–7, **67**, 162, 163, 196
capacity
   compact disc 110, 112
   DVD (digital versatile disc) 112
   hard drive 113
capstan 246
cards
   CF 112
   SD 112
   sound **44**, **214**, 219
cardioid pickup pattern **68**, 77, 162, 163, 194, 205
care of CDs 110–11
career prospects 1, 4–5, 193
   case study 213–14
carpeting, wall 25
cartridge (MiniDisc) **252**
cartridge tape recorders 248–9
cartridges (turntable) 243–4
cart-style CD players **110**
case study
   webcasting 213–14
cassette tape recorders **247–8**, 247–8
cassettes, DAT tape **249**

CBR (constant bit rate) 215
CD players **22**, 109–10, **110**
CD recorders (CD-R) 22, 111, **112**
CD-quality 214, 215
CD-RW (rewritable CD) 111
CDs (compact discs) 6, 110–11
ceiling grid (TV studio) 197
cell 'phone transmission **183**
centralized providers (audio streaming) 218
CF (CompactFlash) cards 112
chain, audio **22**
chairs 26
channel (audio console) 91
channel, input **93**, **94**
channel orientation 127
cheat sheets 162
checklist, location survey **177**
chorusing 51
circumaural headphones **128**, 180
clamping plates **252**
clappers 199
classification of microphones 66
clocks/timers 96, **133**
closed-cushion headphones **128**
close-proximity monitoring 126
clothing, fireproof 185
CNN Radio 5
code, time 203
code of ethics 12–13
coding 41–2, **42**
coding, perceptual 215
coefficient, absorption 25
coil, voice 66, **124**
cold (XLR connector) **130**
combo operation **23**
comments (web) 217
commercials 21, 159–60, **160**
compact discs (CDs) 6, 109–11, **110**, **112**
   player/ recorder 22
CompactFlash (CF) card 112
companders 147
compatibility of media 111
components
   speakers 124
composition, music **202**
compression 29, 111–12, 215
compressors 149
computer maintenance 46
computer monitors 23
computer skills 5
computers and audio consoles 91, **214**
concavity 26
condenser microphones 66–7, **67**, 77, 162, 163, 196
cones (speaker) **124**
connection speed 214

connectors, audio **130–2**, 214
consoles, audio: *see* audio console
constant bit rate (CBR) 215
contact, visual (in-studio) 22, 27–8
content 3–4
continuity (visual media production) 197–8
control, balance 97
control, input volume 92–3, 180
control, remote 22, 96
controls, gain 92–3
controversy 3
convergence 4
converters, analog-to-digital (A/D) 42, 65–6
convexity 26
copy marking 10–11
copyright 12, 219
cords, patch 129
Corian 23
correspondents 2
corroboration 2
counters **23–4**
   anti-static 27
CPU
   system requirements 219
creativity 5, 23–4
cross-fade 97–8, **98**
crossover network **125**
cross-pair miking 71
cue 94–5, **244**
cue sheets **206**
cue talent signal **28**
cue wheels 109
cueing, slip 244–5
curiosity 5
curves, frequency response 31
cut 'n' splice 249
cut signal 28

D/A (digital-to-analog) converters 123
DASH (Digital Audio Stationary Head) 249
DAT tape cassettes **249**
data compression 111–12
DAWs (digital audio workstations) 22, **44**, 44–6
dB (decibel) 28–30, 95, 149
dbx 147
dead air 6, 159, 177
dead spots 182
deadness, studio 25
decay **31**
decibel (dB) 28–30, 95, 149
decision-making **13**
decoration 27
de-essers 150
delay units 147–9, **148**
delivery 5

demographics 3
Denon Professional **110**
depth, stereo 70
designer, sound 1, 193–4
desk stands **76**, 197
devices, signal processing:
   see signal processing equipment
dialogue 194
diaphragms (mic) **66**, 67
diffusion **24**, 24–5
digital audio equipment: *see* audio equipment
digital audio players/recorders 109, **200**
   CDs and CD players/recorders 109–11
   compactflash and other digital recorders 112
   data compression 111–12
   digital distribution networks 114–15
   mp3/portable audio players 113–14
   storage 113
digital audio production 41, 54, **200**
   audio synchronization 46
   the digital process 41–3
   editing 43–4, 46–51
   latency 47
   multitrack 50–4
   roots in analog production 41
   strengths and weaknesses 45–6;
   see also audio console;
   digital audio workstations (DAWs)
Digital Audio Stationary Head (DASH) 249
digital audio workstations (DAWs) 22, **44**, 44–6
digital delay units 147–9
digital distribution networks 114–15
digital input 123
digital media 47
digital microphone 65–6
digital monitor speakers **124**
digital recording 29
digital reverb units 147
digital signal processing (DSP):
   see signal processing equipment
digital versatile disc (DVD) 112
digital-to-analog (D/A) converter 123
dimmers (lighting) 26
diodes 95, 109, 110
direct sound 24
Disc Jockey (DJ) 1, 2–3, **161**, 161–2, **176**

distortion 28–9, 109
distribution networks, digital
114–15
distribution platforms 213–14
   alternative distribution means
      219–22
   copyright and listing sites 219
   encoders 215
   home studio (internet audio
      production) 218–19
   on-demand files and podcasting
      218
   playback software and apps 216
   servers 216
   streaming 214–15
   web pages 214; see also
      internet; radio
DJ (Disc Jockey) 1, 2–3, 161,
   161–2, 176
Dolby 147
double glazing 25
double-glassed windows 25
double-plug (patch cord) 129
double-system sound 199
dovetailing 52
drama 167
dramatic pause 6
dramatic tension 3
driver, dynamic cone 124
drop-out 247
dry signal 143
dryness 25
DSP (digital signal processing):
   see signal processing
   equipment
dubbing, voice 51
DVD (digital versatile disc) 112
dynamic cone driver 124
dynamic microphone 66, 77, 194
dynamic range 42, 109, 143, 149

earbud headphones 128
early sound scattering (ESS) 25
echo 26, 52
editing 1, 49–54, 164, 249–53
   digital 43
editing suite 21
editor 1
   multitrack/two-track 43, 50–4
education, audience 4
effects
   sound 5–6, 199–200
   voice 51–2
   world wide web 148–9
egg carton (sound treatment) 25
electricity, static 27
electrostatic headphones 128
enclosures, speaker 124–5, 125
encoders 215
encoding (digital media) 47, 215
endings 97–8, 98

engineer, sound 1
engineer-assist broadcasting 22–3
enhancing image 160–1
entertainment 2
envelope 30–1, 31
environment
   simulated 147
   studio: see studio environment
equal loudness principle 31
equalization (EQ) 22, 96–7,
   144–6, 185
equalizers 96–7, 145–6
equipment 11, 22; see also audio
   equipment
equipment, field reporting 180
equipment, signal processing:
   see signal processing
   equipment
ergonomics 21, 23
ESS (early sound scattering) 25
ethernet connectors 132
ethics 12, 12–13, 13, 185
Eventide 150
Eventide Harmonizer 151
expander 149
expenses, production 4
extemporaneous speech 5
eye wash 185

fabric softener 27
facilities 11, 45
facts (journalism) 2
fade-in/out 97–8, 98
faders 92
Federal Communications
   Commission (FCC) 11,
   43, 180
feed, satellite/telephone 47
feedback, headphone 128
feedback, microphone 70
feeds, RSS (Really Simple
   Syndication) 219
fees, licensing 219
field mixers 89
field production: see field
   reporting
field reporter 164, 175
field reporting 175, 175–84, 185
   common problems 176–7
   postproduction concerns 185
   provisions 184–5
   signal communication 182–3
   site planning 177–8
   types of field production 175–6
   using equipment 180
   vehicle handling 183–4
field supplies 184
file, sound 48–9, 214
file formats (sound) 49, 111, 217
files, on-demand 218
film 1, 72

filter, blast/pop 74–6, 75
filtering 41–2, 42
filters, audio 146
final decay 31
final mix 205–6
finalization (CD-R) 111
fireproof clothing 185
FireWire connectors 132
First Amendment 11
first-aid kits 185
five-band graphic equalizer 145
flanger 150
flanging 143–4, 150
flares 185
Flash Player (plug-in) 216
flashlights 185
FlashMic, HHB 113
flat-screen monitors 23
floor stands 197
fluorescent lighting 26
FM (frequency modulation) 73,
   220
Foley 204
Foley walker 204
food supplies 185
form, location permission 178
format, compact disc (CD) 110,
   112
format, file 49, 111, 217
Fox News 3
frames, speaker 124
frequency 30
frequency boosting/cancelling 26
frequency modulation (FM) 220
frequency response 31, 42, 66,
   109, 125, 143
full-track (mono) 246
functions, audio console 91
fundamental 30
furniture, adjustable 26
furniture, studio 23, 23–4, 26

gain 144
gain controls 92–3
games, video 199–200, 206
gas masks 185
gates, noise 149
General Public License, GNU 97
generation loss 41
generators, tone 97
geosynchronous satellites 182–3
give mic level signal 28
glare 26
glazing, double 25
glossary 231–42
GNU General Public License 43
goals (production planning) 2–3
Google 2
Grace Digital
   IRC6000 Mondo Wifi Internet
   Radio 216

graphic equalizers 145
grease pencils 249
grid, ceiling (TV studio) 197
group recording 163
groups, interest 3
guide pin/roller 248
guides, tape 245–6

half-track (mono) 246
hand signals 27–8, 28
handling, vehicle (field reporting)
   183, 183–4
hard drive storage 113
hard drives 180
hard hats 185
hardwiring (speaker) 128–9
harmonics 30
harshness 110
hats, hard 185
HD radio 221–2, 221
headphones 22, 128, 127–8,
   180–1, 219
headphones, barefoot 128
headroom 28–9
headsets 197
hearing threshold 149
hear-through headphones 128
helicopters, news 184
hertz (Hz) 30
HHB FlashMic 113
high-impedance 69
hiss 144, 147
hoaxing 11
home page 214
home studios (internet audio
   production) 218–19
hosting, talkshow 165–6
hot (XLR connector) 130
house sound 180
housing, plastic (CDs) 110
hub 248
hum 144, 149, 176–7, 205
humidity 177
humour 5
hyper-cardioid 68, 77
Hz (hertz) 30

iBiquity Digital Corporation 221
IBOC (in-band on-channel) 221
ID, station 160
ideas (production planning) 2
idler arms 246
IDs 46
image enhancement 160–1
imaging, stereo 70, 143
impedance (microphones) 69
importance of sound in visual
   media 193–4
in phase 30
in the mud 95
in-band on-channel (IBOC) 221

indecency 11
indirect sound 24
information 3
information society 2
ink/label layer (CD) 111
input
  audio console 91–3
  channel **93**, **94**
  digital 123
  level 127
  selectors 91–2
  volume control 92–3, 180
inspiration 2
Integrated Services Digital
  Network (ISDN) 132, 180
interaction, on-air (internet) 217
interest groups 3
interface, telephone 133, **134**
interference, multiple-
  microphone 70, 163, 196
internet 2, **214**, 217
  portals 5–6
  webcasting case study 213–14
Internet Service Providers (ISPs)
  216
interview programs 43
interview techniques 166
invasion of privacy 11
ISDN (Integrated Services Digital
  Network) 132, 182
ISPs (Internet Service Providers)
  216
Izotope.com 148–9

jack plug **130**
jacket 184
jingle 160

keyboards 23
kilohertz (kHz) 30
kits, first-aid 185
kitchen-type counters **24**
knock-out tabs **248**

label information (digital media)
  46
language 3
  offensive 11
laser diodes 109
latches, locking (XLR connector)
  130
latency 47
lavaliere microphones **73**, 196–7,
  197
laws 11–12
layers
  CD construction **111**
  magnetic tape **245**
layout, studio 22–3, **23**
leads, pigtail **124**
LEDE (live end/dead end) 25

LEDs (light-emitting diodes) 95
left/right (connectors) **129**
left/right (patch panel) **129**
lenses (CD player) 109–10
LEO (low-earth orbit) satellites 180
level
  line 92, 133
  microphone **28**, 92, 133
  output/input 127
  speaker 133
levels, riding 95
Lexicon 150
libel 11
libraries, music 6
license fees 219
licensing agencies 219
life cycle (sound) 24
light-emitting diodes (LEDs) 95
lighting, fluorescent 26
lighting, visual media **196**
lighting dimmers 26
lightning 177
lights
  on-air/off-air **27**
  recording **27**
limiters 149–50
line level 92
line of sight 182
liners 160
lines, balanced and unbalanced
  132
listener (speaker placement) **126**
listing sites 220
literacy 11
live bouncing 51
live broadcasting 47
live end/dead end (LEDE) 25
live music **176**
live studio 25
liveliness (studio) 25
location permission form **178**
location survey checklist **177**
locks, sound 25–6, **26**
locking latches (XLR connector)
  130
Lo-Fi 149
long-play (LP) 243
looping 159, 203–5
loss, generation 41
loudness 31, 95
low pass filters 146
low-earth orbit (LEO) satellites
  180
LOWER button 144
low-impedance 69
LP (long-play) 243

magnetic tape 41, **245**, **248**
magnets
  mic **66**, **72**
  speaker **124**

maintenance, digital equipment 46
male/female connectors 130–1
malice, actual (libel) 11
Marantz recorder **181**
marking, copy 10–11
Marti RPU 180
masks, gas 185
materials, construction (studio)
  25–6
materials, sound-treatment 25
M-Audio
  FastTrackPro USB audio
    recording interface 219
MD (MiniDisc) 22, 111, 251–3,
  252
media, digital 47
media, visual (sound production):
    see sound production
    (visual media)
media agencies 5
media compatibility 111
medical kits 185
memory cards 112, 180
metallic sound 110
meter, pegging the 95
meters, VU 94–5, **95**, 180
mic level (signal) **28**
microphone 22, 24, 65–6, **65–76**,
    77, 176, **179**, 194
  camera **201**
  feedback 70
  field reporting 179
  hand signals **28**
  impedance and sensitivity 69
  internet audio production 219
  level **28**, 92, 133
  pickup patterns 67–9
  specialist applications and
    accessories 74–7
  position 77, 193–7 **196**
  stands 74–7, **76**, 197, **197**
  stereo and surround sound
    70–2
  techniques 77
  types and classification 66–7
  unwanted effects 69–70
  visual media production 193–7
microphone-to-mouth
    relationship 77
MIDI (musical instrument digital
    interface) 46
midrange speakers 124
mid-side miking 71
miking 70–1, **71**, **163**
MiniDisc (MD) 22, 111, 251–3,
  252
minidramas 159–60
miniphone connectors **131**
mix
  down 53
  final 205–6

mixer 1, 89, 180, **180**, **200**
mixing 91
mobile 'phones 2, **183**
mobile studio 183
modem 214
modes, safe/sync 54
modular architecture (audio
    consoles) 90
modular furniture 23
modulation (radio transmission)
    221, **221**
monaural input 97
monitor amplifiers 93, **94**, 127
monitor speakers 22, 93, **123–4**,
    133, 219
  amplifiers 127
  components 124
  enclosures 124–5
  headphones 127–8
  phase and channel orientation
    127
  sound qualities and placement
    125–6
  types and classification 123–4
  wiring and connections
    128–31
monitors, broadcast 125
monitors, computer 23
monitoring **181**
monitoring techniques 126
mono, full-track/half-track 246
mono signal 53
mono/stereo (phone plugs) 130
morning radio shows 6
mounting (microphone) 73–6
Moving Picture Experts Group
    (MPEG) 218; see also
    MP3 file format; MP4 file
    format
moving-coil microphone 66, 77
MP3 file format 49, 111, 218
  portable audio players
    113–14
MP4 file format 218
M-S stereo miking 70–1, **71**
mud, in the 95
muff, headphone **128**
multi-channel console 90, 96
multi-effects processor **151**
multiplay CD player 110
multiple-microphone interference
    70, 163, 196
multitrack editing 50, **50–4**
music 5–6, 201, **202**
  bed 159
  composition **202**
  libraries 6
  punctuators 160
musical instrument digital
    interface (MIDI) 46
mute switches 97

National Oceanic and Atmospheric Association (NOAA) **165**
National Public Radio (NPR) 3
National Semiconductor 65
near-field monitoring 126
needle 244
needle, pinning the 95
network, crossover **125**
networks, digital distribution 114–15
Neumann KM84 163
news 1, 2, 7, 163–4
   radio **5**
news helicopters **184**
news stations **164**
niche-orientation 3
NOAA (National Oceanic and Atmospheric Association) **165**
noise 22, 28–9, 46, 67–8, 97
   encoders 215
   gates 149
   high-frequency 144
   reduction 144, 147
nondestructive editing 49
nondirectional pickup patterns 67–8, **68**
non-sync sound 199
notch filters 146
NPR (National Public Radio) 3

objectives (production planning) 2–3
obscenity 11
obsolescence 243, 253
octave 31
off-air lights **27**
off-axis **67**
offensive language 11
omnidirectional pickup pattern 67–8, **68**, 77, 178, 194
on-air lights **27**
on-air radio consoles **90**
on-air studios 21
on-axis **67**
on-demand files 218
online data storage 113
operator (speaker placement) **126**
operator, board 1
orientation, channel 127
out of phase 30
output, audio console 91–3, 95–6
output level 127
output selectors 95–6
output volume controls 96
overdriven signal 28
overdubbing 50–1
overhead (bandwidth) 215
overmodulation 95

over-the-air broadcasting 220–1
overtone **30**

pace (speech) 8
pad, pressure **248**
page, home 214
pages, web 213, 214
pain threshold 149
pan knobs 97
pan pots 97
panels, acoustic **25**
panels, patch **128–9**
paperwork 7
parabolic microphones 74, **75**, 77
parametric equalizers 146, **146**
parking 178
particle-board 23
patch
   bays 129
   cords 129
   panels **128–9**
patching (speaker) **128–9**
pause, dramatic 6
payola 11
PDX. abbreviation 21
peaking in the red 95
peaks and valleys 26
pegging the meter 95
pencil, grease 249
penetration **24**
perambulator booms 197
perceptual coding 215
performance releases 7, **10**
performance studios 21
permission forms, location 178
personnel (production planning) 4–5
perspective (visual media production) 197–8
PGM (program) output 95–6
phantom power 67, 163
phase 30
   cancellation 197
phase, speaker 127
phone connectors **131**
phones
   cell/mobile 2, **183**
   satellite **180**
photodiode 110
pickup patterns (microphone) **67–9**, 77
pickup unit, remote (RPU) 182
pigtail leads **124**
pilot/reporter 184
pin, guide **248**
pinch rollers 246
ping-ponging 51
pinning the needle 95
pint-through 247
pitch (speech) 7–8
placement, speaker **126**

Plain Old Telephone Service (POTS) 133
planning, production: *see* production planning
planning, site (field reporting) 177–8
plastic housing (CDs) **110**
plate, clamping **252**
platforms, distribution: *see* distribution platforms
platter 243
PLAY button 48
playback software **214**, 216
play-by-play 165, 177, 182
players, digital: *see* digital audio players/recorders
plosives 66
plug-ins 129, 148
plugola 11
plugs **129–31**
plywood 23
podcasting 218
polar pattern 68–9
polar response patterns 68–9, **69**
pole pieces **72**
polycarbonate layer (CD) **111**
pop filters 74–5, **75**
portable audio players 113–14
portable mixers **180**
position, microphone 77
posters 27
postproduction
   concerns 198–204
   editing 1
   field reporting 185
   visual media 205
potentiometers (pots) 93, 96–7
POTS (Plain Old Telephone Service) 133
pots (potentiometers) 93, 96–7
power, phantom 67, 163
preamplifiers 244
pressure microphones 66
pressure pads **248**
pressure zone microphone (PZM) 73–4, 77
PREVIEW button 146
principle, equal loudness 31
prisms (CD player) 109
privacy, invasion of 11
Pro Tools 144, 214
procedure, recording visual media 198–9
producer 1
production, digital audio: *see* digital audio production
production, field: *see* field reporting
production planning 1, 14
   elements 5–6
   equipment and facilities 11

   ideas and goals 2–3, **4–5**
   importance of voice 7–11
   laws and ethics 11–13
   paperwork 7
   personnel 4–5
   scripting 6–7
   style 3–4
   target audience 2
production situations 159, 167
   announcing 161–2, 163–4
   commercials 159–60
   drama and variety 167
   image enhancement 160–1
   music recording 162–3
   sports/traffic/weather reports 165
   talkshows 165–6
productivity 23–4
professionalism 23–4, 27
program (PGM) output 95–6
promo 21, 160
propagation, wave 29
protective layer (CD) **111**
providers, audio streaming 218
provisions, field reporting **184**, 184–5
proximity effect 69–70
PSA (public service announcement) 21
public service announcement (PSA) 21
punch-in 51
punctuators, music 160
PZM (pressure zone microphone) 73–4, **74**, 77

quantizing 41–2
quarter-track stereo 246
questioning 166

radio 1, 72
   HD 221–2
   internet **216**
   licensing agencies 219
   news anchor **5**
   production elements 5–6
radio, satellite 220
radio console, on-air **90**
radio frequency (RF) microphones 73
radio reporter **2**
Radio-Television Digital News Association (RTDNA) **12**
rain 177
RAISE button 144
random access memory (RAM) 214
range, dynamic 42, 109, 143, 149
rarefaction 29
rate (speech) 8
rate, sampling 41–2

rate, streaming 215
rate, variable (CD player) 109
ratio, signal-to-noise (S/N) 28,
     109, 215
RCA connectors **130**, 214
realistic sound 72
Really Simple Syndication (RSS)
     feeds 219
RECORD button 48
recorder, audio **22**, 180, **181**
recorder, digital: *see* digital audio
     players/recorders
recorder, solid-state **112**
recorders, tape 245–9
recording 162–3
     analog/digital 29
     visual media 194, 198–206
recording lights 27
recordist 1
records, vinyl 41, 243, 244
red, peaking in the 95
reel, take-up/supply 245
reel-to-reel tape recorders 245–7,
     246
reflected sound 24
reflection 24, 24–6
reflection-free zone (RFZ) 25
reflective layer (CD) **111**
reflex, bass **125**
regulated phase microphone 72
release **31**
releases, performance 7, **10**
remote control 22, 96
remote pickup unit (RPU) 182
remote start switches 96
reporter, field 164, **175**; *see also*
     field reporting
reporter, radio **2**
reporter/pilot 184
reports, news/traffic/weather 165
requirements, system
     home studio (internet audio
          production) 218
research 2, 165, 177–8
resistors, variable 92
response, audience 3, 213
response, frequency 31, 42, 66,
     109, 125, 143
reverb 24–25
reverb unit, digital **147**
reverb(eration) 22, 24, 147–9
     route/ring/time 24–25
revolutions per minute (RPM)
     243
rewritable CD (CD-RW) **111**
RF (radio frequency) microphone
     73
RFZ (reflection-free zone) 25
ribbon microphone **72**
riding levels 95
right/left (connectors) 129

ring (phone connectors) **131**
ring, reverb 24–25
rip and read 163
RocknRoller Multi-Cart **200**
roll off, bass 69–70
roller, guide **248**
roller, pinch 246
room tone 185, 201
route, reverb 24–25
routing 91, 129
RPM (revolutions per minute) 243
RPUs (remote pickup units) 182
RSS (Really Simple Syndication)
     feeds 219
RTDNA (Radio-Television Digital
     News Association) 12
rundown sheets 7, 8

S, Dolby 147
safe harbor 11
safe mode 54
sampling 41–2, **42**, 123
sampling rates 41–2
satellite feeds 47
satellite 'phones 180
satellite radio/television 220
save function 49
scattering, early sound (ESS) 25
schedule 9
scratch tracks 203
screenshots
     Adobe Audition 48–50, 145–6,
          148, **150**
     SoundCloud **113**
script, two-column 7, **9**
script supervisor 197
scripting 6–7
SD (Secure Digital) card 112
sealed-box design 124
search engines 2
seating 26
Secure Digital (SD) card 112
segue 98
sel sync (selective
     synchronization) 247
SELECT button 144
selective attention principle 176
selectors
     input 91–2
     output 95–6
self-contained audio consoles 90
Sennheiser Electronic
     Corporation 112, **113**
     MD40 cardioid hand mic **179**
sensitivity 69, 127
servers **214**, 216
SESAC (Society of European
     Stage Authors and
     Composers) 219
sexual obscenity 11
shadow, microphone 196, **196**

sheets
     cheat 162
     cue **206**
     rundown 7, **8**
     spotting **205**
     timing 203, **202**
     track 52–3, 52–3
shellac 243
shield (XLR connector) 130
shock mount (microphone) 74–5,
     76
shotgun microphone **74**, 197
shutter opening (MiniDisc) **252**
sibilance 66
sight, line of 182
signal
     audio/sound 29
     mono/stereo 53
     overdriven 28
     stair-step **42**
     wet/dry 143
signal processing equipment **22**,
     143–4, 150
     audio filters and noise
          reduction 146–7
     common and specialist devices
          150
     dynamic range 149
     equalizers 144–6
     reverb and digital delay 147–9
     software/blackbox signal
          processing 144
signals, hand 27–8, **28**
signal-to-noise (S/N) ratio 28,
     109, 215
simulated environment 147
single-plug (patch cord) 129
single-system sound 199
sit-down design 23
site planning (field reporting)
     177–8
sites, listing 220
slapback echo 52
sleeve (phone connectors) **131**
sleeve (phono plug) 130
sliders 92
slip cueing 244–5
slogan 160, **161**
slow down (signal) 28
slug 7
smartphone 2
SMPTE (Society for Motion
     Picture and Television
     Engineers) 46
society 2
Society for Motion Picture and
     Television Engineers
     (SMPTE) 46
Society of European Stage
     Authors and Composers
     (SESAC) 219

softener, fabric 27
software, playback 214, 216
software signal processing 144
solid-state recorder **112**
solo switches 97
solution, eye wash 185
Sony-Philips Digital (S/PDIF)
     131
sound 29, 29–31
     ambient/atmospheric 185,
          199–200
     direct/indirect 24
     double-/single-system 199
     house 180
     importance of (visual media)
          193–4
     metallic/tinny 110
     reflected 24
     surround 71–2, **72**
     sync/non-sync 199
sound bites 5, 7
sound cards **44**, **214**, 219
sound designers 1, 193–4
sound effects 5–6, 199–200
sound engineers 1
Sound Exchange 219
sound files 48–9, 214
sound locks 25–6, **26**
sound production (visual media)
     193, 206
     continuity and perspective
          197–8
     final mix 205–6
     importance of visual and aural
          elements 193–4
     microphone positioning
          193–7
     postproduction concerns 205
     recording 198–206
     speech 194
sound signals 29
sound transitions 97–8, **98**
SoundCloud **113**
soundproofing 25
source, audio 93
sources 2
source/tape switch 246
spaced pair miking 71
S/PDIF (Sony-Philips Digital)
     131
speaker(s) **214**
     midrange 124
     phase 127
     placement **126**
     systems 124; *see also* monitor
          speaker
speech 5, 8, 194, 214
speed, connection (internet) 214
speed selector switch 243, 244
speed up (signal) 28
splaying (studio walls) 26

splicing
blocks 249
tape 249–50
techniques (analog) 250–1
split pair miking 71
sportscaster 2, 180
sportscasting 165, 181
spots 21, 53, 160, 183
spotting sheets 205
Sprint 180
SR, Dolby 147
stacking 51–2
stair-step signal 42
standby signal 28
standing waves 26
stands, microphone 74–7, 76, 197, 197
stand-up design 23
static electricity 27
station ID 160
stereo 70–1, 71, 246
stereo imaging 70, 143
stereo microphone 70–1, 71, 73
stereo signal 53
stereo synthesizer 150
stereo/mono (phone plugs) 130
stingers 6, 160
stock components, modular 23
stools 26
STOP button 48
storage, data 112–13, 180
storage, physical 200
streaming, audio 214, 214–15
centralized providers 218
streamlined studio 21
stringers 183
studio, home 218–19
studio, mobile 183
studio accessories 123–4, 133-4
balanced and unbalanced lines 132
microphone/line/speaker levels 133
telephone interface 133
wiring and connections 128–31
studio deadness 25
studio environment 21–2, 26, 31
audio chain 22
communications 27–8
construction materials 25–6
dimensions and aesthetics 26–7
layout 22–3
production studio furniture 23–4
sound considerations 24–5, 28–31
style 3–4, 23–4
styluses 243–4
subwoofers 71–2
suite, editing 21
super-cardioid 68, 77

supervisor, script 197
supplies, field 184
supply reels 245
supra-aural headphones 128
surface-mount microphone 73–4
surfaces
acoustic properties 25
surround sound 71–2, 72
survey, location checklist 177
survival bag 184, 184–5
suspense 3
suspension, acoustic 124–5, 125
sustain 31
swallowing the microphone 77
sweepers 160
sweetening 205
switcher, audio routing 129, 129
switches
mute 97
remote start 96
solo 97
source/tape 246
speed selector 243, 244
talk-back 97
symmetry, acoustic 125–6
sync mode 54
sync sound 199
synchronization 46, 247
synthesizer, stereo 150
system requirements
home studio (internet audio production) 219

tab, knock-out 248
takes 52–3
take-up reels 245
talent 24, 28
online interaction 217
talk shows 2, 165–6, 166
talk-back switches 97
tape, magnetic 41, 245, 248
tape
guides 245–6
recorders 245–9
transport 245
target audience 4, 161
teasers 4, 160
technique, interview 166
technique, microphone 77
telephone feeds 47
telephone interfaces 133, 134
television (TV) 1, 72, 220
television reporter 2
television sound production: see sound production (visual media)
Telex 182
temperament 5
template (script) 7
tension, dramatic 3
tension arms 245–6

testimony 2
theme tune 5
three-pin connector 130, 130
three-way speaker system 124
threshold, hearing/pain 149
thunder 177
timbre , 30
time, reverb 24–25
time code 203
timeline (commercials) 160
timers/clocks 96, 133
timing sheet 203, 202
tinny sound 110
tips (phono plug) 130
toeing in 126
Toft ATB16 251
tone 8, 30
tone, room 185, 201
tone arms 243–4
tone generators 97
Toslink 131
touch pad, static 27
track assignment 54
track sheets 52–3, 52–3
tracking, voice 162
tracks
scratch 203
wild 185
traffic reports 165
transducers 65, 123
transitions, sound 97–8, 98
transmission, cell 'phone/wireless broadband 180
transmitter 22
transport, tape 245
tray (CD player) 109
treatment, sound 25
treble 31
tube microphones 72
turntables 243–5, 244
TV (television) 1, 72, 220
booms 23
tweeters 124, 125
two-channel audio consoles 92
two-column scripts 7, 9
two-track editors 43
two-track stereo 246
two-way speaker systems 124

ultra-cardioid 68
unbalanced cables 132
unbalanced wiring schemes 132
unidirectional microphones 68
unity-gain devices 144
universal serial bus (USB) connectors 132
user interfaces 44–5
utility (UTL) output 96

valleys and peaks 26
variable bit rate (VBR) 215

variable rate (CD player) 109
variable resistors 92
variety shows 167, 167
VBR (variable bit rate) 215
vehicle handling (field reporting) 183, 183–4
vented-box design 125
verification 2
versatility 4
vest, bulletproof 185
video games 199–200, 206
vinyl 41, 243, 244
Vinyl (plug-in) 148
virtual audio consoles 91
visual contact (in-studio) 22, 27–8
visual media, sound production: see sound production (visual media)
VO (voice-over) 159, 204
V/O (voicer-actuality) 164
voice, importance of 7–11
voice coils 66, 124
voice dubbing 51
voice effects
multitrack 51–2
voice tracking 162
voice-over (VO) 159, 204
voicer 164
voicer-actuality (V/O) 164
voices 5
volume (speech) 10
volume control, input/output 92–3, 96, 180
VornDick, Bil 163
VU meters 94–5, 95
in the field 180

walker, Foley 204
walking over 162
wall carpeting 25
walla walla 185, 201
water supplies 185
WAV file format 49, 111
wave, standing 26
wave propagation 29
wavelength 30
weather
field reporting 177
weather maps 165
weather reporting 165, 165
web comments 217
web pages 213, 214
webcasting 213–14
well (CD player) 109
wet signal 143
wheel, cue 109
Wi-Fi
internet radio 216
wild tracks 185
windows, double glazed 25

windscreen, microphone 74–5,
    75, 179
wireless
    broadband transmission 180
    headphones 128
    microphone 73, 196–7
wiring schemes, balanced/
    unbalanced 132

WMA file format 218
woofers 124, **125**; *see also*
    subwoofers
workstations, digital audio (DAW)
    22, **44**, 44–6
world wide web effects 148
WORM (write once read many)
    design 111

wow 149, 244
write once read many (WORM)
    111
writing ability 4

XLR connectors 130–1, **130**,
    214
X-Y stereo miking 70–1, **71**

Yamaha 150
    01V96 digital mixer **200**

Zaxcom
    Deva V 10-track digital
        recorder **200**